4300

2000
Zns
2-03

4300

The Particle Odyssey

The Particle Odyssey

A Journey to the Heart of the Matter

Frank Close, Michael Marten, Christine Sutton

OXFORD

UNIVERSITY PRESS

OXFORD
UNIVERSITY PRESS

Great Clarendon Street, Oxford OX2 6DP

Oxford University Press is a department of the University of Oxford.
It furthers the University's objective of excellence in research, scholarship, and education by publishing worldwide in

Oxford New York
Auckland Bangkok Buenos Aires Cape Town Chennai Dar es Salaam Delhi Hong Kong
Istanbul Karachi Kolkata Kuala Lumpur Madrid Melbourne Mexico City Mumbai Nairobi
São Paulo Shanghai Singapore Taipei Tokyo Toronto and an associated company in Berlin

Oxford is a registered trade mark of Oxford University Press
in the UK and in certain other countries

Published in the United States
by Oxford University Press Inc., New York

British Library Cataloguing in Publication Data

Data available

Library of Congress Cataloging in Publication Data

ISBN 0 19 850486 1

10 9 8 7 6 5 4 3 2 1

Art direction: *Richard Adams Associates*
Designed and typeset: *Sam Adams*

Original photography: *David Parker*
Diagrams and illustrations: *Gary Hincks*
Photoshop: *Cesar Pava* and *Paul Gleave/Science Photo Library*

Printed in Italy on acid-free paper:

Contents

Preface

A little over fifteen years ago the three of us teamed up with the aim of producing a book that would show just how visual the world of subatomic particles can be. We brought together classic images of particle tracks in cloud chambers and photographic emulsions, bubble chambers and modern electronic detectors, and we mixed in pictures of leading personalities, from the 1890s to the present day, together with photographs of experiments old and new. The result – *The Particle Explosion* – proved a great success. But subatomic particle physics has had its successes too in the intervening years, and so we have put together a much-requested, new and updated version – *The Particle Odyssey* – with around 250 new pictures and some completely new chapters.

In 1987, when the original book was first published, the particles that carry the weak force, the W and Z bosons, were brand new, and CERN's Large Electron Positron collider (LEP) had not even started up. Now, LEP is no more – decommissioned at the end of 2000, after producing millions of Z particles and thousands of W particles. Elsewhere, the top quark and the tau neutrino have been discovered, completing a pattern of fundamental particles that first began to emerge in the 1960s.

Meanwhile, the century has changed to the twenty-first, and the challenges in particle physics have changed too. The questions have changed from 'what?' to 'why?'; from 'what is matter made of?' to 'why is matter the way it is?'. The explosion of particle discoveries in the 1960s has evolved into an odyssey to explore the underlying relationships and symmetries that give rise to the Universe we observe.

The Particle Odyssey seeks to bring the reader up to date, with images from the LEP collider, new 'portraits' of particles such as the top quark, and pictures of the latest generation of experiments that are asking 'why'? Readers of *The Particle Explosion* will find parts they recognize, but also much that is new. We hope that all our readers – old and new alike – enjoy this new journey into the atom.

Frank Close, Michael Marten, and Christine Sutton
Oxford
January 2002

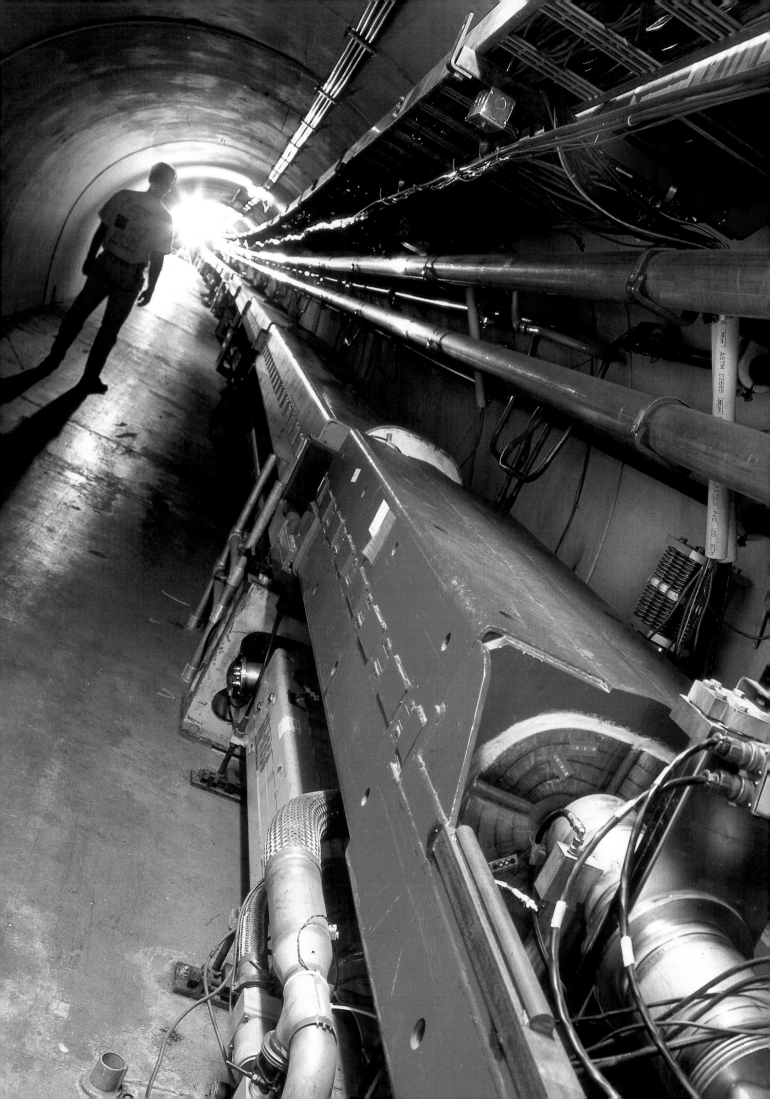

1. The World of Particles

The Executive Lounge at Chicago's O'Hare airport, with its deep pile carpets, soft armchairs, and panoramic view of aircraft manoeuvring, is a temporary oasis for business travellers. The bustle and noise of the concourse disappear once you enter the air-conditioned calm of this living exhibition of state-of-the-art technology. Here you can pause before your flight to enjoy some of corporate America's latest toys. Disregarding the computer screens with their optimistic promises of 'On Time' departures, or the multitudinous channels of world-wide television, you may seek out a glass booth where other travellers' mobile phones will not disturb your business. The booths contain fax machines, modem connections to the Internet for your PC, and optical-fibre links to a mainframe computer should your portable not be up to the task. If you're a television news reporter, you can even make your presentation live through a satellite hook-up.

All of these, and much more to which we give barely a second's thought, are the result of a discovery made more than a hundred years ago by a bowler-hatted, bespectacled Victorian gentleman, Joseph John ('J.J.') Thomson, in Cambridge, England. Every day, among the hordes passing through O'Hare, there are always a few of his modern successors, members of the world-wide network of particle physicists. Take the trio sitting opposite you. They happen to be members of a team whose discoveries have recently completed a chapter in the history of science. They work on an experiment at Fermilab, the 6 km circumference particle accelerator sited 50 km from O'Hare. Their experiment takes place in America, their home universities are in Europe, and their experimental colleagues and collaborators are based in 17 states of the USA, six countries in Europe, plus Canada, China, Korea, and Japan. Their collaboration has enough PhDs to fill a jumbo jet.

The three have been upgraded to Business Class courtesy of their frequent-flyer miles. As particle physicists at large in the twenty-first century, they earn miles so fast that it is hard to unload them and the last thing they want is to take a vacation on yet another flight, even if they could afford the time. For particle physics is big business, the competition global. Managing multimillion dollar budgets and teams of hundreds of PhD researchers, technicians, and engineers is like being head of a major corporation.

Corporate America is power dressed, with sharp suits and crisply ironed shirts. This uniform distinguishes the businessmen from the physicists, who are dressed as ageing undergraduates, with crumpled check shirts open at the neck, casual slacks or jeans, and their notes carried in overweight shoulder bags that bear the logos of recent international conferences in Singapore, Dallas, or Serpukhov. If their dress hadn't proclaimed their profession, the shoulder bags would, as few people other than physicists visit the Serpukhov laboratory near Moscow.

The trio are like missionaries, returning home bearing the latest news and data from their experiment, which in 1995 made headlines with the discovery of the top quark. This fleeting, minuscule fragment of matter had been eagerly sought for more than 15 years; its discovery was the final piece in the story that had begun with Thomson a century earlier.

Six and a half thousand kilometres east of Chicago, a hundred years back in time, Cambridge was a gas-lit stone city of cyclists. Cycling remains today the fastest way around its heart, where international tourists are disgorged from electric buses to gaze at ancient colleges and visit neon-lit superstores with banks of televisions, all tuned to the same satellite station, which turn a news-reader into a choreographed dance of moving

Fig. 1.1 The basic building bricks of the Universe – the fundamental particles of matter – were formed in the initial hot Big Bang. To learn about these elementary constituents, particle physicists reproduce the energetic conditions of the early Universe with machines that accelerate subatomic particles close to the speed of light, through tunnels kilometres long. The machines are monuments to modern technology. Electromagnets guide the particles repeatedly on circular paths through an evacuated 'beam pipe', part of which is just visible in the bottom right corner of the picture. The beam pipe passes through regions of electric field that provide the accelerating power. This view shows the tunnel of the Tevatron at the Fermi National Accelerator Laboratory (Fermilab), near Chicago, as it looked at the time of the discovery of the top quark in 1994–95, when it contained two rings of magnets. The red and blue magnets (the upper ring) form the Main Ring, which has since been dismantled and replaced by an entirely separate machine. The Main Ring was Fermilab's original machine, which started up in 1972, and from 1985 until 1997 accelerated and fed particles into the Tevatron, the ring of yellow magnets just visible below the Main Ring.

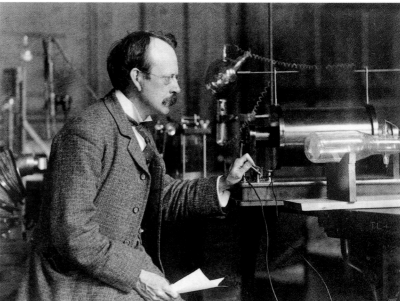

Fig. 1.2 (LEFT) Free School Lane, Cambridge, c. 1890, with the old Cavendish Laboratory, where Thomson discovered the electron.

Fig. 1.3 (RIGHT) Joseph John (J.J.) Thomson gives a lecture demonstration of the kind of tube he used to measure the ratio of electric charge to mass for the cathode rays. His results led him to conclude that the rays consist of minute subatomic particles – electrons.

wallpaper. Here, as everywhere, the city and pace of life have changed in ways that J.J. Thomson never foresaw when, in a laboratory in Free School Lane, he discovered the electron in 1897.

Thomson takes the credit for identifying this workhorse of the modern age and for recognizing that electrons are fundamental constituents of atoms as well as the carriers of electrical current. Like any scientist, he was driven by curiosity. He wanted to determine the nature of the mysterious 'cathode rays', which produced a coloured glow when an electric current passed through a rarefied gas in a glass tube. In his Cambridge laboratory he observed what happened as a narrow beam of cathode rays sped along an evacuated glass tube about 27 cm long to make a glowing green spot at the far end. Using his measurements of how magnetic and electric fields moved the spot, he calculated the properties of the cathode rays and proved that they consisted of particles – electrons.

The electron was the first of what we now know to be fundamental varieties of matter. In the intervening century the list of particles has continually changed as layers of the cosmic onion have been peeled away and deeper layers of reality revealed. Thus nuclei, protons and neutrons, exotic 'strange' particles, and quarks have entered the menu. Throughout, the electron has remained in the list. Today we recognize its fundamentality.

Our best theories require that quarks also are fundamental and that there are six varieties of them, named 'down' and 'up', 'strange' and 'charm', 'bottom' and 'top'. To create the first examples of the top quark, the physicists at Fermilab have had to bring matter and its physical opposite, antimatter, into collision at higher energies than ever before in an underground ring of magnets, 6 km in circumference. The magnets guide protons round in circles as they are accelerated by electric fields; the antimatter equivalents of protons – antiprotons – whirl round the same ring in the opposite direction. As the particles and antiparticles accelerate, their energies increase until eventually they are made to collide head on. Each collision creates a burst of new particles that shoot into giant multilayered detectors surrounding two collision zones. The new particles bear the imprint of events that have happened so swiftly they can never be seen directly. But in 1994–95, the physicists at Fermilab found the 'signatures' expected for the long-sought top quark.

Fermilab stands on enough grassland to support a herd of American buffalo. The offices of its scientists fill ten floors of a graceful cathedral of glass and stone whose atrium soars up to the roof, is grand enough for trees to grow, and sports a dedicated travel bureau. Prairies stretch for hundreds of kilometres to the western plains. Another land of flat earth, the Fens of East Anglia, is home to the grey stone building with gables and bay windows that is the old Cavendish Laboratory in Cambridge. A rabbit warren of staircases connects the corridors of discovery. Doors open onto small rooms where ingenuity has teased from nature those secrets that are just within reach. No buffalo here, no grand entrances; instead Free School Lane is wide enough for pedestrians and Cambridge's ever-present bicycles. On

a misty winter evening today, the illumination can appear hardly more advanced than it would have been in the late nineteenth century. Yet this is where Thomson made his momentous discovery that led to modern particle physics – the science that studies the basic particles and forces and attempts to understand the nature of matter and energy.

Nature has buried its secrets deep but has not entirely hidden them. Clues to the restless agitation within the atomic architecture are all around us: the radioactivity of natural rocks, the static electricity that is released when glass is rubbed by fur, the magnetism within lodestone, sparks in the air, lightning, and countless other clues for those who are prepared to pause and wonder. Such was the arena for J.J. Thomson and much of physics before the twentieth century. Today, Fermilab is looking at matter as it was at the beginning of the Universe, including exotic forms that no longer exist but which seeded the stuff we are made from. In 1897, by contrast, no one knew what stars really are, let alone where the Universe came from.

Fig. 1.4 (ABOVE LEFT) The 6 km circumference ring of the Tevatron at Fermilab is marked out by the lights of a car circling the service road above the underground machine. The land within the circle has been restored to natural prairie by volunteers. The glow of Chicago is visible in the distance.

Fig. 1.5 (ABOVE RIGHT) The atrium of the high-rise main building at Fermilab, which was designed by Robert Wilson, the laboratory's director from 1967 to 1978. Offices of the scientists line the sides of the gracefully symmetric building.

Fig. 1.6 (LEFT) Evidence for the brief existence of the top quark – the heaviest of Nature's building bricks – is captured in this artistic rendition of the aftermath of a proton–antiproton collision in the D0 experiment at Fermilab. The collision has occurred at the centre of the detector, spraying particle tracks (purple and blue) in all directions. Among the particles are an energetic electron, made visible when it deposits its energy, represented by the red blocks to the bottom right, and a ghostly particle known as a neutrino. The neutrino remains invisible, but its direction, marked by the broad pink line to the bottom left, can be calculated from the 'missing energy' it spirits away. The electron, the neutrino, and the two sprays of other particles are together the remnants of the very short-lived top quark.

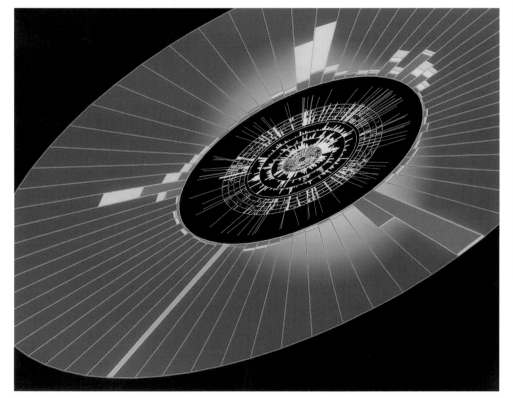

Fig. 1.7 The ethereal beauty of the frozen wastes of Antarctica – location of the AMANDA experiment which detects neutrinos that have traversed the Earth after being created in the atmosphere on the other side of the planet.

Cosmic Explorers

The night is already three months old as the aurora flashes across the sky. It is June at the South Pole. Three thousand metres above sea level, and at a temperature of −70 C, a figure wrapped in a parka and thermal underwear lies on the snow watching the natural display while listening to Tchaikovsky's *1812 Overture* on headphones. The person is a particle physicist, one of a team with an experiment at the South Pole, trying to discover how our Universe came to be. Instead of working at huge man-made accelerators, these researchers make use of the natural accelerators in the cosmos, where electromagnetic forces in space whip into violent motion particles from exploded stars and other exotic events. The moving picture-shows of the aurorae occur when particles from the Sun are trapped by the magnetic arms of the Earth and hit the atmosphere. When higher-energy particles from more distant sources smash into the atmosphere the result is an equally dramatic but invisible rain of particles that cascade to Earth. These messengers from the stars show scientists on Earth what subatomic matter is like 'out there'. They have revealed a Universe that is far richer and more mysterious than anyone imagined a hundred years ago.

The particle physicists at the South Pole are working with AMANDA – the 'Antarctic Muon and Neutrino Detector Array'. This is a telescope, but a telescope that is a far cry from the more familiar structures with lenses or mirrors. Buried under a kilometre of ice, its purpose is to detect not light, but high-energy cosmic neutrinos from our own or nearby galaxies. Neutrinos are mysterious particles that are associated with radioactive phenomena; they have little mass, no electric charge, and are as near to nothing as you can imagine. They travel straight through the Earth as freely as a bullet through a bank of fog. However, they are so numerous in the cosmos at large that they have a significant influence on events in the Universe. They roam the Universe as leftovers of its creation, they are emitted by the processes that fuel the Sun and other stars, and they spill out in huge numbers from colossal stellar explosions.

Neutrinos are very shy and to capture them scientists need to think big. They interact so feebly with other matter they are all but invisible. A telescope for neutrinos must contain enough matter for there to be some chance that occasionally one of the millions of neutrinos passing

Fig. 1.8 AMANDA consists of an array of nearly 1000 light-sensitive phototubes held in the ice 1500–2000 m below the surface at the South Pole. The phototubes detect faint light (Čerenkov radiation) emitted as charged particles produced in the rare interactions of neutrinos pass through the ice. The phototubes are attached to cables and lowered into holes drilled in the ice by a jet of hot water. The drill tower is clearly seen here, together with the 'heater room' – the large dark building near the centre – where the pressurized water is heated before it is pumped down the hole.

through will hit an atom and cause an observable effect. To detect high-energy neutrinos from cosmic sources requires a cubic kilometre or so of matter, and to build this in a customized laboratory would cost an unrealistic amount. So the ingenious idea with AMANDA is to use the natural detector that the Antarctic ice provides. When a neutrino hits an atom in ice, its interaction can give rise to a brief, faint flash of blue light, which can be detected if the ice is clear enough.

In the Antarctic, the ice a kilometre below the surface condensed from snow that fell more than ten thousand years ago, soon after the last Ice Age. Down here the pressure has squeezed out all the air bubbles and the ice is as clear as diamond – so pure that the light flashes caused by neutrinos can travel undimmed for more than a hundred metres to be detected by sensitive devices known as photomultipliers. These 'eyes' are special tubes that convert the faint light to an electric current, which then goes to equipment on the surface that records what has happened.

In AMANDA, photomultipliers are attached at intervals to long cables, which are dropped into holes in the ice up to 2.4 km deep. The holes are made with a special drill that sprays out hot water, rather like a large shower-head. This scalding blast melts its way straight down into the ice, with gravity as its engine. The 'strings' of photomultipliers are then lowered down the holes to sit in the columns of warm water. After a few days the water freezes, trapping the tubes in the ice-pack. From then on they record data continuously.

A full-scale, kilometre-sized version of AMANDA has still to be built, but the tubes so far deployed in the Antarctic ice can detect neutrinos that have travelled right through the

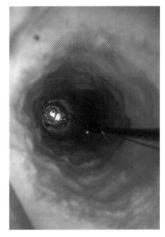

Fig. 1.13 Victor Hess (1883–1964), centre, around the time of his pioneering balloon flight of 7 August 1912, in which he found that levels of radiation became greater at high altitudes. This led him to conclude that the radiation came from outer space. He had discovered cosmic rays.

Earth after being created by cosmic rays interacting in the atmosphere over the North Pole (see Fig. 11.17, p. 216). The full size will be necessary to pinpoint neutrinos from distant cosmic sources, but the next time a star in our Galaxy dies and explodes as a supernova, the existing AMANDA will really come into its own. The associated burst of neutrinos will fly through the Earth and send flashes of blue light through the Antarctic ice. Meanwhile, the scientists can only wait while AMANDA keeps watch.

More than 80 years before the arrival of AMANDA's first contingent of particle physicists, Roald Amundsen was the first person to reach the South Pole, in December 1911, followed a month later by Robert Scott's fateful expedition. This was the heroic era of Antarctic exploration. Several thousand kilometres away, the First World War was soon to change the shape of Europe. On the River Elbe, just south of the German border, the Bohemian town which today lies in the Czech Republic and is known as Ústí nad Labem was then called Aussig and was in the Austro-Hungarian empire.

It is here, in the dawn of 7 August 1912, that Austrian physicist Victor Hess is preparing for what will prove to be a historic balloon flight. On previous flights he has found that radiation detected above the Earth does not diminish as it should if it were due to the Earth's natural radioactivity; indeed, by 2000 m the radiation begins to increase. He has come to the conclusion that the radiation must originate in outer space. The Sun seems an obvious source, but has already been ruled out, as a flight during a solar eclipse on 17 April showed no reduction in the radiation. To confirm that the radiation indeed comes from outer space, Hess has decided to go as high as the technology of the time allows.

Thus it is that around 6 am on this August morning Hess, together with a pilot and a meteorological observer, each with his own oxygen cylinder, climbs aboard the tiny basket slung beneath the balloon. The basket is cramped. There is a small bench to sit on, assorted instruments and baggage, and about 800 kg of ballast in 52 sacks, hung so they can be emptied by cutting a string (in order to avoid unnecessary physical strain at great altitude). After casting off ten sacks of ballast they ascend to 1500 m. At 7.30 am they cross the German border near Peterswald, and by 8.30 am (and 20 ballast sacks lighter) they are 3000 m high. At 9.15 am they are 4000 m above Elstra in eastern Saxony.

It is now freezing cold and measurements of the radiation are exhausting. Hess takes some oxygen to stay alert. By 11 am they are at more than 5000 m and Hess, despite the oxygen, is so weak that he is able to complete only two of the three planned measurements. But that is enough. Although there are still 12 sacks of ballast, which if dispensed could enable them to rise even higher, they decide to come down, and land about 50 km east of Berlin around midday. They collect the equipment and return to Vienna by overnight train.

The scientific results from this pioneering ascent proved to be a great success. Hess discovered that the radiation had become more and more intense the higher they had risen: at 4000 m the radiation was half as strong again as on the ground and at 5000 m more than twice as strong. The conclusion was that the radiation was invading the atmosphere from outer space. With this historic balloon adventure in 1912, Hess had discovered the existence of cosmic rays.

Soon scientists were going up high mountains, laden with equipment to capture the rays and find what they consist of. The cosmic rays have proved to be particles with energies far higher than anything previously known, and they revealed exotic forms of matter never before seen on Earth. The challenge of understanding the message of the rays led physicists to build high-energy particle accelerators in order to reproduce their effects in the laboratory – and so gave rise to modern particle physics.

Particle Physics Now

The form and state of matter today on the cool Earth is the frozen end-product of creation: the early Universe, we now know, was a cauldron of heat and ephemeral varieties of matter that have been long gone. Nonetheless, fifteen thousand million years after that epoch there remain hints of the profound history, hidden from our immediate senses.

Matter as we know it today is made of atoms, which are so small that up to a million could fit into the width of a single human hair. Once thought to be the ultimate seeds of everything, today we know that atoms are themselves made of yet smaller pieces. Their basic constituents were created within the first seconds of the Big Bang. Several thousand years would elapse before the ferment of the Big Bang had subsided to the more quiescent conditions where these particles combined to make atoms. The cool conditions in which atoms exist today are enormously far removed from the intense heat of the Big Bang.

The inner labyrinths of an atom are as remote from daily experience as are the hearts of stars, but to observe the atomic constituents we have to reproduce in the laboratory the intense heat of stars. This is the world of high-energy particle accelerators, which create feeble imitations of the Big Bang in small volumes of a few atomic dimensions.

Particle physicists today have a rich subatomic world to explore. They have discovered hundreds of new varieties of particle. There are pions and kaons, omegas and psis, 'strange' particles and 'charmed' ones. The members of this subatomic 'zoo' have been named with apparent disregard for logic. Many particles are called after letters of the Greek alphabet, and physicists habitually refer to them simply by the Greek letters. The pion, for example, is π.

If the particles are akin to the letters of nature's alphabet – the building blocks from which all else is made – then the analogue of grammar is the set of natural forces that choreograph the cosmos. Particle physicists recognize four basic forces at work that make things the way they are. Gravity causes apples to fall to Earth, and controls the motions of the planets and galaxies. The electromagnetic force affects compass needles and glues atoms to one another to make solids, liquids, and gases, such as human flesh and blood and the air we breathe. Two further forces, known as the strong force and the weak force, control the structure of atomic nuclei. The strong force binds quarks together to form neutrons and protons, which in turn form the nuclei of atoms. The weak force underlies certain forms of radioactivity and also regulates how the Sun burns, the source of all life on Earth.

Fig. 1.14 Richard Feynman (1918–1988), one of the greatest physicists of the twentieth century, gives a lecture at CERN, the European centre for research in nuclear physics near Geneva. In 1965, the year this photograph was taken, he shared the Nobel prize for physics with Sin-Itiro Tomonaga and Julian Schwinger, for work on quantum electrodynamics, or QED, the theory that describes the electromagnetic interactions of subatomic particles. Theorists such as Feynman play an important role in organizing the discoveries of particle physics experiments into theories, which in turn may predict new phenomena to be discovered.

Fig. 1.15 The control room is the nerve centre of a particle accelerator. In this image, banks of monitors show the status of key components in the various machines at the Stanford Linear Accelerator Center in California. The machine crew is in charge, ensuring that the accelerators deliver their beams as smoothly as in an industrial process.

We exist not least because these four forces have the varied properties that make them appear so different in the world about us. Yet theorists conjecture that in the initial heat of the Big Bang all four forces might have been as one, only to split apart as the Universe cooled so that their unity is now obscured. The search for such a 'unification' of forces has become an important strand in the fabric of particle physics. Indeed, it carries a significance beyond particle physics itself, for it is a search for the physics of the Big Bang. One of the unexpected developments in particle physics has been the way that it has become increasingly intertwined with astrophysics and cosmology. This work concerns some of the major questions posed by the very existence of the Universe. How did it all begin? Why does it have the form and structure it has? Will it continue expanding forever or will it eventually begin to contract?

These theoretical constructs are not a modern analogue of ancient theological debates concerning the number of angels on the head of a pin. Theories survive or fall by experimental tests. There is a symbiosis between two breeds of particle physicist: the experimenter and the theorist.

The theorist organizes what has been discovered into a theory, which may predict the existence of new particles. Part of what the experimenter does is to search for the predicted particles, but there is much more than this. A great stimulus to experimenters is the possibility that they will discover something totally unpredicted, which the theorist must then explain in a modified or entirely new theory. It is a measure of the growth of the science that the time is long gone when individual physicists could lay claim to have both experiment and theory at their fingertips. Now specialization is the order of the day, though theorists and experimenters still need to appreciate the subtleties of the other's craft as they feed off each other's work.

Another characteristic of modern particle physics is its internationalism. A typical experiment today involves hundreds of people. It is not something that a single institution can develop, build, and operate. The largest current experiment at CERN, the European particle physics laboratory on the outskirts of Geneva, involves more than 200 institutions, not only from Europe, but also from North and South America, Asia, Africa, and Australia. CERN is itself a multinational effort funded by 20 European nations. Enter the canteen there and you are immersed in a multilingual babble. Furthermore, CERN has forged links with its counterpart in Eastern Europe – JINR, at Dubna 100 km north of Moscow – and more recently has established important relationships with North America, Japan, and India, *en route* to becoming a veritable United Nations of Physics.

CERN, Fermilab, and laboratories like them, provide accelerators where scientists come to perform their experiments. These scientists are, however, only a part of the whole. There are also engineers who maintain the accelerators and keep them working. 'Driving' a particle accelerator is like flying a spacecraft. The 'bridge' is the accelerator control room, consisting of rows of computer monitors. While the particles whirl around several kilometres of beam pipe at almost the speed of light, nothing much seems to be happening. Two or three people may be drinking coffee, consulting a computer display, or telephoning someone at the experiments that the machine is feeding.

The automatic pilot is in control. The path of the particles is programmed. The constant adjustments of accelerating units and magnets, of coolants and vacuum pumps and electricity supply, are all controlled by the computers, which teams of experts have spent hours programming. The people in the control room have little to do, except to make periodic checks. But there are moments of high stress, as when the pilot prepares to land the spacecraft. For example, the machine physicists at CERN and Fermilab can prepare beams of antimatter, which survive only so long as they are kept out of the way of the matter that is all around them. It may take a whole day to prepare the beam, accumulating enough antimatter particles to be of use for the experimenters. Then the controllers must pilot the beam correctly so that it eventually arrives at the experimental apparatus. One push of the wrong button at the wrong moment and all will be lost. A whole day could be needed to put it right again.

Why do particle physicists need to accelerate particles such as electrons and protons to high energies? In some instances, the energy can assist in materializing additional particles, in accordance with Albert Einstein's famous equivalence of mass and energy: $E = mc^2$. An extreme example is when matter and antimatter mutually annihilate into pure energy, which can rematerialize as new, different particles. In this way, particle physicists have been able to create particles and forms of matter that do not occur naturally here on Earth, but which may be commonplace in more violent parts of the Universe.

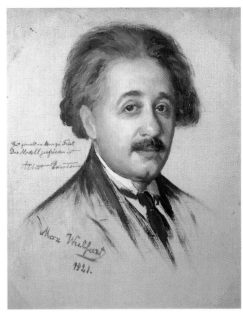

Fig. 1.16 Albert Einstein (1879–1955). He aptly summed up the problems experimental particle physicists face when he described detecting particles as 'shooting sparrows in the dark'.

Creating extreme conditions, hotter than in any star, akin to the early Universe, is only part of the challenge. It would be useless if we were unable to see what happens and record the results. The particles created in today's high-energy collisions can be smaller than 10^{-16} cm across – smaller relative to a grain of sand than a grain of sand is to our distance to the Sun. And not only are these particles triflingly small, they live for only a few hundredths of millionths of a second, or less. Recording these tiny and ephemeral pieces of matter is the job of the detectors.

Detectors come in a variety of types and sizes, but today most are huge, multilayered pieces of apparatus. Despite their differences, they all rely on the same basic principles. They never reveal the particles directly; instead they make visible the effects that the particles have on their surroundings.

Much as an animal leaves tracks in the snow, or a jet plane forms trails of condensation across the sky, electrically charged particles leave trails as they gradually lose energy when they travel through a material, be it a gas, a liquid, or a solid. The art of particle detection is to sense this deposited energy in some manner that can be recorded. Then, in the way that measurements of the footprints of our ancestors can reveal something about their size and the way they walked, the information recorded can reveal details of a particle's nature, such as its mass and its electric charge. All the techniques described in later chapters rely on this same principle, from the simple photographic emulsions of the 1930s and 1940s to the metre-long gas-filled chambers, criss-crossed by thousands of wires, of the 1980s, and the barrels of silicon wafers of the twenty-first century.

Modern detectors are hybrid devices consisting of many subdetectors – scintillation counters, drift chambers, Čerenkov counters, silicon strips – whose job is to measure the paths, angles, curvatures, velocities, and energies of the particles created in a particle collision. The many subdetectors are sandwiched together, sometimes in a series one behind the other (in a fixed-target experiment), sometimes in a kind of Swiss roll wrapped around a beam pipe (in a collider experiment). And every part of the detector has hundreds of cables running from it, each of which goes to a particular place in the control system.

A typical detector at a modern particle physics laboratory is a major undertaking. It will take 5–10 years to design and build, it may operate for another 5–10 years, and its results will continue to be analysed for a further 2–4 years. Someone involved in the project from beginning to end may spend up to 25 years on this one detector. It is not something that a

Fig. 1.17 This view of one end of the H1 experiment at the DESY laboratory in Hamburg shows the complexity of modern particle physics detection. H1 is like a huge Swiss roll – a cylinder of layers of different particle detectors, each with a specific task. Each of these detectors produces electrical signals that contain information about the path of a particle, the energy it deposited, and the time it passed through. And each of these signals must pass through cables to the electronics and computer processors (see Fig. 12.14, p. 228) that piece together the information, ultimately to reveal the particles created in the high-energy collision of an electron with a proton at the heart of the apparatus.

handful of individuals can set up on a laboratory bench. It requires computer experts, draughtsmen, engineers, and technicians, as well as hundreds of physicists from a large number of institutions.

The images the particles create have always played an important role in particle physics. In earlier days, much of the data were actually recorded in photographic form – in pictures of tracks through cloud chambers and bubble chambers, or even directly in the emulsion of special photographic film. Many of these images have a peculiar aesthetic appeal, resembling abstract art. Even at the subatomic level nature presents images of itself that reflect our own imaginings.

The essential clue to understanding the images of particle physics is that they show the *tracks* of the particles, not the particles themselves. What a pion, for instance, really looks like remains a mystery, but its passage through a substance can be recorded. Particle physicists have become as adept at interpreting the types of track left by different particles as early hunters were at interpreting the tracks of animals.

Most of the subatomic zoo of particles have brief lives, less than a billionth of a second. But this is often long enough for a particle to leave a measurable track. Relatively long-lived particles leave long tracks, which can pass right through a detector. Shorter-lived particles, on the other hand, usually decay visibly, giving birth to two or more new particles. These decays are often easily identified in images: a single track turns into several tracks.

Relativity plays a vital role in studying these ephemeral particles. An energetic particle with a lifetime of only one hundred millionth of a second – 10^{-8} seconds – before it breaks up into other particles, can in fact travel several metres before it does so, thanks to an effect in Einstein's special relativity called 'time dilation'.

This means that the faster a particle is travelling through space, the slower time elapses for the particle than for the laboratory-fixed experimenter who sees it fly past. The faster its speed, the greater is its time dilation; for a particle travelling at nearly the speed of light, time almost stands still. It is like the twin who ages less in a high-speed rocket than the sibling who stays at home. In this way, short-lived particles, such as pions and kaons, can be produced in high-speed beams that survive long enough to be useful in experiments.

A Journey to the Start of Time

It is some fifteen thousand million years since the Big Bang, four thousand million since life first began on Earth, yet only in the past hundred years have we discovered what our Universe is made of. But as the twenty-first century begins, our questions are turning from 'what' to 'why'. Why is there anything rather than nothing? Why do the fundamental particles have the masses they have? Why do the forces have their special strengths and properties? The range of experiments that are seeking the answers is extensive, in scope, style, size, and also geographically – the Sun never sets on particle physics!

Whereas in 1897 J.J. Thomson discovered the electron all by himself, using apparatus that was about 27 cm long, by 1997 physicists at CERN were speeding electrons around a ring of magnets that was 27 km in circumference. That is a measure of how the magnitude of science and technology has grown in a century. Now, as the new century begins, the most ambitious experiment in the history of physics is being prepared at CERN. The apparatus involves a new particle accelerator – the Large Hadron Collider or LHC – which will swing two counter-rotating beams of protons around the 27 km tunnel that previously housed the electron accelerator. The protons will pack a greater punch than the electrons, thereby probing deeper into the Big Bang than has been possible before. Huge detectors will catch the debris of millions of collisions, the raw material to analyse for answers to the questions that intrigue today's physicists.

To voyage to the start of time you have to build all the pieces for yourself: there is no customized 'Big Bang apparatus' for sale in the scientific catalogues. Protons

Fig. 1.18 (OPPOSITE) The tracks of many charged particles are made visible in this image from the NA35 experiment at CERN, Geneva. The particles emerge from the collision of an oxygen ion with an atomic nucleus in a lead target at the lower edge of the image. Tiny luminous streamers reveal their tracks as they pass through an electrified gas and curve under the influence of a magnetic field, positive particles bending one way, negative particles the other. Most of the particles are very energetic, so their paths curve only slightly, but at least one particle has a much lower energy, and it curls round several times in the detector, mimicking the shell of an ammonite.

Fig. 1.19 The 27 km long tunnel of the Large Hadron Collider (LHC), as it will appear in 2006 when it begins to collide together beams of protons at higher energies than ever before. The two counter-rotating beams will be guided by magnets within this pipe-like structure, which is designed to keep the magnets at their frigid operating temperature, only 1.9 degrees above absolute zero.

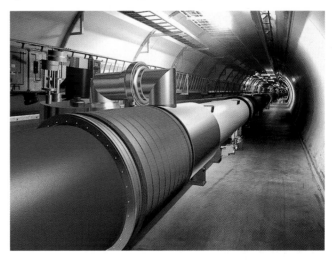

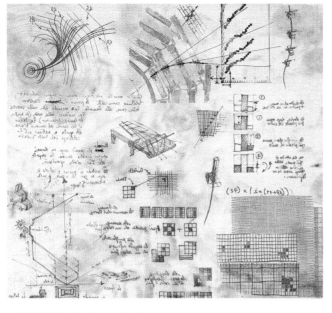

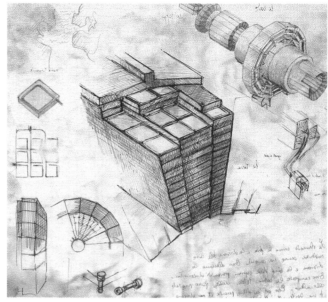

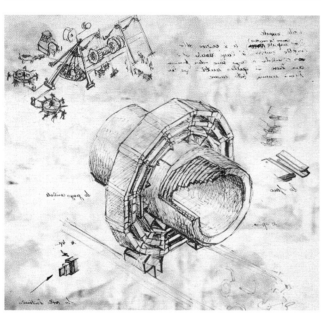

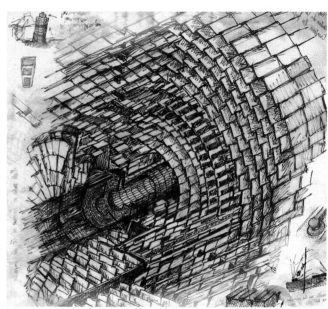

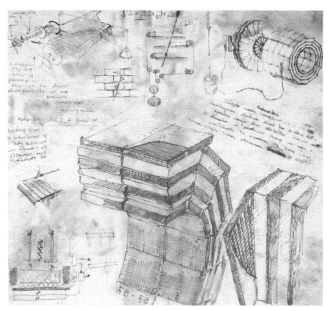

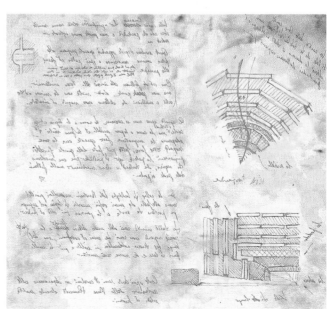

stripped out of hydrogen gas will provide the particle beams of the LHC. Ores dug from the ground are melted, the metals extracted and alloyed to make magnets capable of guiding the beams of protons at more than 99.999999% of the speed of light – so fast that they will make over 10 000 circuits of the 27 km ring every second. Sand provides the raw materials for the nervous system of the ubiquitous computer chips that will orchestrate the enterprise. Speeding beneath Swiss vineyards, the protons will cross the international border into France, scurry under the statue of Voltaire in the town where he spent his final years, rush beneath fields, forests, and villages, until they smash head on into protons that have been doing the same but in the opposite direction. Each collision will in effect create momentarily, in a small volume, temperatures not known since the first moments of the Universe.

Years ago, particle accelerators were known as 'atom smashers'. Today's accelerators, such as those at CERN, Fermilab, and a handful of similar laboratories around the world, might be better termed chronoscopes – time machines that are using pieces of atoms to mimic the condition of the new-born Universe. From such experiments we are on the threshold of discovering how matter came to be, and are even set to answer profound questions such as why there is any material Universe at all.

This book is the story of how a century of discovery and invention has brought us to our modern understanding of the subatomic particles and the nature of the material Universe. It is also a showcase of particle imagery, from early cloud chamber and emulsion photographs to the latest computer displays. These pictures show that the subatomic world is real and accessible; they also have their own peculiar beauty.

The Particle Odyssey is both a voyage through time and a journey to the heart of matter. Chapters 2, 4, 6, and 8 describe the history of particle physics over the past century, and the techniques developed to generate and study the particles. Chapters 3, 5, 7, and 9 provide individual portraits of all the major particles discovered by these techniques. Chapters 10 and 11 describe the questions that are absorbing particle physicists today and the experiments they hope will help to provide answers. Finally, Chapter 12 takes a look at how techniques and discoveries of particle physics have been put to work in society, from diagnostic scans in medicine to the invention of the World Wide Web.

Fig. 1.20 (OPPOSITE) Sketches by physicist Sergio Cittolin in the style of Leonardo da Vinci, complete with mirror writing, show aspects of the various component parts of the CMS detector, which is being built to record head-on proton collisions at the LHC. Like most experiments at colliding-beam machines, CMS (for Compact Muon Solenoid) will consist of different detector layers surrounding the central beam pipe. Clockwise from top left the illustrations show ideas for 'triggering' to sift out the tiny proportion of interesting collisions; sections of the 'hadron calorimeter' to measure the energy of particles such as protons; the layers of detectors to reveal the tracks of charged particles; the cover for the experiment's technical proposal; the outer layers to detect the penetrating particles known as muons; and the location of the cylindrical magnet within a segment of the outer detector layers.

Fig. 1.21 One of the tasks of the experiments at the LHC will be to search for clues to the origin of mass – one of the fundamental properties of particles. The most favoured theory involves a new particle – the Higgs particle – which is thought to interact with all other particles to give them their mass. This image shows how evidence for the Higgs particle might appear in the CMS detector. The various coloured dots and lines show the simulated tracks of the many particles produced in the head-on collision of two protons at the centre of the detector. Four particles, however, shoot out at large angles to the others, towards the top left and bottom right of the image. These are the tracks of penetrating muons, which have been produced in the decay of the Higgs particle created in the initial collision.

2. Voyage into the Atom

Take a deep breath! You have just inhaled oxygen atoms that have already been breathed by every person who ever lived. At some time or other your body has contained atoms that were once part of Moses or Isaac Newton. The oxygen mixes with carbon atoms in your lungs and you exhale carbon dioxide molecules. Chemistry is at work. Plants will rearrange these atoms, converting carbon dioxide back to oxygen, and at some future date our descendants will breathe some in.

If atoms could speak, what a tale they would tell. Some of the carbon atoms in the ink on this page may have once been part of a dinosaur. Their atomic nuclei may have arrived in cosmic rays, having been fused from hydrogen and helium in distant, extinct stars. But whatever their various histories may be, one thing is certain: most of their fundamental constituents – the electrons and quarks – have existed since the first split second of the Big Bang. Atoms are the complex end-products of creation.

At the end of the nineteenth century, the existence of atoms was little more than hypothesis. Today the reality of these tiny bundles of matter is accepted as indisputable. We know of many different types of atoms, one for each different chemical element – from hydrogen to uranium – that occurs naturally on Earth; and we have created and characterized in the laboratory atoms of at least 15 other elements heavier than uranium. We know that the atoms of all these elements are combinations of electrons, protons, and neutrons (except for atoms of the lightest element, hydrogen, which usually consist of a single electron and a proton, but no neutron). We understand the structure and behaviour of these atomic constituents in great detail, and how atoms link together to form molecules and complex organic and inorganic chemicals.

One of the most surprising features of atoms is that they contain an enormous amount of empty space in which the lightweight negatively charged electrons gyrate. By contrast, the massive positively charged protons and neutral neutrons are tightly bunched in a dense, central nucleus which is smaller relative to the atom's electron cloud than the hole is relative to a 500 m fairway on a golf course. The number of protons in the nucleus identifies the element. For example, hydrogen, the lightest element, has one proton; uranium, the heaviest naturally occurring element, has 92. The negative charge on each electron exactly balances the positive charge of each proton, so if the number of surrounding electrons exactly equals the number of protons, the atom will be electrically neutral overall.

The choreographer of the electronic dance around the nucleus – the ultimate controller of the atom – is the electromagnetic force. It binds the negatively charged electrons to the positively charged nucleus according to the rules of quantum mechanics, the theory developed in the 1920s that has proved triumphantly successful in describing and predicting subatomic events and processes. Quantum theory recognizes an inescapable limit in observing the subatomic world, which is enshrined in Werner Heisenberg's famous Uncertainty Principle. The precise path of any individual electron around a nucleus can never be known – the more we try to pin it down, the more it eludes us, like a subatomic will-o'-the-wisp. However, the average paths of millions of electrons in a million atoms can be described statistically with great accuracy. So, quantum theory replaces certainty with probability. Physicists sometimes speak of electrons forming a 'cloud' around the nucleus, but it would be more accurate to describe them as producing a blur, like the spokes of a rapidly revolving bicycle wheel. We cannot distinguish their individual motions, only the

Fig. 2.1 While we cannot distinguish the individual electrons in atoms, we can observe the average effects of their motions. This scanning tunnelling microscope image shows standing wave patterns of electrons in the surface of copper, caused by scattering from the ring of 48 iron atoms. The ring has a diameter of about 14 nanometres (0.000 014 mm).

generalized effect of repeated motions.

This understanding of the basic structure of the atom transformed the twentieth century. The exploration of the atomic nucleus led to the development of nuclear power and also nuclear weapons. The detailed understanding of the behaviour of electrons around the nucleus revolutionized the chemical industry and created electronics. This chapter describes the journey that scientists in the early twentieth century took deep into the atom; how the discovery of X-rays in 1895 led to the accidental discovery of radioactivity; and how that in turn led to a new view of the nature of the atom and the birth of nuclear physics.

X-rays and Radioactivity

In the latter part of the nineteenth century, the industrial revolution brought a new standard of living to Europe and North America. Machines performed tasks that had previously involved dirty, even dangerous manual labour. Nature was being tamed and exploited with the aid of science. At the same time, the development of new technologies provided the opportunity to extend the domains of scientific investigation, in particular into the nature of electricity.

Understanding electricity was one of the great scientific challenges of the nineteenth century, but its origins and properties were still largely a mystery. In the course of their investigations scientists passed electricity through all manner of substances, including gases. Earlier in the century, Michael Faraday at the Royal Institution in London had studied the beautiful glow that appears when an electric current flows through a gas at low pressure – a phenomenon common today in mercury and sodium streetlights. By the 1880s, new improved vacuum pumps, one of the many inventions of the nineteenth-century boom, enabled other scientists to follow up these investigations more thoroughly. The basic equipment was a glass tube with metal electrodes fitted at each end and a pump to remove most of the gas.

When an electric current passed between the electrodes, an eerie glow appeared in the rarefied gas left in the tube. As investigators pumped out more and more gas, they found that although the gas ceased to glow, the current continued to flow and a luminous spot appeared on the wall of the tube opposite the negative electrode – the cathode. Objects placed in the tube would cast shadows in this glow, showing that a stream of rays must

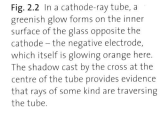

Fig. 2.2 In a cathode-ray tube, a greenish glow forms on the inner surface of the glass opposite the cathode – the negative electrode, which itself is glowing orange here. The shadow cast by the cross at the centre of the tube provides evidence that rays of some kind are traversing the tube.

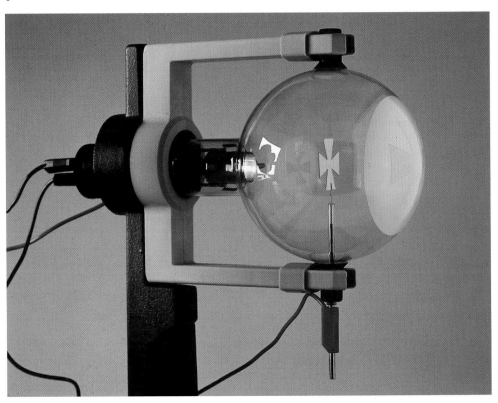

emanate from the cathode, causing the glass to glow only where they struck it. These emanations became known as 'cathode rays'.

One of the many people to study the strange lights in the cathode-ray tubes was Wilhelm Röntgen, who worked in Würzburg in Germany. In 1895, he inadvertently left some unexposed photographic plates, tightly wrapped, near his tube. Later, upon taking the plates out for use, he found that they were fogged. Moreover, when he repeated the sequence of events, he found the same results: the wrapped plates, unexposed to light, always became fogged when left near the cathode-ray tube.

One night as he was leaving his laboratory, Röntgen remembered that he had forgotten to switch off the tube. Returning to the room in the dark he noticed a glow coming from a sheet of paper on a nearby table. The paper was coated with barium platinocyanide, a substance known to glow in a strong light – but there was no light, and the cathode-ray tube was covered by thin black cardboard!

Röntgen realized that the cause of the glow must be the same as that of the fogged photographic plates: invisible rays of some unknown type must be coming from the cathode-ray tube. He called them 'X-rays'. He soon discovered their most startling property – their ability to penetrate as easily through many objects as ordinary light passes through glass. We now know that X-rays are light with a very short wavelength. Materials that are opaque to the longer wavelengths of visible light can easily transmit the shorter wavelength X-rays. The rays can for instance pass through skin and tissue, casting a shadow only when they meet more solid bone.

For his discovery of X-rays, Röntgen received the first Nobel prize for physics in 1901. By that time, popular magazines had seized on the bizarre photographs showing the inside of things, revealing a world previously unseen. The prudish Victorians even worried whether ladies could be seen naked beneath their layers of petticoats! For the scientists, however, X-rays provided a fascinating new phenomenon to investigate, and it led inadvertently to another discovery that was to have even more far-reaching consequences.

An early question concerned the origin of X-rays: were they unique to cathode-ray tubes, or were they emitted by all fluorescent materials – materials that glow on exposure to a strong source of light, such as the Sun? One person to investigate this was Henri Becquerel, a professor at the École Polytechnique and the Museum of Natural History in Paris, who came from a family of distinguished scientists. Several years earlier, while helping his father with an experiment involving a uranium salt, Becquerel had noticed that the crystals would

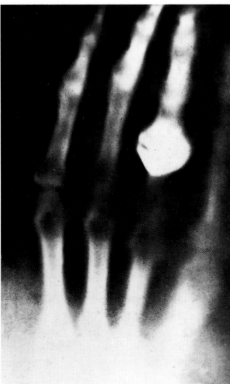

Fig. 2.3 (LEFT) Wilhelm Röntgen (1845–1923).

Fig. 2.4 (RIGHT) Röntgen's first X-ray photograph of a human shows the hand of his wife with the ring she was wearing.

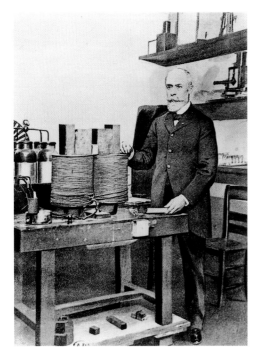

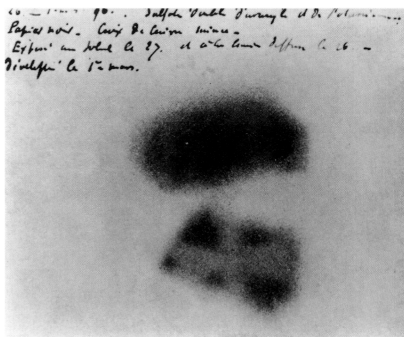

glow for some time after they had been removed from sunlight. He decided to find out if the same sample of salt, which he had inherited, emitted X-rays. All that was needed was to wrap a photographic plate in black paper, put a piece of the uranium salt on top, lay it in the sunlight for a while, and then develop the plate. In addition, he placed a metal cut-out pattern between the salt and the paper-wrapped plate. This would produce a shadow on the plate and leave no doubt as to the cause of any image produced.

On 26 February 1896, Becquerel prepared his experiment, but clouds shut out the Sun. So he put everything away in a drawer just as it was: plates wrapped in black paper; a metal cut-out on top; and finally the all-important uranium salt – impotent without the sunlight to stimulate it into action. Or so Becquerel thought.

The Sun did not come out for three days and on 1 March Becquerel decided to develop the plate anyway – presumably to prove that without the Sun there was no effect. He was astonished to find instead a very clear image. The uranium salt evidently gave out invisible rays even in pitch darkness.

Becquerel soon found that the rays emanated from the uranium in the salt, and he formed images from samples of pure uranium metal. He also found that the rays differed from Röntgen's X-rays in two crucial ways. First, the uranium rays did not penetrate materials. Second, uranium and its compounds emitted the rays spontaneously. Day and night, for weeks on end – and we now know for millennia – the uranium gave out its invisible rays. The X-rays, on the other hand, were produced instantaneously when cathode rays struck a material, such as the glass at the end of the cathode-ray tube.

Becquerel had discovered the phenomenon that was soon to become known as 'radioactivity'. Today the word conjures up a multitude of images, from the frightening fall-out of atomic bombs and the hazards associated with nuclear power stations on the one hand to treatments for cancer on the other. It became a byword of the twentieth century, and one that continues to arouse suspicion in many people. Yet radioactivity is a natural process, happening constantly all around us and even within us; and not only is it natural, it is also essential. Without radioactivity stars would not shine and the ingredients from which we are built would never have been formed. Moreover, it provides us with a window onto the fundamental nature of all matter – a window that scientists began to open soon after Becquerel's discovery.

One of these scientists was Marie Curie. As Manya Sklodowska she came from her native Poland to study in Paris in 1891. Her life was hard. She earned the money to rent a small room in an attic by cleaning apparatus at the Sorbonne and giving lessons. She was top student at the university and soon after graduating in 1895 she married Pierre Curie, a professor of physics, and started work in her husband's laboratory. When the Curies heard

of Becquerel's discoveries, Marie decided to investigate the new kind of radiation for her doctoral thesis. In particular, she wanted to know if uranium was the only element that emitted the rays and to quantify the amounts of radiation emitted by different substances.

Marie tested a vast number of materials and found effects from only one element apart from uranium – thorium. However, she also found to her surprise that raw, impure uranium ores showed more radioactivity than she could explain in terms of the uranium they contained. She suspected that the raw materials must contain something over and above uranium, a more powerful emitter yet. From one tonne of the uranium ore known as pitchblende, the Curies managed to extract a few grams of the culprits during 1898. Two new radioactive elements emerged: polonium, named after Marie's homeland, and radium – the most powerful radioactive substance known, which emits a million times more intensely than uranium.

Although Becquerel discovered radioactivity and the Curies isolated radium, it was Ernest Rutherford who honed their findings into a scientific tool, eventually using the new

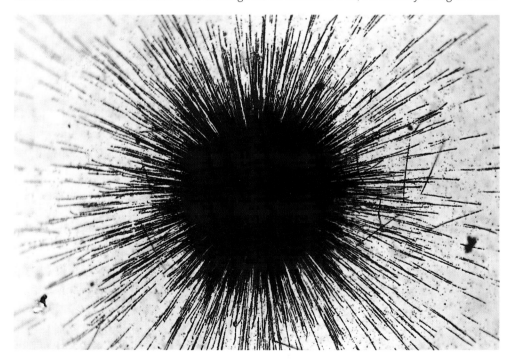

Fig. 2.9 Alpha particles shoot out from a speck of radium salt on the surface of a photographic plate covered with a special emulsion. The electrically charged alphas leave tracks in the emulsion, which appear as dark lines on the negative image formed on the developed plate. (The central blob is about a tenth of a millimetre across.)

Fig. 2.10 An early portrait of Ernest Rutherford (1871–1937).

radiations to bombard atoms and probe their inner secrets. His researches into the new phenomena began while he was still a young research student at Cambridge, where he had arrived in 1895 from his native New Zealand. At first Rutherford worked with J.J. Thomson, investigating the way that X-rays 'electrify' air, making it a good conductor of electricity at normal pressures. The two physicists found evidence that X-rays split the air into equal numbers of positively and negatively charged atomic particles, or 'ions'. Later, once he had established the existence of the electron, Thomson began to think of the positive ions as atoms with one or more of their electrons missing. This is indeed the case. X-rays and charged particles, such as electrons themselves, knock electrons out of atoms. In the case of an energetic charged particle, it loses a little of its energy at a time, creating a trail of ion–electron pairs in its wake. It is through making this ionization visible that we are able to 'see' the particle trails in many of the images in this book.

After the discovery of radioactivity, Rutherford turned to studying how the radiation from uranium could also 'ionize' air. He soon became more interested in the radiation itself, and began to use the ionization of gases as a means of studying radioactivity, rather than the other way about. The instrument that Rutherford generally used for these investigations was an electrometer. The details of the operation vary from one design to another, but the basic principle is to measure the deflection of a charged metallic strip in an electric field. If the air around the strip becomes ionized, the charge leaks away – a current flows – and the strip moves. Rutherford could measure the rate of leakage, and hence the amount of ionization, by timing the movement: the faster the leakage, the more the ionization and the stronger the radiation.

In the course of his studies, Rutherford covered a sample of uranium with sheets of aluminium foil, which absorbed the radiation. As he gradually increased the thickness of foil, at first he found that less and less radiation penetrated. This much he expected – the radiation is progressively absorbed. But more surprisingly, as he increased the thickness beyond about a hundredth of a millimetre, he discovered that the radiation maintained its intensity. Only when he had added several millimetres of foil did he find that the remaining radiation was absorbed. Rutherford concluded that there were in fact two types of radiation. One of these, which he named 'alpha', was absorbed very quickly; the other, which he called 'beta', was a lower-intensity, more penetrating radiation. He later looked at the radiation from thorium, and found an additional, extremely penetrating radiation. This became known as gamma radiation. The three emissions – alpha, beta, and gamma – are all quite different, as Rutherford and others soon discovered, but historically they all became known as rays or radiation (the words are interchangeable).

Fig. 2.11 The gold-leaf electroscope (left) was one of the earliest instruments used in studying electrical phenomena. It consists basically of a box with a window. A metal rod, which passes through an insulating collar in the top of the box, has two thin sheets of gold foil attached. When the rod is electrically charged, the two gold leaves acquire the same charge and repel each other. If the air in the box is ionized (for example by radiation), the charge on the leaves becomes gradually neutralized and the leaves collapse together. More advanced instruments, which allow amounts of electricity to be measured accurately, are called electrometers. The device shown here (right) is of the kind designed by Theodor Wulf (see p. 50). It contains two metallized quartz fibres held under tension, which repel each other when charged. The degree of separation between the wires can be measured by using the microscope attachment.

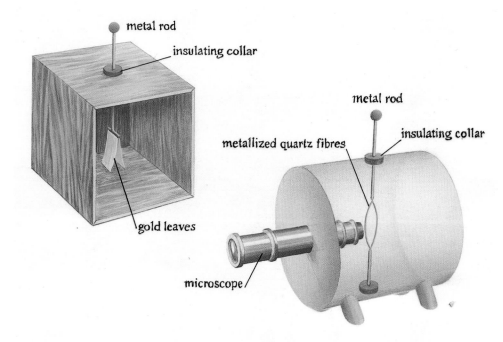

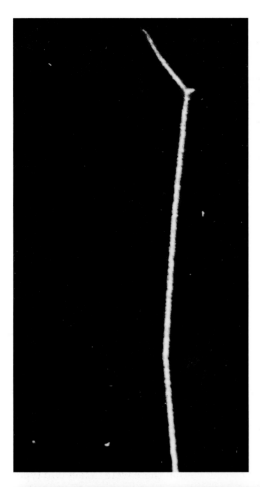

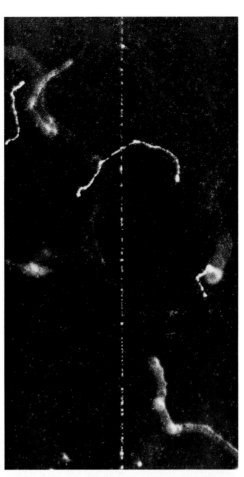

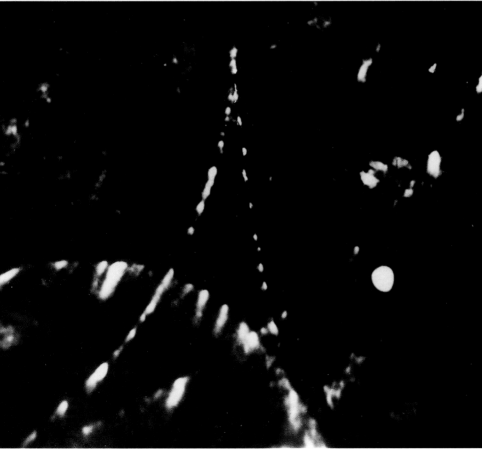

Figs. 2.12–2.14 Rutherford discovered that radioactive materials emit three distinct types of radiation – alpha, beta, and gamma – which produce different characteristic tracks in a cloud chamber (see p. 30).

Fig. 2.12 (TOP LEFT) The final portion of the track of an alpha particle. The track changes direction where the alpha collides with atoms in the air inside the chamber. Finally, close to the end, the track becomes fainter as the positively charged alpha particle captures electrons, losing its charge and hence its ability to ionize.

Fig. 2.13 (TOP RIGHT) Electrons – beta rays – have a much smaller mass than alpha particles and so have far higher velocities for the same energy. This means that fast electrons do not lose energy so readily in ionizing the atoms they pass. Here we see the intermittent track of a fast beta-ray electron. (The short thick tracks are not caused by the beta ray; they are due to other electrons knocked from atoms in the gas filling the chamber by invisible X-rays. Their tracks are thicker because they are moving more slowly than the beta ray and are therefore more ionizing; and they wiggle about because they are frequently knocked aside in elastic collisions with electrons in the atoms of the gas.)

Fig. 2.14 (BELOW) Gamma rays are non-ionizing and therefore leave no tracks in a cloud chamber. However, they can convert into equal amounts of matter and antimatter if they have high enough energy. Here an invisible gamma ray from a radioactive source materializes near the top of the picture as an electron and an antielectron (a positron); being oppositely charged, they curl away in opposite directions in the cloud chamber's magnetic field.

The First Particle

Fig. 2.15 Joseph John (J.J.) Thomson (1856–1940).

The trail that had led to the discovery of radioactivity had begun with the mystery of cathode rays, and while it had uncovered several new kinds of invisible radiation, it revealed little about the nature of the cathode rays themselves. By the mid-1890s there were two schools of thought as to what cathode rays might be – wave-like vibrations or energetic charged particles. In 1895 in Paris, Jean Perrin found that a magnet would deflect the fluorescent spot at the end of the tube. This indicated that the rays must also be deflected by the magnet, and the direction of deflection could be explained if the rays carried negative electric charge. But there was a general reluctance to believe that the rays could be a new type of charged particle. Then, in 1897, J.J. Thomson, professor of physics at Cambridge University, performed a series of experiments that were to prove conclusively that the cathode rays are indeed streams of particles.

Thomson had found that he could deflect the rays by electric fields as well as by magnetic fields. He was able to do this because he could produce a better vacuum than other investigators; the residual gas in a poor vacuum is sufficient to conduct electricity, so a static electric field cannot be maintained. By measuring the motion of the rays in both magnetic and electric fields Thomson came to a remarkable conclusion: the rays must consist of negatively charged particles with a mass approximately two thousand times less than the mass of a hydrogen atom, the lightest atom in the Universe. But since atoms at the time were considered indivisible, nothing lighter than a hydrogen atom was expected to exist.

Thomson obtained the same results irrespective of the material of the cathode or the gas in the tube. So he concluded that the cathode rays were 'matter in a new state, a state in which the subdivision of matter is carried very much further...this matter being the substance from which all the chemical elements are built up'. The new particles became known as electrons, and in 1906 Thomson was rewarded with the Nobel prize.

It is electrons that carry the electric current across a cathode-ray tube and give rise to the eerie glow. Electrons from the cathode gain energy in the electric field along the tube. They can pass this energy on to atoms in collisions in the rarefied gas in the tube, and these 'excited' atoms then divest themselves of the extra energy by emitting light: the gas glows. Once the gas pressure is low enough, however, the electrons can travel along the tube without colliding at all, so the main glow disappears and the cathode rays leave only a fluorescent spot where they strike the opposite end of the tube.

Later, other experimenters were able to show that the beta rays emitted in radioactivity are also electrons. The electrons of beta rays are indistinguishable from those found in atoms, but their origin is different. We now know that they are ejected from the nucleus within the atom, and that they move so fast they penetrate sheets of lead a millimetre thick. For some time physicists thought that the beta-ray electrons were actually present within the nucleus, but this is not so. Beta-ray electrons are created by changes within the nucleus and immediately ejected; they are no more part of the nucleus than a dog's bark is part of the animal.

Thomson's revelations provided the first evidence that atoms are not like featureless billiard balls but have a complicated inner structure. However, his discovery also raised a question. If atoms contain negatively charged electrons, then there must also exist positive charges to render the atoms neutral overall. Where are these positive charges in the atom? How can we ever hope to look inside minute atoms and see them? The tool to answer all these questions was radioactivity, and the man who used it to such advantage was Ernest Rutherford, who had been Thomson's student in Cambridge.

Rutherford and the Atom

In 1898, Rutherford left Cambridge for McGill University in Montreal, where he continued his research into the radioactivity from uranium and thorium. While studying thorium, he found that the amount of radiation produced seemed to vary and be very sensitive to currents of air. After a detailed series of experiments, he came to the conclusion that the thorium was emitting something that was also radioactive. He referred to the unknown substance as an 'emanation' and found that it remained radioactive for only a short time, quite unlike thorium or uranium. Rutherford was convinced that the emanation was a gas, but he needed the assistance of a chemist to analyse it properly. To this end he enlisted the talents of the young Frederick Soddy, newly arrived at McGill from Oxford.

Rutherford the physicist and Soddy the chemist together made a formidable team. In a series of detailed investigations they found conclusive proof that the emanation was not only a gas, but was chemically quite different from thorium and more akin to the unreactive 'inert' gases, such as argon. It was in fact a new element, which is now known as radon. With this discovery, that the element thorium could produce a different element, radon, Rutherford and Soddy had found the first amazing evidence of the transmutation of one element into another. This was alchemy at work, but naturally.

Still more surprises were in store. Further work by Rutherford and Soddy showed that there are several steps in the transmutation of thorium to radon. At each stage one element turns into another, spitting out radiation, mainly alpha rays. By 1902, Rutherford was able to demonstrate that the alpha rays must be fragments of matter. He showed that a strong magnetic field bends the paths of alpha rays and concluded that the rays must consist of positively charged particles. Here was proof that heavy atoms such as thorium change from one type to a slightly lighter type by ejecting tiny atomic fragments.

In 1907, Rutherford returned to England from Montreal to become professor of physics at Manchester University. There he attracted all sorts of people to work with him. His origins in New Zealand freed him from the class-consciousness of Edwardian England. The brilliant and the best came from the north of England, and others joined from overseas. In Rutherford's team we find the makings of the international research group – the norm today but a very new idea in the first decade of the twentieth century.

Among those in Rutherford's team at Manchester was a young German, Hans Geiger, who is famed today for the 'Geiger counter', which he was to develop in the 1920s. Using a forerunner of that counter, he and Rutherford were able to proceed a step further towards discovering the nature of the alpha rays. Rutherford suspected that the positive particles he had detected at McGill were positive ions – atoms with electrons knocked out – of helium, the second lightest element after hydrogen. This idea was strengthened by the discovery of helium gas in association with radium. To settle the question, Rutherford needed to be able to detect alpha particles one at a time, so that he could measure their exact charge and mass.

Fig. 2.16 Frederick Soddy (1877–1956).

Fig. 2.17 Rutherford (right) and Hans Geiger in their laboratory at Manchester University.

The apparatus Rutherford and Geiger used to tackle this problem was located in a cellar in the physics department at Manchester. The key feature of their detector was that it could greatly amplify the tiny amount of ionization caused by the passage of a single alpha particle. It consisted of a brass tube some 60 cm long, with a thin wire passing along the centre, which was pumped out to a low pressure. The wire and tube had 1000 volts applied between them. The voltage set up an electric field, which became much stronger nearer the wire. When an alpha particle passed through the rarefied gas, the ions created were attracted towards the wire. Nearer the wire, where the field became stronger, the ions would move faster and in turn ionize more gas, amplifying the initial effect. One ion could produce thousands of ions, which would all end up at the central wire. There the ions would produce a pulse of electric charge large enough to be detected by a sensitive electrometer connected to the wire.

Geiger and Rutherford used their device – which nowadays we would call a 'proportional counter' – to count individual alpha particles coming down a narrow tube from a thin film of radium. From this they could calculate how many alpha particles were radiated by the whole sample of radium, and compare the answer with the total charge emitted, as measured with the electrometer. The calculations showed that the charge of an alpha particle is twice that of the hydrogen ion, and this in turn indicated almost certainly that alphas are helium ions.

To prove that single alpha particles were entering the wire detector, Rutherford, with characteristic thoroughness, looked for a different technique. The answer came in a letter from Otto Hahn, who had worked with Rutherford in Canada, but was now in Berlin. Hahn described how a colleague, Eric Regener, had been detecting alpha particles by letting them hit a screen coated with zinc sulphide. When a particle struck, the screen emitted a flash of light – a phenomenon known as 'scintillation'. Inspired by this work, Rutherford and Geiger built improved zinc sulphide screens and were astonished to find the technique every bit as good as the electrical methods they had been using. By combining the two techniques they were able to prove that they were detecting individual alpha particles.

Later in the same year, 1908, Rutherford confirmed that alpha particles are indeed helium ions by collecting some in a tube, and allowing them to neutralize themselves by picking up electrons from their surroundings. In this way he collected atoms of a gas, which he could stimulate into emitting light, in the same manner as a sodium lamp. The spectrum of this light provided a fingerprint that identified the gas as helium without a doubt. Rutherford announced this result in his speech when he was awarded the Nobel prize in 1908 – not for physics, but for chemistry – for his work with Soddy on transmutations.

Today we know that alpha particles are the *nuclei* of helium atoms. There was however no way that Rutherford could make this last step in 1908 as the idea of the nucleus was still unknown. That was to be his next dramatic contribution: deducing what the inside of the atom looks like.

At McGill, Rutherford had noticed that when a beam of alpha particles passed through thin sheets of mica, they produced a fuzzy image on a photographic plate. The alphas were apparently being scattered by the mica and deflected from their line of flight. This was a surprise because the alphas were moving at 15 000 kilometres per second, or one-twentieth the speed of light, and had an enormous energy for their size. Strong electric or magnetic fields could deflect the alphas a little, but nothing like as much as when they passed through a few micrometres (millionths of a metre) of mica. This suggested that there must be unimaginably powerful forces at work within the atoms of the mica sheet.

In 1909, Rutherford assigned to Ernest Marsden, a young student of Geiger's, the task of discovering if any alphas were deflected through very large angles. Marsden used gold leaf rather than mica, and a scintillating screen to detect the scattered alphas. He placed the screen not behind the gold foil, but to the side, next to the radioactive source. In this way he could detect alphas reflected back through large angles.

The effort in counting the scintillations was enormous. At each angle individual flashes were observed through a low-power microscope focused on the screen. The flashes were faint and sparse and could be counted by eye only in a darkened room. This placed a great strain both on the observer's eyes and on his powers of concentration. Work would continue for only a few minutes at a stretch, so Rutherford, the master, assisted Marsden, the student. One watched and one recorded for a few minutes only and then they changed places.

Fig. 2.18 Ernest Marsden (1889–1970) in 1911.

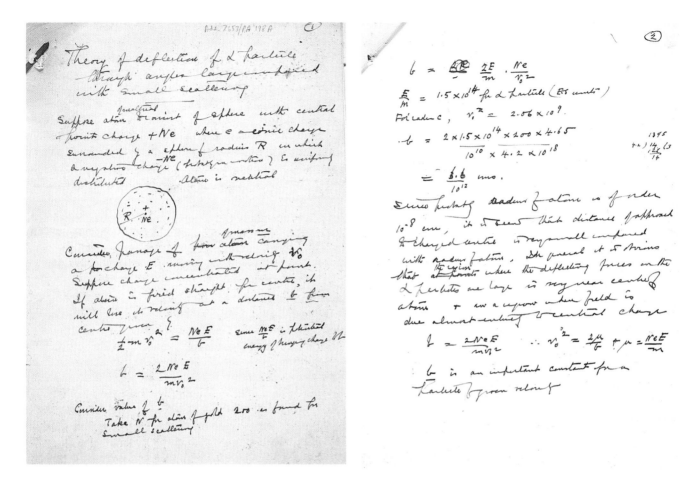

To the amazement of Rutherford and Geiger, Marsden quickly discovered that about 1 in 8000 alphas bounced right back towards the source, reflected back by more than 90°. This was an incredible result. Alpha particles, which were hardly affected at all by the strongest electrical forces then known, could be turned right round by a thin gold sheet only a few hundreds of atoms thick! No wonder that in later life Rutherford exclaimed, 'It was as though you had fired a 15-inch shell at a piece of tissue paper and it had bounced straight back and hit you.'

At first neither Marsden, Geiger, nor even Rutherford could understand these results at all. Geiger took up some earlier work again, and Marsden left the team for a while to do some research on the atmosphere at a meteorological station; but Rutherford kept on puzzling, his normal output of scientific papers falling dramatically. Then late in 1910, with the aid of a very simple calculation, Rutherford at last saw the meaning of the results. The key was that he knew the energy of the incoming alphas. He also knew that each alpha particle carries a double dose of positive charge. The positive charge within the gold atoms must repel the approaching alphas, slowing them and deflecting them. The closer the alphas approach the positive charge in the atom, the more they are deflected, until in extreme cases they come to a halt and are turned round in their tracks.

Rutherford could calculate just how close to the positive charge the alphas should get, and the result astounded him. On rare occasions the alpha particles come to within 10^{-12} cm of the atom's centre, one ten thousandth of the atom's radius, before they are turned back. This means that the positive charge is found only at the very centre of the atom, not distributed throughout the atom as Thomson, for one, had surmised.

Rutherford had discovered that atoms consist of a compact positively charged nucleus, around which circulate the negatively charged electrons at a relatively large distance. The nucleus occupies less than one thousand million millionths of the atomic volume, but it contains almost all the atom's mass. If an atom were the size of the Earth, then the nucleus would be the size of a football stadium. The atom's volume is mostly empty space.

Fig. 2.19 These pages from Rutherford's notebook show where he calculates how close an alpha particle must approach the positive charge within an atom if it is to be turned back completely in its tracks. The answer, 6.6 x 10^{-12} cm, astonished Rutherford for it showed that the positive charge is concentrated in a tiny core deep within the atom. This is the moment the atomic nucleus was 'discovered'.

Inside the Nucleus

Fig. 2.20 Alpha particles of the same energy have the same range, and radioactive materials emit alphas at one or more specific energies. Here the majority of alphas from a source of thorium C' (polonium-212) travel 8.6 cm in a cloud chamber filled with air before stopping, while a single higher-energy alpha travels 11.5 cm.

Fig. 2.21 Four high-energy protons cross a bubble chamber (see p. 92) but the one on the right collides with a nucleus in the liquid and knocks it to the right. The angle between the paths of the two particles is 90°, indicating that they have the same mass – in other words, the nucleus is also a single proton, so the liquid in the chamber is liquid hydrogen. (The angle appears less than 90° because of the perspective from which the photograph is taken.)

Rutherford's discovery that the positive charge of an atom is concentrated in a central nucleus raised the question of what precisely carries this charge. The negative charge of the atom is carried by the tiny electrons; are similar objects responsible for the positive charge?

The experiments at Manchester had penetrated the atom to reveal the nucleus, but they had not probed the nucleus itself. Rutherford realized that gold nuclei have a relatively large positive charge, in fact nearly 40 times greater than the positive charge of an alpha particle. This means that an approaching alpha particle begins to feel a strong repulsive force long before it reaches the nucleus at the heart of the gold atom. However, in light atoms, with smaller nuclei and fewer positive charges, the repulsive force should be less powerful, and an alpha particle should make a closer approach to the nucleus.

So Rutherford and Marsden turned to firing alpha particles through hydrogen gas, the lightest element of all. They expected that the alphas would all come to a halt at more or less the same distance from the radioactive source, as they were all emitted with the same energy. Beyond this distance – the 'range' of alphas in hydrogen – the particles should no longer penetrate the gas to strike a zinc sulphide screen.

However, Marsden and Rutherford found that as they moved the screen beyond the range of the alphas a few scintillations *did* occur, up to four times further through the gas. A magnetic field deflected the culprits, and the direction of the deflection showed that they must be positively charged particles. Rutherford argued that the new particles – which he called 'H particles' – could be nothing other than the nuclei of hydrogen, knocked out from atoms in the gas by the energetic alphas. The hydrogen nuclei, each carrying only a single positive charge compared with the double charge of the alphas (helium nuclei), could travel four times as far through the gas.

Marsden later noticed similar long-range particles when he was measuring the distances alpha particles travel in air, and he wondered if they too could be H particles. However, this was in 1914 and the First World War interrupted his work. Geiger had returned to Germany, Marsden departed to become a professor in New Zealand, and many of the students went off to the war. Rutherford became involved in work on submarine detection for the Board of Inventions and Research, although he was able to continue with a little of his own research.

By 1917 Rutherford decided that Marsden had indeed seen H particles, chipped out of nitrogen atoms in the air in the detector. In similar experiments, Rutherford had used alpha particles to knock H particles – hydrogen nuclei – out of atoms of six different elements: boron, fluorine, sodium, aluminium, phosphorus, and nitrogen. He concluded that hydrogen nuclei must form part of the nuclei of all elements, and he named the particles 'protons', as they were the first nuclear building bricks to be discovered.

Rutherford had found the carriers of the positive charge in the nucleus, but puzzles remained. The nucleus also contains most of the atom's mass – about 99.95% – so the protons in the nucleus should presumably account for all this mass. A nucleus with twice the charge of another should have twice the mass. But this is not so. Nuclei have at least double the mass expected from the number of protons suggested by the total charge.

To account for this discrepancy, Rutherford speculated in 1920 that there are electrically neutral particles within nuclei – 'neutrons'. But he was alone in this idea. The picture that most physicists accepted was of a nucleus containing protons and *electrons*. The theory was that the nucleus contains twice as many protons as there are electrons in remote orbits; half the protons are neutralized by these electrons, while the other half are neutralized by electrons *inside* the nucleus. The phenomenon of beta

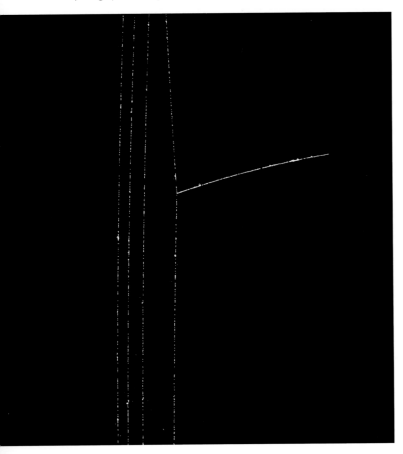

decay, in which electrons are emitted from the nucleus, gave strong support to this notion.

The first indications that Rutherford might be correct came in 1930 in experiments by Walther Bothe and Herbert Becker in Germany, though they did not realize the significance of their work. They bombarded beryllium with alpha particles from polonium and observed the emission of an extremely penetrating neutral radiation, which they assumed to be gamma rays.

This work was soon followed up by Irène Curie, daughter of Marie and Pierre Curie, and her husband Frédéric Joliot. They found the same neutral radiation, and observed that it had the power to knock protons out of paraffin wax – a substance rich in hydrogen. Like Bothe and Becker, however, they believed the radiation to be gamma rays, although they were surprised how readily it could scatter heavy protons. The Joliot-Curies published their results in January 1932, and their paper had an immediate impact at the Cavendish Laboratory in Cambridge.

Rutherford had returned to Cambridge as Cavendish Professor in 1919, on the retirement of his old master J.J. Thomson. There he began increasingly to direct the researches of younger scientists, rather than do experiments himself. One of these was James Chadwick, who had worked with Rutherford in Manchester and then with Geiger in Berlin, before being interned in Germany during the First World War. In 1919, he had rejoined Rutherford at Cambridge, and in the following years, among other research, made several unsuccessful attempts to search for neutrons. As soon as he heard of the Joliot-Curies' results early in 1932, Chadwick realized that the neutral radiation from beryllium was not gamma radiation at all, but neutrons.

To prove this, Chadwick allowed the neutral rays to collide with a variety of gases – hydrogen, helium, and nitrogen. In this way he could observe the differing amounts by which the atomic nuclei in the various gases recoiled, and so calculate the mass of the individual neutral projectiles. He found that they had more or less exactly the same mass as the proton; gamma rays, by contrast, have no mass. This made it clear that nuclei contain not only positively charged protons but also neutral neutrons. Chadwick was rewarded with the Nobel prize in 1935 for his discovery of the neutron – just one of several Nobel prizes that Rutherford's group in Cambridge earned in the course of revealing atomic structure.

Fig. 2.22 James Chadwick (1891–1974) in 1935.

Fig. 2.23 Apparatus Chadwick used in his discovery of the neutron. Alpha particles from a polonium source, at the right end, bombarded a beryllium target at the left end. A penetrating, neutral radiation – neutrons – emerged from the beryllium.

Splitting the Atom

The insidious nature of neutrons had at first been hidden from the physicists' view, but it was soon to lead directly to the most well-known – and contentious – phenomenon associated with the atomic nucleus. Unlike alpha particles, neutrons can readily penetrate the nucleus; being neutral, they are not repelled by its positive charge. Late in 1938, Otto Hahn and Fritz Strassman in Germany discovered that if they directed neutrons at uranium, the particles split the uranium nucleus in two. This process of nuclear fission not only releases energy, it also frees further neutrons, which can in turn trigger the fission of neighbouring nuclei, leading to the possibility of a chain reaction. When the process is properly controlled, we have the release of useful nuclear energy; when uncontrolled, the chain reaction will multiply catastrophically, and we have one of the most destructive weapons the human race has invented – the atomic bomb.

Nuclear fission is the phenomenon that most people associate with the term 'splitting the atom'. Less well known is that in 1932, the *annus mirabilis* at the Cavendish Laboratory, Rutherford's group had split the atom in a less dramatic way, soon after Chadwick's discovery of the neutron. This achievement was the culmination of a quest to break into the nucleus that had driven Rutherford for years, but the story begins even earlier, in September 1894, when a young researcher at Cambridge first developed a device that was to make nuclear reactions visible.

Charles Wilson was working in the meteorological observatory on the summit of Ben Nevis and became fascinated by the beauty of coronas – coloured rays around the Sun – and glories, where the Sun glows around shadows in the mist. Back at the Cavendish Laboratory, he decided to investigate these phenomena more closely. To do so he needed a ready-made mist, so he built a glass chamber fitted with a piston and filled with water vapour. When he withdrew the piston quickly, the sudden expansion cooled the gas so that a mist formed in the cold damp atmosphere.

In the course of his investigations, Wilson found that if he made repeated small expansions of the chamber – without allowing in fresh air – the mist would disappear. He could explain this because he knew the droplets in the mist formed on specks of dust; on repeated expansions the droplets would slowly sink to the bottom of the chamber and so remove the dust, leaving nothing on which further droplets could form. The surprise came when, having cleared the chamber in this way, Wilson made large expansions. He found that these always produced a thin mist no matter how many times he expanded the chamber. But what could the droplets be forming around? Wilson surmised that they were condensing on the electrically charged particles, or ions, known even in the 1890s to cause

Fig. 2.24 Charles (C.T.R.) Wilson (1869–1959).

Fig. 2.25 Wilson's first cloud chamber. The chamber itself is the squat glass cylinder at the top left of the picture; the coils are where an electric field was applied to clear away stray ions between expansions. Below the chamber is a cylinder containing the piston. The glass bulb to the right was pumped out to a low pressure. When a valve between the bulb and the chamber beneath the piston was opened, air would rush into the bulb, causing the piston to fall and the air in the glass cylinder to expand suddenly. Water vapour in the air would then condense out on any ions present, so making the ionized tracks of particles visible.

conductivity in the atmosphere.

Wilson was soon able to test his theory with the newly discovered X-rays – and confirmed that their ionization of the air in his chamber caused an immense increase in condensation. However he soon abandoned the experiments with his 'cloud chamber', and turned instead to work on atmospheric electricity. He did not return to his device until 1910 when, using alpha and beta radiation, he saw for the first time the tracks of individual particles, which he described as 'little wisps and threads of clouds'. Cloud drops formed instantly around the ions produced by the radiation and when illuminated the tracks stood out like the dust motes in a sunbeam.

Wilson's cloud chamber provided the first visible records of the motion of particles smaller than an atom, and he was rewarded with the Nobel prize in 1927. The technique, meanwhile, had been seized upon by a man at Cambridge with a passion for gadgets. Patrick Blackett had adapted Wilson's basic idea and devised a chamber that expanded automatically every 10–15 seconds and took a picture on ordinary cine film.

Between 1921 and 1924, Blackett obtained more than 23 000 photographs of alpha particles bombarding nitrogen in a cloud chamber. While in many photographs the alphas shot through the gas without interruption, there were several where a nitrogen nucleus had deflected the alpha particle as they bounced off each other like billiard balls. But most exciting of all were eight precious examples, which were quite different. In each of these, the track of a proton was clearly visible, more lightly ionized than the alpha track because of the proton's smaller charge. Also visible was a short stubby track, similar to that of a nitrogen nucleus; but there was no sign of the recoiling alpha particle. The conclusion was that the alpha had become bound to the nitrogen to make a form of oxygen, leaving the lone proton to continue on its way. The alpha particle had modified the nitrogen – nuclear transmutation had been captured on film.

To be caught by the nitrogen nucleus, the alpha particle must have forced its way through the electric field surrounding the nucleus. The highest-energy alpha particles are only just powerful enough to do this. Rutherford noticed, however, that faster alpha particles are more penetrating than slower, less energetic ones, and he became interested in somehow making alphas travel even faster, at greater energies than nature provides, to create a tool that would probe deep into the atomic nucleus.

At first, Rutherford's goal seemed out of reach. Charged particles are accelerated by an electric field – this is what happens to electrons in the cathode-ray tube. But to achieve energies similar to those of alpha particles from radium, for example, would require electric fields of several million volts, far beyond the technology of the 1920s. However, in 1928 a paper by the Russian theorist George Gamow arrived at the Cavendish Laboratory, and

Fig. 2.26 One of the first photographs of the tracks of ionizing particles in a cloud chamber, obtained by Wilson early in 1911. The tracks are due to alpha particles.

Fig. 2.27 (BELOW LEFT) Patrick Blackett (1897–1974).

Fig. 2.28 (BELOW RIGHT) An example of a nuclear transmutation induced by an alpha particle in Blackett's cloud chamber in 1925. Alpha particles travel up the picture. Most continue for the full length of their range, but the one on the far left interacts with a nitrogen nucleus in the air in the chamber. The alpha is captured, and a nucleus of a heavy isotope of oxygen – ^{17}O – is formed, accompanied by a proton. The proton shoots off to the right, leaving a faint track; the recoiling oxygen nucleus leaves the thick track to the left, and collides again before halting.

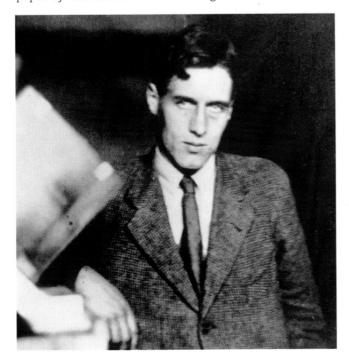

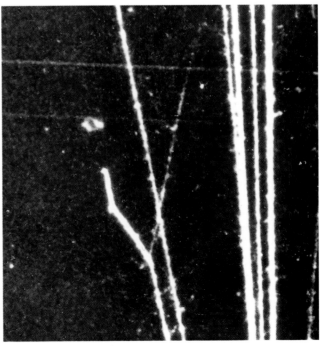

there it was read by John Cockcroft, who realized immediately that there was an alternative way to achieve Rutherford's dream.

Gamow showed that the indeterminate nature of quantum theory implied that even at relatively low energies alpha particles could sometimes penetrate the repulsive field around the nucleus. Cockcroft realized that Gamow's paper meant that millions of volts were not necessary to produce nuclear transmutations artificially; lower voltages would do provided you had an intense enough beam of alphas, or protons for that matter. He calculated that protons accelerated by 300 000 volts (300 kilovolts, or 300 kV) could penetrate the nuclei of boron at least 1 in every 1000 times. With Rutherford's backing and a grant of £1000 from Cambridge University, Cockcroft set about building a 300 kV proton accelerator. He was joined by Ernest Walton, a young man from Dublin who had already tried various schemes for accelerating particles.

To produce their protons, Cockcroft and Walton used an electric discharge in hydrogen gas to strip electrons from the atoms. The lone protons (hydrogen nuclei) laid bare in this way were attracted towards a negatively charged plate, and drawn off through a small hole. Beyond the hole another negative electrode drew the protons towards the far end of an evacuated glass tube, where they were directed onto a suitable target. The voltage between the ends of the tube provided the electric field to accelerate the protons, just as the electrons are accelerated in a cathode-ray tube.

Initially, with a system working at 280 kV, Cockcroft and Walton found no evidence for nuclear transmutations when their accelerated protons struck the target. So they began to try higher voltages using a voltage multiplying system that Cockcroft had designed, which could take their apparatus up to 800 kV. Late in April 1932 they achieved their goal. When they directed their high-energy protons onto a target of lithium they saw at once bright scintillations on a screen set up to detect any charged particles produced in nuclear reactions. The inference was that the high-velocity proton had been absorbed by a lithium nucleus and that the new compound had then split into two alpha particles. The three protons and four neutrons of the lithium nucleus and the additional accelerated proton had rearranged themselves into two helium nuclei – alpha particles – each with two protons and two neutrons. Rutherford described the sight as 'the most beautiful in the world'.

The technology of 'atom splitting' – the application of our knowledge about the atomic nucleus – has provided us with nuclear power and weaponry. It has also led to the development of artificial radioactive species, which are used as tracers in medical diagnosis, and to cancer therapy using beams of particles from accelerators. The science of nuclei has also progressed, broadening into two complementary branches. One branch – nuclear physics – deals with the structure of the nucleus, and the complex behaviour of the conglomeration of protons and neutrons that form it. The other branch – elementary particle physics – involves the quest to understand the deeper structure of matter, the simplest building blocks with which the Universe began, and the fundamental forces that mould them together. This is the story that unfolds in the following chapters.

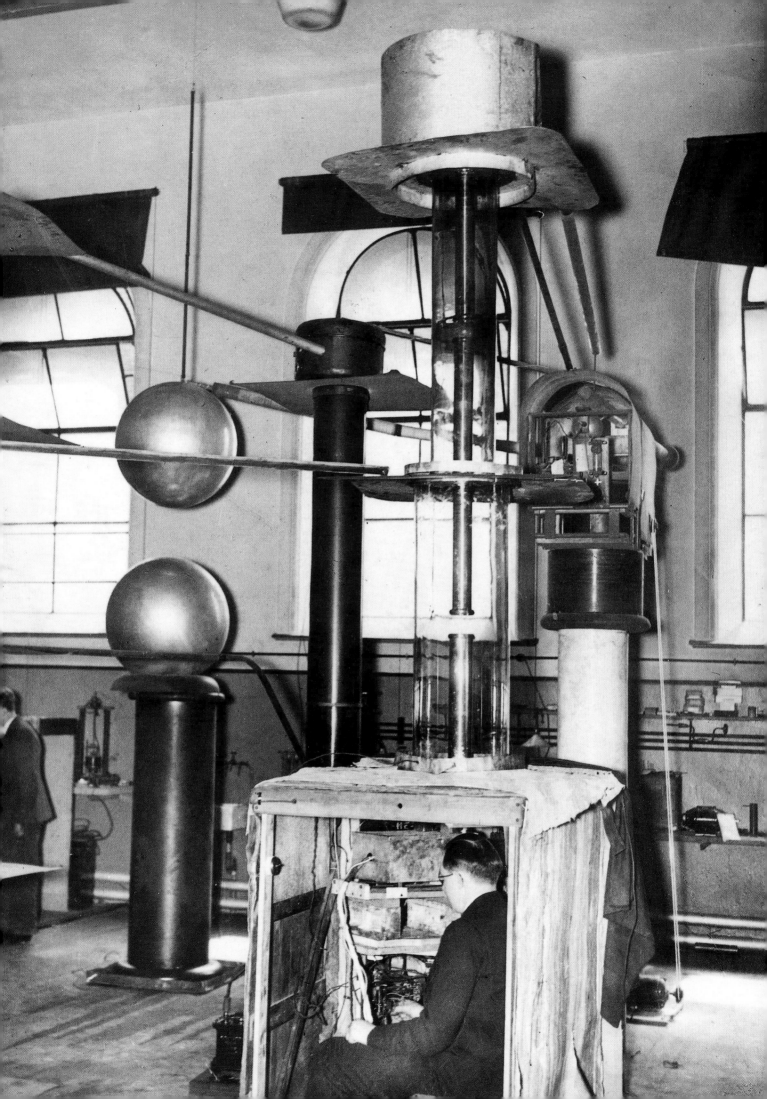

3. The Structure of the Atom

The idea that matter is made of invisibly small particles was first put forward in the fifth century BC by the Greek philosopher Leucippus. He called them *atomos*, meaning indivisible, to indicate that they were the elementary constituents of matter. By the end of the nineteenth century AD, atoms had become reality, as the building blocks of the chemical elements. But it was also becoming clear that the atoms were not indivisible and by 1932, physicists had discovered three basic constituents of atoms – the electron, the proton, and the neutron. They had also come to appreciate that the electromagnetic force, which binds the electrically charged constituents together, is carried by the photon – the particle of familiar light. The photon plays a crucial role in atoms; it is in a sense the 'mortar' that holds together the 'bricks' in building the matter of the everyday world.

The atoms and their inhabitants are unimaginably small, so how can we ever hope to see them, and picture what the atom looks like? The experiments in Manchester, where high-energy alpha particles were turned back in their paths by thin gold foil, led to the inescapable conclusion that the atom resembles a solar system in miniature. Modern physicists still use this picture as a kind of verbal and visual metaphor, but while this is a useful way of imagining the atom, it does not explain what twentieth-century physics revealed about the behaviour of electrons in atoms.

The Danish physicist Niels Bohr first developed the theoretical aspects of the 'planetary' atom in detail. By 1913 he had formulated a detailed description of electrons in atoms which explained the spectra of light emitted by certain atoms, notably hydrogen, with great success. But Bohr's model was only the beginning of a revolution in physics in the 1920s, which led to a theory of subatomic behaviour that continues to this day to underlie our understanding of the behaviour of atoms and their constituents. This theory is quantum mechanics. At the level of the constituents, it provides the basic ground rules to explain the behaviour of the particles – how they interact electromagnetically by 'exchanging' photons in a game of 'quantum catch'. And at the level of atoms, it yields the modern, intrinsically fuzzy picture of electron 'clouds' in an atom, which show not where an electron is exactly, but where the electron is most likely to be.

This does not mean that individual electrons – or protons or neutrons – cannot be studied or in some sense 'seen'. The early experiments that initially revealed these particles often used detectors that produced no lasting visual image. However, the detectors did provide accurate information about the quantities, energies, directions, and so forth, of the invisible particles, and this allowed experimenters to build up a picture of the unseen world. Rutherford used electrometers in much of his early research on radioactivity before turning to scintillation techniques, which allowed charged particles to be counted as they hit a special screen. This was how he discovered the nucleus at the heart of the atom.

In the 1920s, Rutherford became a champion of the cloud chamber, the first device to show the ionized tracks of individual charged particles. Later, in the 1940s, physicists brought to perfection another technique, which records the tracks of a charged particle directly in special photographic emulsion (see pp. 59–61). Once developed, the layers of emulsion provide a record of the particle's track. Visual techniques like these do not show us the particles themselves, but they do create 'footprints' – like those of an animal in the snow – which allow us to recognize different particles. In this chapter we meet the inhabitants of the atom, and discover how varied their footprints are.

Fig. 3.1 (OPPOSITE) An 'electron tree' produced in a block of plastic (15 cm square by 2.5 cm thick) by a beam of electrons. The electrons initially penetrate about 0.5 cm into the block and stop. As the number of electrons builds up, however, their mutual electric repulsion begins to force them apart. If the beam is switched off before this happens, a subsequent small tap to the block with a metal punch releases the electrons suddenly and they shoot out, rather like lightning, leaving a pattern of tracks that becomes 'frozen' in the plastic. (The colour has been added.)

Fig. 3.2 Niels Bohr (1885–1962) in 1920.

In human terms, electrons are the most important of all subatomic particles. Electrons in the outer reaches of atoms give structure to the Universe: they are the means by which atoms are bonded together to form molecules and, eventually, large aggregates of matter such as ourselves and everything around us. The achievements of modern chemistry and biochemistry, from the invention of plastics to the synthesis of new drugs and the manipulations of genetic engineering, are ultimately based on our detailed understanding of the behaviour of electrons in atoms.

Remove electrons from atoms, set them in motion, and you have an electric current: a stream of electrons flowing through a substance capable of conducting them, such as the copper of electric wiring. Electrons not only provide us with electricity, they are the basis of electronics – from televisions to microchips. Electronics, in fact, is defined as the applied science of the controlled motion of electrons.

The electron was discovered by J.J. Thomson in his experiments on cathode rays in 1897, as described in Chapter 2. But 40 years elapsed before physicists came to understand the behaviour of electrons in detail. We now know that electrons are stable, lightweight, negatively charged elementary particles. They appear to be stable in the sense that when unperturbed they can live forever; unlike most of the particles described in this book, they do not transmute spontaneously into other particles. They also seem to be elementary, in that they do not consist of smaller entities, in contrast to the other constituents of atoms – protons and neutrons – which are now known to be built from smaller particles called quarks. All the evidence suggests that electrons are truly indivisible, and are at least as small as physicists can currently measure – less than 10^{-18} m across.

An important characteristic of the electron is its lightness. It weighs in at 9.1×10^{-31} kilograms. To understand how light this is, consider the difference between the weight of

Figs. 3.3–3.4 Images of electrons knocked from atoms at different speeds. Like many pictures that follow, these black-and-white cloud chamber photographs have been coloured to identify the tracks of particular particles.

Fig. 3.3 (LEFT) The track of an alpha particle (blue) in a cloud chamber shows wispy side shoots (red) due to electrons. Only when the alpha comes very close to an atomic electron does the electron acquire sufficient velocity to form a short trail. Such energetic electrons, which in this image travel as far as 2.9 mm, are known as 'delta rays'.

Fig. 3.4 (RIGHT) A high-energy cosmic ray particle (green) shoots into a cloud chamber and knocks an electron (red) out of the gas, giving it enough energy for the electron to leave a long track as it curls under the influence of a magnetic field. The track of the cosmic ray is about 7 cm long.

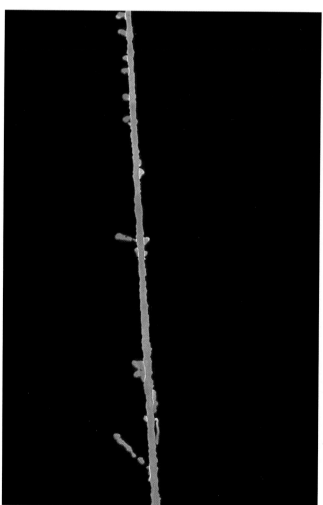 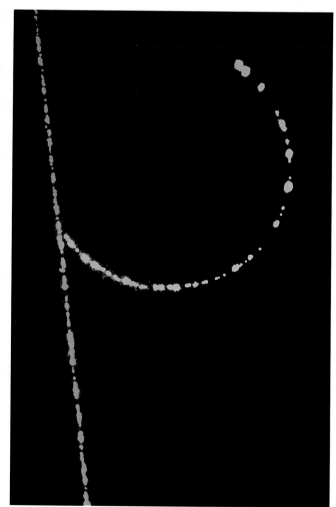

a small bag of sugar (1 kg or 2.2 lb) and 15 million planets the size of Earth. The same ratio exists between the weight of an electron and the bag of sugar. For physicists, however, the vital statistic is that an electron weighs 1836 times less than a proton. Partly because they are so light, electrons can easily be deflected from their paths or knocked out of their positions in atoms. In Fig. 3.3, an alpha particle emitted in a radioactive decay is seen knocking electrons from atoms of the gas that fills a cloud chamber. The electrons leave short squiggly tracks as they scatter off other atoms.

The paths of electrons are also easily bent by magnetic fields. Figure 3.4 shows a single electron being knocked out of an atom by the passage of a cosmic ray. The electron's track is longer than in Fig. 3.3 because it has been given more energy in the collision; it is curved because the chamber is subject to a magnetic field. In Fig. 3.5, a powerful magnetic field causes the track of an energetic electron to curl round many times. This cloud chamber photograph shows the coiled track that is typical of electrons, and which we will see in many other images in this book.

A single electron carries the smallest quantity of electric charge ever detected in isolation – 1.6×10^{-19} coulombs. This is a trifling amount, but its impact is tremendous. The strength of the electrical force between electrons and protons in an atom is comparable to the gravitational force between bulk matter: the electrical forces within a microgram speck of dust are equal to the Earth's gravitational pull on an object weighing several hundred tonnes. If there were a sufficient excess of electrons on matter in the Sun and Earth, the Sun would repel the Earth and send it spiralling out of the Solar System.

Electrons exist both on their own, as free particles, and as constituents of atoms, and they can change from one role to the other and back. An electron forming part of a carbon atom in the skin of your wrist could be knocked out of position by a passing cosmic ray and become part of the tiny electric current in your digital wristwatch, and then in turn become part of an oxygen atom in the air you breathe as you raise your arm to look at the time.

Fig. 3.5 Under the influence of a magnetic field, an electron in a cloud chamber spirals around some 36 times, producing a track about 10 m long. The electron starts its life at the left of the picture, where it has been created together with an antielectron, or positron, by an invisible gamma ray. The electron's spiral moves slowly across the page due to a slight variation in the magnetic field. Notice how the spiral becomes significantly tighter about half way across the picture; this is because the electron has lost energy by radiating a photon.

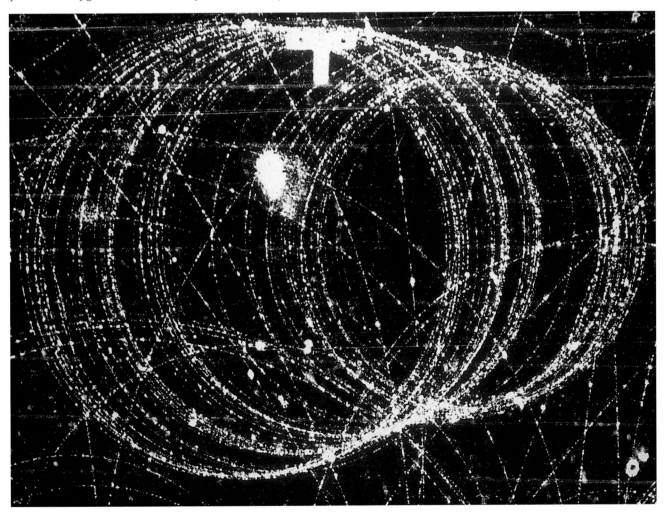

Fig. 3.6 Electric currents – the familiar manifestation of the motion of electrons – become visible in a 'plasma sphere' filled with gas. Electricity discharges from the terminal at the centre as electrons move away towards an outer glass sphere (unseen), colliding with atoms in the intervening gas. In the collisions, the 'current' electrons lose energy to electrons in the gas atoms. These 'excited' atoms soon lose their extra energy, radiating it as light to make glowing coloured streamers along the paths of the electric currents.

Figure 3.6 shows the visible effect of the continual interactions between electrons in currents and electrons in atoms. Tiny electric currents dance between the central region and the outer glass wall of a 'plasma sphere', making the intervening gas glow eerily. The electrons in the currents knock electrons from atoms in the gas; other electrons move in to occupy the empty locations in the atoms, and lose energy in the form of light as they do so.

In hot and violent parts of the Universe – inside stars or in the superhot shells of gas expanding after a supernova explosion – the particles of matter tend to disintegrate into their constituent parts and to be in constant flux. Stable atoms are rarely or never formed. Nuclei stripped bare of electrons mingle with clouds of free electrons, in a state of matter known as 'plasma'. In comparison, conditions on Earth are cold and electrons and nuclei are 'frozen' into the relative stability of atoms.

Where did electrons come from in the first place? Physicists believe that most of them were created out of intense radiation in the first instants after the Big Bang, when the Universe was still very hot. They were not created alone, but together with their antimatter equivalents – positively charged electrons or 'positrons' (see pp. 66–68). Such 'electron–positron pairs' are still produced all the time out of intense radiation, both naturally and in physicists' laboratories. Electrons are also created in the beta-decay form of radioactivity. Indeed, the so-called 'beta rays' emitted when atomic nuclei decay in this way are nothing other than electrons, no different from the electrons in an electric current or in the atoms that make up our bodies.

For over a century, electrons have proved to be extremely useful tools in exploring the nature of matter, because they are electrically charged and lightweight and therefore easy to accelerate. An electron in an electric field experiences a force and is accelerated; this is precisely the effect that impelled the cathode rays across the glass tubes that Faraday, Thomson, and others used. Nowadays we see a similar effect every time we watch television. Electric fields inside the television tube accelerate the electrons towards the screen, where they stimulate atoms to emit flashes of light. Millions of electrons in coordinated bursts produce the visual image on the screen. A similar process is at work in an electron microscope, which records the patterns of energetic electrons scattered from the specimen in its field of view. With bigger 'tubes' and stronger electric fields, physicists can use electrons to probe still further into matter, firing electrons right at the heart of the atom – the nucleus.

The Nucleus

Strip an atom of its electrons and you are left with its nucleus, a compact bundle that occupies only a thousand million millionth of the atom's volume but which provides 99.9% of its mass. The nucleus also provides the positive electric charge that balances the negative charge of the electrons to make the atom neutral overall. The electrons in the outer reaches of an atom determine its external relations – how it bonds with other atoms – but it is the nucleus that determines the atom's nature.

The nucleus is more than just a core. It is an entirely new level of reality, where the forces of electromagnetism and gravity that govern atoms, molecules, and larger conglomerations of matter play only a minor role. In the nucleus, different forces are at work, which are unfamiliar in the wider world: the 'weak' force governs beta radioactivity, and the 'strong' force holds the constituents of the nucleus together.

All nuclei except for the hydrogen nucleus, which consists of a single proton, are compound entities built from both protons and neutrons. This complex inner structure is illustrated in Fig. 3.7, which shows what happens when the nucleus of a sulphur atom (16 protons and 16 neutrons) collides head on with a nucleus in photographic emulsion. The two nuclei shatter into a multitude of fragments, although only those that are electrically charged produce tracks in the picture.

Fig. 3.7 In this colour-coded positive print of a cosmic ray collision in photographic emulsion, a cosmic ray sulphur nucleus (red) collides with a nucleus in the emulsion, shattering into a multitude of fragments. The electrically charged fragments leave visible tracks in the emulsion. They include a fluorine nucleus (green) and other nuclear fragments (blue) as well as several short-lived particles (yellow) known as pions (see p. 73). The track of the incoming sulphur nucleus is about 0.11 mm long.

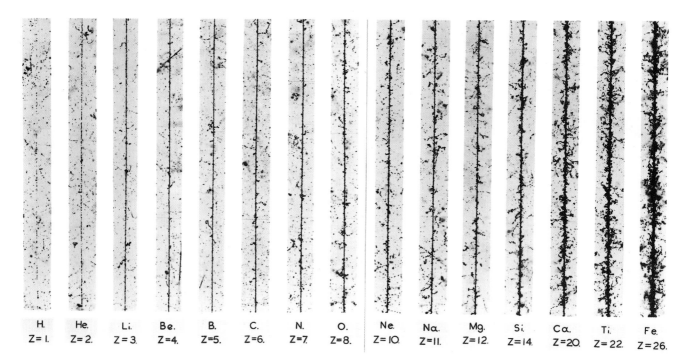

H.	He.	Li.	Be.	B.	C.	N.	O.	Ne.	Na.	Mg.	Si.	Ca.	Ti.	Fe.
Z=1.	Z=2.	Z=3.	Z=4.	Z=5.	Z=6.	Z=7.	Z=8.	Z=10.	Z=11.	Z=12.	Z=14.	Z=20.	Z=22.	Z=26.

Fig. 3.8 The more protons a nucleus contains, the higher its positive charge and the greater its ability to ionize; hence the denser its trail in a detector. Here are the tracks of 15 different nuclei, captured in emulsions exposed to the primary cosmic radiation high in the atmosphere. Hydrogen (H), with a single proton and therefore only one unit of positive charge, leaves a barely visible track. But because the ionization depends on the square of the charge, the track of lithium (Li), with only three protons, is nine times stronger and stands out clearly. Iron (Fe) has 26 protons, enough to produce a track 676 times stronger than hydrogen. The wispy side-strands are due to delta rays – electrons knocked from atoms in the emulsion, which have enough energy to produce their own short tracks as in Fig. 3.3 (p. 36).

Positive protons contribute both electric charge and mass to the atomic nucleus, whereas the neutral neutrons contribute mass alone. Thus the number of protons determines the charge of an atom, while its mass depends on the sum of its protons and neutrons. Each proton carries one unit of electric charge. When we look at the Periodic Table of the elements, we find that they are ranked according to the number of protons in their nuclei, from hydrogen and helium, with one and two protons respectively, to the 'transuranic' elements, which go beyond 100 protons. By 2001, the list had extended to element 116, the heaviest so far.

The nuclei of heavy elements, with a large amount of electrical charge, exert a correspondingly large electrical influence on matter. When such nuclei pass through a photographic emulsion, for example, these powerful electrical forces ionize atoms and leave behind a trail of sensitized emulsion, which then appears on the developed image. Thus the greater the charge of the nucleus, the thicker the track that it leaves in the emulsion.

This effect is seen in Fig. 3.8, which shows the tracks of 15 nuclei, from hydrogen to iron. With its single proton, hydrogen is all but invisible; the 26 protons of the iron nucleus, on the other hand, leave a thick track liberally bedecked by the curlicues of electrons knocked out of atoms in the emulsion as the iron nucleus ploughs through them.

The atom, as we have seen, is held together by the electromagnetic force: the positive nucleus attracts the negative electrons. The more protons there are in the nucleus, the greater its positive charge, and the more electrons it is able to attract. But what holds the nucleus itself together? Why do the protons, all with the same electric charge, not repel each other and cause the nucleus to disintegrate? The answer lies with the strong force, which holds the nucleus together despite the mutual electromagnetic repulsion of its protons. Within the nucleus the strong attraction is over 100 times more powerful than the electromagnetic disruption.

The strong force does not distinguish between protons and neutrons. In other words, neutrons and protons attract one another with the same strength as either one attracts its own kind. The neutrons, having no electrical charge, are not subject to electromagnetic repulsion, so extra neutrons help to stabilize nuclei. They provide sources of strong attraction for the protons and help them to fight the disruptive electrical force.

This is why nuclei tend to have more neutrons than protons. Too great an excess of neutrons, however, will destabilize a nucleus. If such a nucleus is created – in high-energy collisions, for instance – then it becomes more stable by the process of beta decay, in which a neutron converts into a proton, at the same time emitting an electron and a particle known as a neutrino. For a particular element, characterized by a specific number of protons, there may be several relatively stable nuclei with different numbers of neutrons.

These are called isotopes. One example is uranium, which occurs in nature mainly as uranium-238; its 92 protons are accompanied by 146 neutrons, making 238 'nucleons' in total. Less than 1% of naturally occurring uranium is uranium-235, which with its 143 neutrons, is the one responsible for the chain reactions in nuclear reactors and bombs.

Protons and neutrons in nuclei arrange themselves in 'shells', rather as electrons do in atoms, and the two systems follow very similar rules. In atoms, the shell with the lowest energy is full when it contains two electrons. In nuclei, it is full when it contains two neutrons and two protons. This makes an exceptionally stable configuration; in fact, it is the nucleus of helium – the alpha particle. The stability of alpha particles accounts for their appearance in radioactive decays, particularly of heavy elements such as uranium and thorium. These elements become lighter, or more stable, by shedding protons and neutrons in alpha-particle clusters.

The alpha particles emitted by natural radioactivity were used in the early part of the twentieth century as tools to probe the structure of nuclei, as Chapter 2 describes. Figures 3.9–3.11 show some classic pictures of alpha-particle scattering, made visible through tracks in a cloud chamber.

Figure 3.9 shows what happens when an alpha particle collides with the nucleus of one of the hydrogen atoms filling a cloud chamber. With its two protons and two neutrons, the alpha is four times as massive as the single proton of the hydrogen nucleus. As a result the proton is propelled forwards, while the alpha is only slightly deflected. The angle between them is less than 90°. This all accords with experience if you have ever played shove-halfpenny: a big coin knocks a small one forwards. (Note also that the proton, with its single unit of electric charge, causes less ionization and so produces a less dense track than the doubly charged alpha.)

What happens when identical coins collide, or two billiard balls (assuming that they are not given any spin or 'side')? They move off at 90° to one another. We see this in Fig. 3.10, where the cloud chamber is filled with helium instead of hydrogen. An alpha particle entering from below bounces off a helium nucleus, and the outgoing tracks produce a fork with an angle of 90°. This is because helium nuclei have the same mass as alphas; indeed, experiments of this kind helped to show that alpha particles are nothing more than the nuclei of helium atoms. (The tracks in fact appear to diverge at a little less than 90° because they are being viewed at a slight angle.)

In Fig. 3.11, the cloud chamber contains nitrogen. A nitrogen nucleus consists of seven protons and seven neutrons, and when an alpha particle entering from below collides with one head on, it is turned back in its tracks, at the same time transferring most of its energy to the nitrogen nucleus, which moves a short distance forwards. The angle between their tracks is 142°.

Figs. 3.9–3.11 Patrick Blackett obtained clear evidence for the way that alpha particles scatter from nuclei of differing mass in a series of classic cloud chamber experiments in the late 1920s. The photographs have been coloured to identify particular tracks, with interacting alpha particles in yellow and non-interacting ones in green.
Fig. 3.9 (LEFT) An alpha particle collides with a proton in the hydrogen gas filling a cloud chamber. The proton (red) shoots off to the right, while the heavier alpha is deflected only slightly from its previous direction, so that the angle between the tracks is noticeably less than 90°.
Fig. 3.10 (CENTRE) An alpha particle bounces off a nucleus in helium gas in a cloud chamber, the two particles moving off at right angles to each other. The 90° angle reveals that alpha particles and helium nuclei have the same mass; they are one and the same thing. (Note that the angle appears less than 90° because the tracks have been photographed at an angle.)
Fig. 3.11 (RIGHT) An alpha particle bounces back from a much heavier nucleus in the cloud chamber gas, this time nitrogen. The alpha gives much of its energy to the nitrogen, which moves onwards, leaving a thick track (red) owing to its relatively high positive charge from seven protons. The deflected alpha also leaves a thick track because it is now moving much more slowly and therefore produces more ionization. Here the angle between the tracks is much greater than 90°.

The Proton and the Neutron

By the 1930s, Rutherford and his team at the Cavendish Laboratory had shown that there are two types of particle that build up atomic nuclei: protons and neutrons. Electrically they are quite different – the proton is positively charged, while the neutron is neutral. But in other respects, the two particles are almost indistinguishable, and they are often regarded as charged and neutral versions of the same basic particle – the 'nucleon'. We now know that the nucleons are not elementary in the sense that electrons are; they are themselves built from more basic particles – 'quarks'.

This structure is reflected in the size of the nucleons, which are about 10^{-15} m across, at least 1000 times bigger than electrons or quarks. For many purposes, especially in studies of the behaviour of complex nuclei, it is useful still to regard the proton and neutron as the basic constituents of the nucleus, just as it is useful to consider atoms as simple objects when studying large-scale molecular behaviour.

In some senses the proton is more basic than the neutron, for free neutrons ultimately decay to protons – indeed, many of the particles that will be introduced later in this book decay ultimately to protons. With a mass of 1.6726×10^{-27} kg, the proton is the lightest member of a family of particles – the baryons – built from three quarks. It carries a positive electric charge, which so precisely balances the negative charge of the electron that atoms and bulk matter are normally electrically neutral. This precise equality of charge, which is essential for the stability of matter, is profound because the electron and proton appear to be quite different forms of matter and yet conspire in this delicate way.

Fig. 3.12 A proton (yellow) enters a bubble chamber from bottom right and scatters from other protons (also yellow) in the liquid hydrogen filling the chamber. Each time the angle between the scattered particles is close to 90°, revealing the equality of their masses. The tracks of particles not involved in this game of subatomic billiards are left uncoloured (grey); the spirals are electrons knocked from hydrogen atoms in the liquid.

Fig. 3.13 The tracks of a 1.6 MeV proton (red) and a 7 MeV alpha particle (yellow) curl round in a cloud chamber in a magnetic field. The curvature is proportional to charge and inversely proportional to momentum (the product of mass and velocity). So the alpha particle, with double the charge of the proton but four times the mass, curls less in the magnetic field. The particles were produced in collisions of a beam of 90 MeV neutrons from the 4.6 m (184 inch) cyclotron at the Lawrence Berkeley Laboratory. The thin blue horizontal lines are wires, which produce an electric field to remove old ions, while the thick black rectangle at the bottom of the picture is part of the structure of the cloud chamber.

Protons are the nuclei of the simplest chemical element – hydrogen. This is made clear in Fig. 3.12, where an energetic proton enters a detector called a bubble chamber (see p. 92) which contains liquid hydrogen. The proton eventually scatters from a hydrogen nucleus (a proton) in a 'billiard ball' collision, and the two particles move off at 90°, revealing their equal masses. (Once again, the viewing angle makes this look like less than 90°.) The 90° scattering angle occurs again and again as more and more protons are set in motion. It is also easy to see the difference between a proton and an alpha particle – the nucleus of helium. Both types of particle are positively charged and bend the same way in a magnetic field (Fig. 3.13), but although the alpha has double the charge of the proton, the proton has only a quarter of the alpha's mass. Consequently, the proton bends more easily in the magnetic field.

Until recently, physicists believed that protons are completely stable and live forever. The latest theories, however, suggest that they may decay after an enormously long period of time. The question is, how long? We know that they must live on average for more than 10^{17} years, otherwise our bodies would be highly radioactive. This is because particle lifetimes are only an average, and a human body contains so many protons – roughly 10^{27} – that many would decay even during a 70 year lifespan. Subtle experiments on proton decay show that protons must in fact live for at least 10^{32} years, which is 10^{22} times longer than the estimated age of the Universe.

The availability, stability, and electrical nature of protons makes them a favourite tool for particle physicists. Intense beams of protons can be accelerated and fired into matter and used to study nuclear behaviour under extreme conditions. This complements experiments using electrons because protons interact by both the electrical and strong forces, whereas electrons do not experience the latter.

Neutrons are about 0.1% heavier than protons; they weigh 1.6749×10^{-27} kg. The mass of a neutron in fact exceeds the total mass of a proton and an electron, and this is sufficient in certain circumstances to make neutrons unstable. An isolated neutron decays into a proton and an electron, on average after about 15 minutes. This is the basic process of beta radioactivity. In some nuclei, neutrons also decay this way, but in others, subtle nuclear effects tip the scales and enable the neutrons to survive, leading to a stable nucleus. Whereas protons leave visible trails, neutrons are like H.G. Wells's invisible man, who gave away his presence indirectly – by jostling the visible crowd. If an invisible neutron bumps into a proton and sets it in motion, we can detect the trail of the proton. An example is in Fig. 3.14, where a single neutron has struck a proton in a paraffin sheet. The proton shoots out across the cloud chamber. It is clear that something bulky has come in – protons do not

Fig. 3.14 One of the first photographs of a proton recoiling after being struck by a neutron, taken by Frédéric Joliot and Irène Curie in 1932. An invisible neutron, knocked out of beryllium by an alpha particle, enters a cloud chamber from below. It strikes a sheet of paraffin wax across the chamber (the white horizontal line) and knocks out a proton, which shoots up the picture. The small gap at the start of the proton's track, above the paraffin sheet, occurred because drops did not condense on ions created near the paraffin.

spontaneously fly off without cause.

Figure 3.15 shows what happens when an intense beam of artificially energized neutrons, coming from the bottom of the picture, enters a cloud chamber. The hydrogen in the chamber has in this case been mixed with a little alcohol, which is basically a mixture of carbon (six protons and six neutrons) and oxygen (eight protons and eight neutrons). These nucleons like to clump together as alpha particles, making three alphas in the case of carbon, and four for oxygen. The energetic neutrons entering the chamber can break up these 'alpha clusters', forming the series of four-pronged stars which are seen one above the other in this colour-coded picture.

The neutron's neutrality can be an advantage. Whereas protons are repelled initially by the electrical forces surrounding nuclei, neutrons feel no such disruption. Slow neutrons can freely approach and enter nuclei, modifying their internal structure and creating new isotopes. This is the key to several technologies, such as the production of special radioactive isotopes for medicine.

Another consequence of the neutron's penetrating powers is its ability to split a uranium-235 nucleus into two fragments, releasing nuclear energy and two or three additional neutrons in the process. These neutrons can in their turn split further nuclei of uranium-235, releasing more energy and more neutrons. In a sufficiently large lump of uranium-235, a chain reaction will occur in which the multiplying neutrons cause the fission of ever more nuclei, leading to an explosive release of energy. This is how the atomic bomb works.

A single fission of uranium (with no chain reaction!) has been captured on film in Fig. 3.16. Neutrons enter a cloud chamber containing a thin uranium-coated foil running down the centre of the picture. Many protons and nuclei recoil. But there are also thick tracks due to the two fragments of the divided nucleus – the two 'fission products' – which leave the foil in opposite directions. Their high electrical charge produces strong tracks, which end in characteristic branches where the fission fragments hit nuclei in the chamber's gas.

Fig. 3.15 (OPPOSITE) A narrow beam of 90 MeV neutrons, produced by the 4.6 m (184 inch) cyclotron at the Lawrence Berkeley Laboratory, enters a cloud chamber from the bottom. The chamber is filled with hydrogen gas saturated with a mixture of ethyl alcohol and water. While the neutrons are invisible, their effects certainly are not. Neutron-induced transmutations of oxygen and carbon in the alcohol produce three 'stars', one above the other. Tracks not involved in these interactions have been removed; those remaining have been coloured according to which star they belong to.

Fig. 3.16 Neutrons entering a cloud chamber knock on many protons and nuclei, leaving short tracks. But one neutron has induced fission in a uranium nucleus in a thin layer coating a gold foil down the centre of the chamber. The two bulky, heavily ionizing fission fragments move out sideways, producing long tracks with short branches towards their ends. These are due to nuclei from the chamber gas, which have been knocked by the fission fragments. The chamber has a diameter of 25 cm.

The Photon

Our senses perceive the visible light and warm rays from the Sun as a continuous stream of radiation. So it comes as no surprise to learn that light and heat (infrared), and all the other forms of electromagnetic radiation, can be described in terms of vibrating waves – oscillations of intertwined electric and magnetic fields. Yet the quantum theories of the twentieth century have provided us with another view of light, as a staccato burst of particles of zero mass called photons. This is true all the way across the electromagnetic spectrum, from high-frequency (short-wavelength) gamma rays through to low-frequency (long-wavelength) radio waves. Frequency and wavelength describe waves of light, but photons are described in terms of their energy and momentum. Gamma rays consist of high-energy photons, radio waves of low-energy photons.

Still more surprising is that photons are responsible for transmitting electromagnetic forces. The electric forces that hold atoms together are in effect 'carried' by photons, which flit back and forth between the atom's electric charges. These photons within the atom are transitory entities, existing on timescales too fine for us to perceive with our slow, macroscopic senses. Physicists call them 'virtual' photons. This description of electromagnetic forces in terms of photons is a consequence of quantum theory. First defined by the British physicist Paul Dirac in 1928, it is now formulated in the theory known as quantum electrodynamics, or QED, which was worked out in the late 1940s by two Americans, Richard Feynman and Julian Schwinger, and Sin-Itiro Tomonaga in Japan. In 1965, these three theorists shared the Nobel prize for their work. Figure 3.17 shows examples of the diagrams that Feynman invented as an aid to calculating detailed electromagnetic effects through the exchange of photons between charged particles. The predictions of QED have been verified many times and to a precision of better than one part in a billion. So precise and powerful is QED that it has become the template for subsequent theories of other fundamental forces, such as the weak and strong forces.

The photon belongs to a class of elementary particles, distinct both from the electron and its relatives, and from the quarks that make up nucleons and other subatomic particles. It is an example of what is now called a 'gauge boson'. It acts as a mediator and force-carrier between particles. And just as the photon 'carries' the electromagnetic force, so other gauge bosons carry the other fundamental forces – the W and Z particles in the case of the weak force, the gluons in the case of the strong force between quarks, and the hypothetical graviton in the case of gravity.

If we shake an atom, by heating it or firing electrons or other charged particles at it, we can shake the photons loose from the atom's electric fields. When this happens, photons

Fig. 3.17 Richard Feynman invented diagrams like these to help in calculating the electromagnetic interactions of charged particles. The lines and vertices in the diagrams are a stylized shorthand for detailed mathematical expressions describing the behaviour of the particles. In general, charged particles are indicated by straight lines, photons by wiggly lines. The top diagram could represent a muon (on the left) kicking an electron (right) out of an atom by exchanging a photon (wiggly line). This is the kind of process that has occurred in Fig. 3.4 (p. 36). The centre diagram shows an electron and a positron (the two straight lines on the left) annihilating at A and producing a photon (wiggly line), which rematerializes at B as new forms of matter and antimatter, such as an electron and a positron or a muon and an antimuon. The bottom diagram illustrates an electron emitting a photon at A, absorbing a second photon at B, and then reabsorbing the first photon at C. This kind of process makes a measurable contribution to the behaviour of an electron in a magnetic field.

Fig. 3.18 Photons of sunlight of specific energy (and therefore wavelength) are absorbed by elements in the Sun's atmosphere, and so do not reach Earth. When the Sun's spectrum is photographed, these missing wavelengths show up as black 'absorption lines'. This image shows the complete spectrum across visible wavelengths, divided into 50 slices. Each slice covers a range of 6 nanometres (nm) wavelength, so the complete image spans from 400 nm (red) to 700 nm (blue).

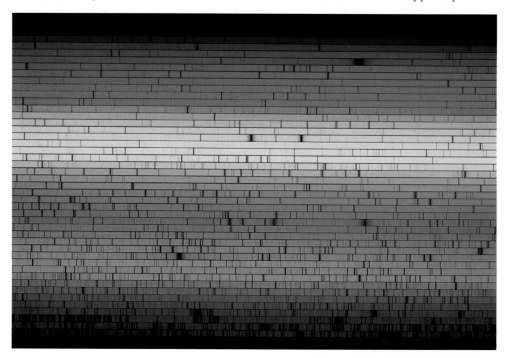

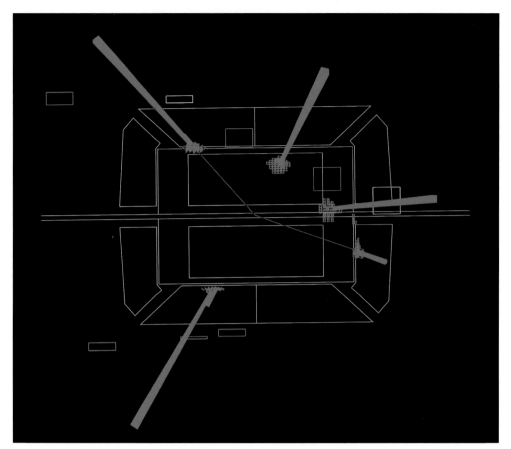

Fig. 3.19 Three high-energy photons create the bright red 'towers' in this computer display from the L3 detector at the Large Electron–Positron (LEP) collider at CERN. An electron and an antielectron (a positron) have travelled unseen in opposite directions along the beam pipe (depicted across the centre of the display) and collided at the centre of the detector, which is seen here in schematic outline. A new electron and positron have formed, which leave tracks in the central part of the detector and then deposit all their energy, represented by the blue 'towers', in the 'electromagnetic calorimeter' – the part of the detector designed to stop photons and electrons. The three photons by contrast are electrically neutral and leave no tracks, but are made visible through their energy deposits, which are represented by the red 'towers'.

emerge with energies characteristic of the parent atom, reflecting the pattern of electron energy levels unique to each specific element. And because the wavelength (colour) of light is related to its energy, the resulting spectrum is an autograph of that element. Thus the light from sodium vapour glows yellow, that from neon is red, while copper colours a flame green. It was in attempting to explain the broad spectrum of radiation emitted from objects, that the German physicist Max Planck realized in 1900 that the radiation must be bundled into 'quanta', which we now call photons.

Atoms do not only emit photons, they can also absorb them. When a photon encounters an atom, the photon can give up its energy to an electron and raise it to a higher energy level. Absorption of this kind happens only if the energy of the photon exactly matches one of the energy steps within the atom, but it provides an important means of analysing the elements in a material. This is how various elements were first discovered in the outer regions of the Sun. Photons streaming out from the Sun cover a whole range of energies, some of which match the energy steps of particular elements. These elements absorb the photons and as a result we find dark shadows at these wavelengths in the Sun's spectrum, as in Fig. 3.18.

The energy of photons in the visible range can raise electrons from one energy level to another within atoms. But photons can also knock electrons out of atoms entirely. This phenomenon is known as the photoelectric effect, and working out the theory won Albert Einstein the Nobel prize for physics in 1921. Nowadays the effect is exploited in many modern processes and gadgets such as solar cells and 'electric eyes'. The essential feature is that the photons liberate electrons, which can then flow and carry an electric current.

These devices detect photons at the relatively low energies of visible light. Figure 3.19 shows photons with much higher energies. An electron and an antielectron (a positron) have entered unseen in the beam pipe to collide at the heart of the detector. They annihilate in a burst of energy, which immediately rematerializes as a new electron and positron. These two new particles leave tracks in the detector and also deposit energy, shown by the blue 'towers'. Also unleashed in the violence of the interaction are three gamma-ray photons. Being electrically neutral, they leave no tracks, and reveal themselves only by their energy deposits, represented by the red 'towers'.

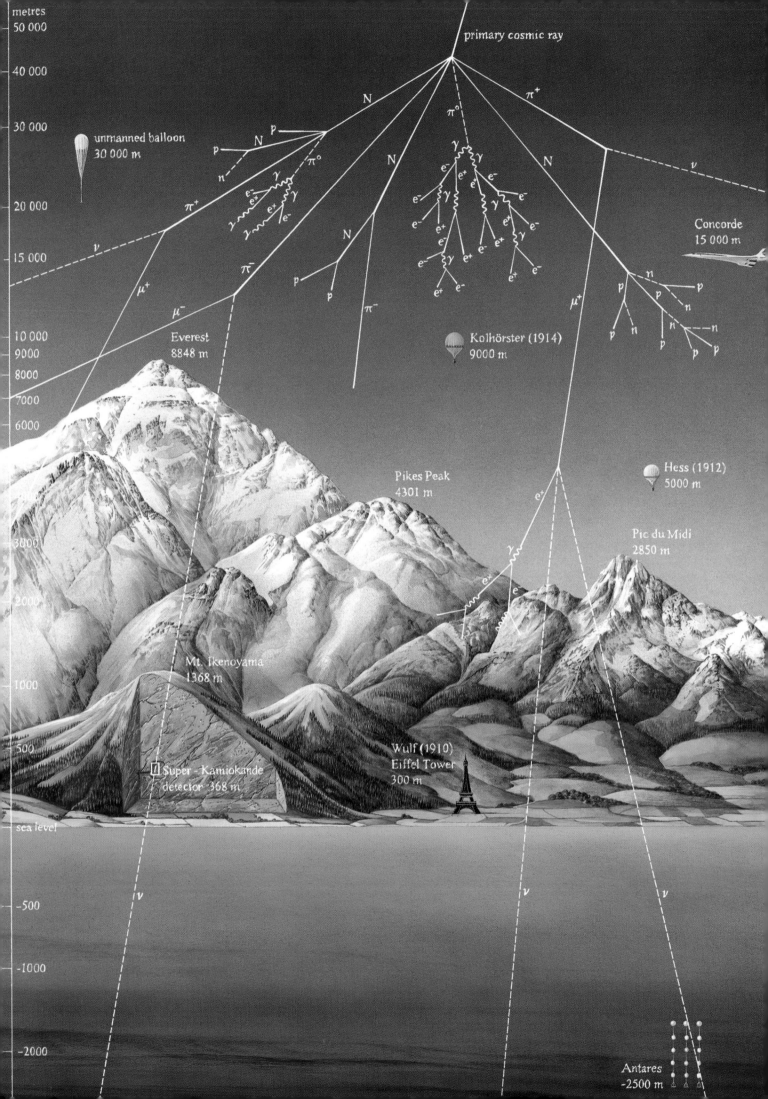

4. The Extraterrestrials

Thousands of metres above the Earth's surface, the outer atmosphere experiences a continuous bombardment. The 'artillery' comes in two forms: photons and other subatomic particles. The photons cover the whole spectrum of electromagnetic radiation, from radio waves through visible light to gamma rays. Most of the other particles are energetic nuclei, that is atoms stripped of all their electrons. The astronauts on board the Apollo spacecraft on their way to the Moon saw the effects of such energetic nuclei as they tried to sleep. Occasionally a heavy nucleus would strike the retina in the closed eye of one of the dozing men, causing him to see a tiny flash of light.

The high-energy rain of particles has become known as cosmic radiation, but in many respects it is quite different from the alpha and beta radiation emitted by radioactive nuclei, described in Chapter 2. Cosmic ray particles have much higher energies than alpha and beta rays, and are thinly spread. These two factors in particular have made the nature of cosmic rays much more difficult to discover.

The Apollo astronauts were out in space, exposed directly to the cosmic rays, but here on Earth we are shielded by the atmosphere. When the high-energy primary radiation from space hits the atoms and molecules of the upper atmosphere, it generates showers of subatomic particles, which form a rain of secondary radiation. Most of this is absorbed in the atmosphere before reaching the Earth's surface, leaving only a thin 'drizzle' of radiation to pass harmlessly through our bodies. We now know that this drizzle consists mainly of electrons, particles called muons (which were discovered in studies of cosmic rays), and neutral particles called neutrinos. Roughly 20 'primary' cosmic rays per square centimetre arrive each second at the top of the atmosphere. At sea level there is about one charged 'secondary' cosmic ray per square centimetre per minute. By comparison, a gram of a radioactive substance such as radium emits thousands of millions of particles every second.

Cosmic rays have very high energies, rising to 10 million million times the maximum energy of the radiation from radioactive sources. However, the highest-energy particles are very rare, with only a few arriving in a square kilometre over a period of 10 years. The majority of cosmic rays have energies up to 1000 times more than those produced by radioactivity. Most of these particles come from within our own Galaxy, the Milky Way, and are accelerated by many different processes.

The origin of ultra-high-energy cosmic rays is much more mysterious. Though they are few and far between, they produce characteristic showers of millions of particles that spread over several square kilometres when they reach the Earth's surface. Studies of these rare 'extensive air showers' show that the primary cosmic rays seem to come from all directions, so their source, either inside or outside our Galaxy, remains a puzzle that continues to challenge researchers (see pp. 217–219).

Though we have a relatively poor understanding of where cosmic rays come from and how they reach the Earth, we do know in great detail what the radiation consists of and what it does to the atmosphere. Figure 3.8 (p. 40) shows the tracks of different kinds of nuclei in the cosmic radiation, caught in photographic emulsions flown high in the atmosphere by balloon. The interactions of this primary radiation initiate a rain of secondary radiation, which develops through the atmosphere as Fig. 4.1 illustrates, and which can penetrate below ground.

Fig. 4.1 A continuous rain of cosmic rays – energetic photons, protons, and other atomic nuclei – enters the Earth's atmosphere from outer space. This primary radiation collides with the nuclei of atoms in the upper atmosphere and produces showers of secondary particles. They include protons (p), neutrons (n), light nuclei (N), and many charged and neutral pions (π). The pions are relatively short-lived and decay in flight. The neutral pions (π^0) decay to pairs of gamma-ray photons (γ), which in turn produce pairs of electrons (e$^-$) and positrons (e$^+$). If they have sufficient energy, these electrons and positrons can radiate more photons, which produce further electron–positron pairs, creating an avalanche of particles that can reach down to sea-level. The charged pions that are not absorbed by nuclei in the atmosphere also decay, transmuting into penetrating muons (μ) and neutrinos (ν) which can continue their journey far underground.

The exploration of cosmic rays began at ground level, and progressed gradually up through the atmosphere. Theodor Wulf only climbed the Eiffel Tower, but Victor Hess rose to 5000 m in his balloons, and Werner Kolhörster to 9000 m. Modern unmanned balloons can reach the edge of the atmosphere at 30 000 m or more. In the 1930s, mountain-climbing physicists established a cosmic ray observatory on the Pic du Midi, in the French Pyrenees. In modern experiments, the penetrating muons and neutrinos are detected underground in experiments such as Super-Kamiokande in the Kamioka mine under Mt Ikenoyama in Japan. In other experiments, natural volumes of water (or ice) are put to use to detect neutrinos, as in the ANTARES detector 2500 m below the surface of the Mediterranean.

Many scientists studied the cosmic radiation in great detail after its discovery in 1912. Their investigations progressed in parallel with the studies of the atom and the nucleus described in Chapter 2. The work involved similar experimental techniques, and indeed sometimes the same inquisitive minds. And there were surprises in store. By the end of the 1940s, the cosmic radiation had revealed unexpected particles of matter, different from the electrons, protons, and neutrons of atoms and nuclei. The discovery of these particles inspired the building of modern accelerator laboratories where we can produce high-energy particles to order – though still none as energetic as the highest-energy cosmic ray particles.

Modern particle physics has thus grown out of the early studies both of radioactivity and of cosmic rays. Many of the cosmic ray physicists of the 1930s and 1940s were later to work at accelerators, bringing with them an armoury of techniques which, even today, underlie the complex technology of experiments in particle physics. The cosmic rays had in their turn been discovered by intrepid scientific explorers who went up in balloons at the turn of the century, to answer some of the questions raised in the study of radioactivity.

The Discovery of Cosmic Rays

Fig. 4.2 Victor Hess (1883–1964) in 1936, the year he won the Nobel prize for physics for his discovery of cosmic rays.

Radioactivity attracted everyone's attention soon after its discovery. The radiation from radioactive bodies was easy to detect, for it splits the molecules of the air into positive and negative ions, and makes the air electrically conducting. In this way, the radiation reveals its presence in electrometers (see Fig. 2.11, p. 22). But a puzzling phenomenon soon became apparent: even when no radioactive source was present, electrometers would indicate the presence of some other 'radiation' that ionized the air.

Armed with electrometers, scientists looked all over for the tell-tale indications of the mysterious emanations. The rays showed up everywhere, even out at sea, far from the radioactivity of rocks. But the most peculiar thing was that however much the researchers shielded their detectors, some radiation still penetrated. The rays from radioactive materials could not breach the shielding, so it seemed that another source of unknown rays of immense penetrating power must exist. But where?

The first clues emerged in 1910 when Father Theodor Wulf, a Jesuit priest, went up the Eiffel Tower and measured more radiation than he expected. Wulf guessed that the rays might have an extraterrestrial origin and he proposed going up in balloons to great heights to test this idea. But the spirit of adventure must have deserted him, since he seems to have been reluctant to do this himself! The risky exercise was undertaken instead by others, notably the Austrian, Victor Hess. During 1911–12 he made ten ascents in balloons complete with detecting apparatus, reaching heights of over 5000 m (see p. 6). These experiments showed that the intensity of the radiation increases rapidly above 1000 m, being some three to five times greater at 5000 m altitude than at sea level. Hess concluded that there must be a powerful radiation originating in outer space, entering the Earth's atmosphere, and diminishing as it passes through the air towards the ground.

Hess discovered the cosmic rays with the aid of instruments that required the personal attendance of the experimenter to watch and record the results. Robert Millikan's group at the California Institute of Technology (Caltech) in the mid-1920s developed an electrometer whose readings could be recorded on a moving film without need for anyone to be present. This extended the

observational possibilities enormously. Unmanned balloons could take the recording equipment to very high altitudes and, at the other extreme, great depths of water could be plumbed for the presence of the rays.

At first Millikan had not believed Hess's claims that the radiation came from outer space. However, he made extensive investigations of his own and in 1926 he changed his mind, even going so far as to claim credit for the discovery himself! As the cosmic rays penetrate matter so easily, and gamma rays are the most penetrating form of radioactivity, Millikan believed, along with many others, that the cosmic rays are ultra-high-energy gamma radiation. He proposed that the primary gamma rays came from nuclear reactions occurring out in space, in which heavier elements were being synthesized from lighter elements, ultimately from the lightest of all, hydrogen. He referred to the cosmic rays as the 'birth cries' of newly born matter.

Hess was awarded the Nobel Prize in 1936, and he is generally acknowledged as discovering the cosmic radiation. Millikan is honoured through the evocative name 'cosmic rays', which he coined and which is commonly used today. Wulf, whose ambitions reached only as high as the Eiffel Tower, is all but forgotten.

Once everyone had accepted the existence of the cosmic rays, there remained the problem of finding out exactly what they are. Hess had correctly identified outer space as the source of the rays. But the intensity of the cosmic radiation is so low, even at high altitudes, that with his relatively simple instruments Hess could do no more than recognize that cosmic rays exist; he could not identify their content. Also, having high energy and therefore high velocities, the cosmic rays are less ionizing than the lower-energy rays from radioactive sources. The more energetic cosmic ray particles pass atoms too quickly to have much effect, and so knock electrons out of fewer atoms; cosmic rays are not only thinly spread, they are also elusive. In addition, there is the problem of not knowing precisely from which direction the next cosmic ray will come. This is utterly different from radioactivity. Whereas a radioactive source set at the end of a collimating tube produces a narrow, well-defined beam of radiation, cosmic rays arrive at the Earth from all directions.

A big step forward came in 1928, when Hans Geiger and Walther Müller, at the Physics

Fig. 4.3 Robert Millikan (1868–1953), right, together with Carl Anderson on Pikes Peak in Colorado. Millikan gave 'cosmic rays' their name, but it was Anderson who discovered two new particles in studies of cosmic rays – the positron and the muon. With a summit at 4300 m served by a rack-railway, Pikes Peak was an ideal high-altitude site for investigating cosmic rays, and it was in experiments here that Anderson and Seth Neddermeyer discovered the muon in 1936.

Fig. 4.4 A basic Geiger–Müller tube, or Geiger counter, consists of a metal cylinder, plugged at both ends, with a wire held at a high positive voltage (around 1000 volts) running along its axis. The tube is filled with a gas at low pressure. A charged particle passing through the gas will ionize atoms, releasing electrons which are attracted by the positive wire. As the electrons approach the wire they are accelerated in the electric field and become energetic enough to knock out more electrons, so an 'avalanche' develops. The field around the wire is so intense that the avalanche propagates along the length of the wire, generating a large electric pulse that can activate a loudspeaker to produce the familiar 'click' associated with modern Geiger counters.

In 1928, Bothe and Kolhörster used two of the new Geiger–Müller tubes, one vertically above the other, to show that individual cosmic rays could penetrate a gold block 4 cm thick. A simultaneous discharge of the two electrometers attached to the tubes revealed the passage of a single cosmic ray.

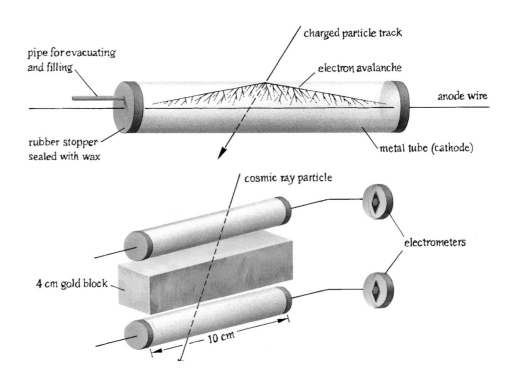

Institute in Kiel, developed what we now call the 'Geiger counter'. This was an improved version of the cylindrical counter with a wire down the centre that Geiger and Rutherford had used in 1908 to count alpha particles (see p. 26). With the Geiger–Müller counter, the electric field at the wire is so high that a single electron anywhere in the counter can trigger an avalanche of ionization. As many as 10 thousand million electrons would be released along the whole length of the wire. The tiniest amount of ionization would produce a signal from the new Geiger–Müller tube.

This ability to detect low levels of ionization makes the Geiger counter ideal for studying the high-energy cosmic rays. But the tubes form a still more powerful tool when two or more are used together. Soon after Geiger and Müller had demonstrated their new detector, Walther Bothe and Werner Kolhörster, both in Geiger's old laboratory in Berlin, used two tubes one above the other to make a 'telescope' – a device to define the path of a cosmic ray. With this they found the first conclusive evidence about the nature of the cosmic rays.

The common belief at that time, that cosmic rays are high-energy gamma radiation, implied that there should be charged particles associated with the cosmic rays, because the rays would tend to knock electrons out of atoms in the atmosphere. The Geiger counter

Fig. 4.5 (LEFT) Walther Bothe (1891–1957), in 1936. He invented the coincidence method in 1924, using a forerunner of the Geiger counter, and was rewarded with the Nobel prize for physics in 1954 for his work with this technique.

Fig. 4.6 (RIGHT) Werner Kolhörster (1887–1946) in 1928.

Fig. 4.7 Bruno Rossi (1905–1993) working in his laboratory at the Physics Institute of the University of Florence, around 1930. The horizontal tubes just visible in the centre of the picture are Geiger–Müller counters. Ranks of dry cells (batteries) provided the high voltage for the counters.

telescope was the perfect tool to test these ideas.

Bothe and Kolhörster connected the two Geiger counters to electrometers and immediately observed simultaneous deflections of the fibre needles in the two electrometers. The surprise was that there were so many 'coincidences', each of which implied the passage of a cosmic ray through both tubes. A gamma ray fires a Geiger counter only if it first knocks an electron out of an atom; it is in fact the electron that triggers the counter. So the observation of coincident signals suggested that a cosmic gamma ray had either fortuitously produced two separate electrons – which was very unlikely – or that a single electron had fired both counters.

To test the latter possibility, Bothe and Kolhörster inserted material between the counters, to absorb electrons knocked from atoms. But they found that 75% of the rays passing through the telescope were impervious even to gold blocks 4 cm thick! In fact, the particles triggering the Geiger counters were as penetrating as the cosmic radiation. The researchers were forced to conclude that these particles *are* the cosmic radiation; that the cosmic rays are highly penetrating electrically charged particles, and not gamma rays as had been supposed.

These results inspired, among others, Bruno Rossi at the University of Florence. He saw a way of using electronic valves – the predecessors of today's transistors – to register coincident pulses from the Geiger counters. This did away with the cumbersome arrangement of electrometers viewed by a self-winding camera, which Bothe and Kolhörster had used. Rossi used his technique to detect coincidences between counters that were not in a vertical line, but instead were arranged in a triangle so that a single particle could not traverse all three counters. With a shield of lead over them, Rossi detected many coincident signals from all the counters. This showed for the first time the production of showers of secondary particles. Rossi went on to perform many key experiments in the detailed study of cosmic rays. More important still, his coincidence circuitry forms the basis of all the electronic counter experiments that today record the creation of many particles in man-made high-energy collisions.

Results from 'coincidence experiments' with Geiger counters, particularly those by Rossi, showed just how penetrating the cosmic rays are. They can pass through metre-thick lead plates and have even been detected deep underground, thousands of metres below the surface of the Earth. But because of their high energies and consequent low ionizations, identifying the exact nature of the particles present in the rays was nearly impossible; that is, until cloud chambers were used to study the cosmic radiation. The technique provided beautiful images of the cosmic ray tracks, and the real excitement began.

The First New Particles

In 1923, a young physicist called Dmitry Skobeltzyn began to investigate gamma rays in his father's laboratory in Leningrad. Gamma rays from a radioactive source knock electrons from atoms. Skobeltzyn hoped to detect the tracks of such electrons in a cloud chamber, but he encountered a problem: the gamma rays also knocked electrons out of the walls of the chamber, and these interfered with his measurements of the electrons produced from the gas in his chamber. To overcome this, Skobeltzyn decided to place the chamber between the poles of a large magnet, so that the magnetic field would deflect the electrons away from the chamber walls.

A magnetic field exerts a force on an electrically charged particle so that its path becomes curved. The curvature depends on the strength of the field, and on the particle's momentum – the product of its mass and its velocity. The curvature is greater for slow or lightweight particles (low momentum) than for fast or heavy ones (high momentum). In this way, the magnetic field can distinguish different particles.

On some of the photographs he took in 1927–28, Skobeltzyn noticed a few tracks that were almost straight. These indicated values of momentum and energy much greater than for electrons from any source known at the time. Skobeltzyn supposed that the tracks were due to fast-moving electrons knocked from atoms by cosmic gamma rays. (This was the year before Bothe and Kolhörster set up their cosmic ray 'telescope' and discovered that the cosmic rays actually are charged particles.) Although he did not realize it then, Skobeltzyn had become the first person to observe directly the tracks of the cosmic rays themselves.

Skobeltzyn did not follow up his discovery, but two years later, in 1930 at Caltech, Robert Millikan instructed one of his research students, Carl Anderson, to build a cloud chamber to study the energies of the cosmic ray particles. With the assistance of engineers at the nearby aeronautical laboratory, Anderson built a powerful water-cooled electromagnet, which could produce fields more than 10 times stronger than Skobeltzyn had used. His first dramatic results showed that the cosmic rays contain both positively and negatively charged particles in about equal numbers.

Millikan still believed that the cosmic ray particles were electrons knocked from atoms by primary gamma rays, so this observation came as something of a surprise. He insisted that the positive particles must be protons, also knocked from atoms by the high-energy gammas. To produce tracks of similar curvature to the electrons, the much heavier protons would have to be moving much more slowly. However, few of the tracks had the dense

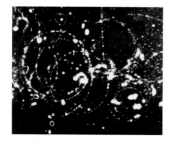

Fig. 4.9 One of the first photographs showing the track of a cosmic ray in Skobeltzyn's cloud chamber. He was studying the tracks of electrons recoiling under the influence of energetic gamma rays when he decided to subject the chamber to a magnetic field, so as to deflect unwanted tracks emanating from the walls of the chamber. In 1927 he noticed a few straight tracks, such as the vertical one at the centre of this picture, amid the curling tracks of the recoiling electrons. These straight tracks must have been made by very energetic particles, otherwise they would have been deflected by the magnetic field; it seemed they must be due to cosmic ray particles.

Fig. 4.10 Carl Anderson (1905–1991) working with the electromagnet for his cloud chamber in the Aeronautics Laboratory at Caltech, where a generator could supply 600 kW to power the magnet. The two coils of the magnet are wound with copper tubing, and cooled by tap water.

ionization expected for slow particles. Anderson, on the other hand, thought that the tracks could be due to electrons moving upwards through the chamber, rather than downward-moving positive particles. But Millikan did not like that idea at all, and stuck to his belief that the tracks were due to protons.

To settle the debate, Anderson inserted a lead plate across the chamber. Particles traversing the plate would lose energy and be curved more by the magnetic field on emerging into the chamber again. In this way he could tell the directions of the particles, and so be certain of their charges. The modification soon revealed that both Millikan and Anderson had been wrong. Anderson found a beautiful example of a positive particle that was clearly much lighter than a proton; its ionization and curvature suggested a mass similar to the electron's. He had discovered the positron – the 'antielectron' predicted by theorist Paul Dirac (see pp. 66–68). This was the first observation of an 'elementary' particle that does not reside inside the atom.

Positrons are not contained within atoms, so where do the positrons in the cosmic rays come from? Anderson was uncertain and it was an experiment by two physicists in Britain in the same year that provided the answer – positrons are created by the cosmic radiation itself. Patrick Blackett and Giuseppe Occhialini, in the Cavendish Laboratory at Cambridge University, had developed an improved version of the cloud chamber, which was to provide an exciting new window on nature.

Up to this time, cosmic ray research with cloud chambers was in some senses rather hit and miss. The cloud chamber was expanded at some random instant and more often than not no cosmic ray happened conveniently to pass through. How could this success rate – about 1 in 20 – be improved upon? Blackett had been working on cloud chambers, producing images of nuclear transmutations (see p. 31), when he was joined by Occhialini, a young Italian physicist. Together they set about devising a way whereby cosmic rays would announce their presence. Occhialini had been a student in Florence under Rossi, and was able to bring Rossi's work on coincident signals from Geiger counters to blend with Blackett's gift for gadgets and expertise with cloud chambers.

Their idea was brilliantly simple. Put one Geiger counter above and another below the cloud chamber, then if both fire simultaneously it is very likely that a cosmic ray has passed through them and, by implication, through the chamber. Blackett and Occhialini connected the Geiger counters up to a relay mechanism so that the electrical impulse from their coincident discharges actuated the expansion of the cloud chamber and a flash of light to allow the tracks to be captured on film. Notice how it is the knowledge of the passage of the cosmic ray that is crucial – not an instantaneous knowledge of its presence. By the time the

Fig. 4.11 The first evidence for the positron, obtained in a cloud chamber photograph taken by Anderson. The particle must be moving up the picture, because it loses energy as it crosses a 6 mm thick lead plate at the centre, and curls round more in the top half of the chamber. The anticlockwise direction in which the track bends shows that the particle is positive, but its track is too faint for it to be due to a proton or an alpha particle. Anderson had earlier evidence for tracks from particles that appeared to be positive, because they were curling the opposite way to electrons, but which were too light to be protons. With hindsight we can say that they were positrons, but at the time Anderson believed them to be electrons moving upwards through the chamber, which would curl the opposite way in the magnetic field to electrons moving down. Millikan, Anderson's professor, argued that they must be protons. As Anderson has since recalled, 'Curiously enough, despite the strong admonitions of Dr Millikan that upward-moving cosmic ray particles were rare, this indeed was an example of one of those very rare upward-moving cosmic ray particles.'

triggering signal from the Geiger counters is formed, the ray has already passed through the chamber. But the ionized track remains and the all-important droplets form on the ions. The cosmic rays are like jet-planes: their trails remain for a time and show where they passed.

Instead of 1 cosmic ray in 20 or more photographs, now the success rate was 4 out of 5! Blackett and Occhialini took their first photographs using this method during June of 1932, and then accumulated nearly a thousand photographs of cosmic rays during the late autumn of 1932.

Anderson was the first to report the observation of a positron, but Blackett and Occhialini's experiments confirmed its existence without a doubt. Many of their pictures showed as many as 20 particle tracks diverging from some point in a copper plate just above the chamber, like water from a shower. The strong magnetic field throughout the chamber curved the tracks, showing roughly half the particles to be negatively charged and half positively charged. The pictures provided dramatic evidence that positrons are produced in the collisions of cosmic rays; Anderson's particle was not a peculiar 'extraterrestrial' object that entered the atmosphere with the primary cosmic radiation.

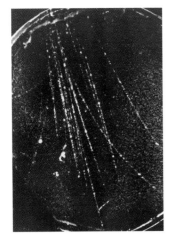

Fig. 4.13 An image from the cloud chamber used by Blackett and Occhialini shows a 'shower' of electrons and positrons produced by a cosmic ray that has interacted in the wall of the chamber. The particles are clearly of opposite charge, bending in opposite directions in the chamber's magnetic field.

The explanation of the 'showers' begins with an energetic electron entering the copper plate; there the electric fields of the positive charges of the copper nuclei cause the electron to radiate photons – gamma rays. Provided they have sufficient energy, these gamma rays can in turn produce pairs of electrons and positrons, again under the influence of the nuclear electric fields. These are the electrons and positrons that Blackett and Occhialini saw, spawned from gamma radiation, itself produced by the cosmic rays in the copper plate above the chamber. Albert Einstein's equation, $E = mc^2$, implies that energy (E) can be converted into mass (m) – radiation into matter. Blackett and Occhialini had for the first time captured the process on film.

So, by the early 1930s, it was clear that cosmic rays, at least near ground level, contain electrons and positrons. But this did not seem to be the whole story. Among the tracks

Fig. 4.14 Carl Anderson (1905–1991), left, and Seth Neddermeyer (1907–1988).

showing up in the cloud chambers were many that occurred with positive or with negative charge, but which were far more penetrating than electrons and positrons and which did not create showers. At Caltech, Anderson and his colleague Seth Neddermeyer at first favoured the idea that there might be two types of electron; they referred to the penetrating particles as 'green' electrons and the shower-producing particles as 'red' electrons. By 1936, however, they had convinced themselves that the penetrating particles were something new. In November of that year, they presented publicly their evidence that the particles have a mass between that of the electron and the proton.

The new particles were originally named 'mesotrons' from the Greek for 'middle', though this was later shortened to mesons. And as with the positron, there already existed a theory that could account for the mesotron – or so it seemed. In 1935 Hideki Yukawa, a Japanese theorist working in Kyoto, had built a theory of the powerful forces that bind the atomic nucleus. One of its consequences was that a new particle should exist, weighing some 250 times more than the mass of an electron and about 1/7 as much as a proton. Yukawa's work was not widely known outside Japan, but when Anderson and the others announced a new particle with a mass only 20% away from his prediction, Yukawa claimed the particle as his own. Moreover, Robert Oppenheimer and Robert Serber at Berkeley in California had made the same connection, and promoted Yukawa's theory in the West.

Fig. 4.15 Niels Bohr, left, Hideki Yukawa (1907–1981), Mrs Yukawa, and Robert Oppenheimer (1904–1967).

But they were wrong to do so. Yukawa's prediction was fulfilled a decade later, with the discovery of the particle now known as the pi-meson, or pion (see pp. 72–73). The original 'mesotron' observed by Anderson and Neddermeyer, which is now known as the muon, is something quite different and uncalled for. It remained an enigma for over half a century before the first hints of its place in the scheme of things began to emerge (see pp. 69–71).

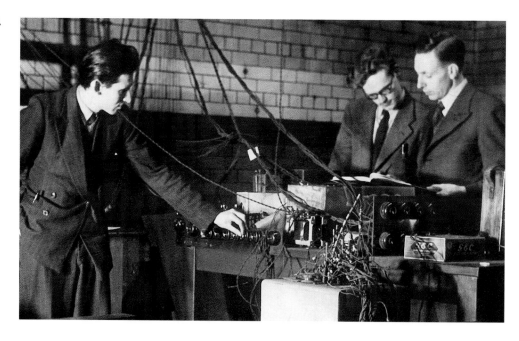

Fig. 4.16 George Rochester (b. 1908), right, in 1944 in the Cosmic Ray Laboratory, Manchester University, with Douglas Broadbent, and Lajos Janossy (1912–1978), far left.

Strange Particles

The discoveries of the positron and the muon were the first of a series showing that Earth-bound physics had sensed only a small corner of nature's rich pageant. By 1950 cloud chambers exposed to cosmic rays had revealed yet more new particles, inexplicable by the existing theories. Their discovery owed much to the improvements in experimental techniques immediately following the Second World War.

In 1937, Patrick Blackett moved to Manchester University from Birkbeck College, London, where he had been a professor since leaving Cambridge in 1933. He immediately began to gather about him a strong team of cosmic ray physicists, but their research was short-lived, being brought virtually to a standstill by the Second World War. Blackett became science adviser to the RAF and anti-aircraft defence and later director of the British naval operations research and anti-U-boat strategy. Many of his recently gathered team went off to other duties, but a few remained to carry on the training of physics students. So a little research on the cloud chamber studies of cosmic rays continued at Manchester, with Blackett's permission and the proviso that it cost nothing! George Rochester and the Hungarian Lajos Janossy spent their spare time investigating a particular interest of Janossy's – the so-called 'penetrating showers', in which extremely energetic particles enter a cloud chamber and produce a cascade of many tracks.

After the war Janossy moved to Dublin. Blackett, however, was sufficiently impressed by the wartime studies to encourage Rochester, together with Clifford Butler, to continue the work, this time with a cloud chamber in a magnetic field. (Electromagnets use electricity, so it had been out of the question to employ a magnetic field for the research during the war!) When Blackett came to Manchester he had brought with him the large electromagnet that in 1935 he had built especially for work with counter-controlled cloud chambers. Rochester and Butler set about building an entirely new cloud chamber, which they placed in Blackett's magnet. They arranged Geiger counters between sheets of lead above and below the

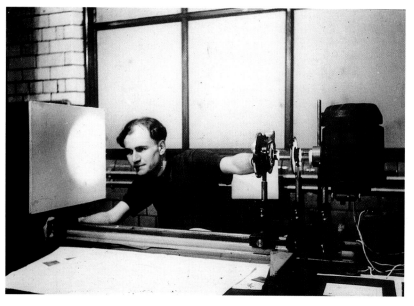

Fig. 4.17 Clifford Butler (1922–1999), in about 1947, with a system for projecting photographs from a cloud chamber. The image of a particular track was projected through a prism onto the white screen, and the prism was turned until the track became straight on the projected image. The curvature of the track could then be read from the position of the prism, which was calibrated in terms of radius of curvature.

chamber, so that it would expand only when a penetrating shower had occurred.

The pictures collected with this arrangement during 1946 and 1947 revealed a major surprise – the first examples of the so-called 'strange' particles. Cosmic rays hit the lead above the chamber and created penetrating showers. Among the many tracks were two curious pronged, or V-shaped, patterns. Rochester and Butler pointed out that both these 'vees' could be explained in terms of the spontaneous decay of new unstable particles. In one case the new particle had to be neutral (Fig 5.10, p. 74), in the other a charged particle was called for; and both particles would have to weigh about half as much as a proton.

Such a discovery was totally unexpected and, on such sparse evidence, somewhat controversial. As two years passed and no more examples turned up, the tension rose for Rochester and Butler. To improve their prospects of observing penetrating showers, Blackett gave Butler the task of setting up a magnet and cloud chamber at altitude.

After some deliberations, Blackett agreed to the choice of the 2850 m high Pic du Midi de Bigorre in the French Pyrenees. An astronomical observatory already existed at the summit, and by November 1949 a team from Manchester had successfully installed Blackett's 11 tonne magnet there. Then, over on White Mountain in California, Carl Anderson reported that he and Eugene Cowan were obtaining one of the 'vees' per day – a total of 28 in all. He said that 'to interpret our photos we require the same remarkable conclusion as Rochester and Butler: spontaneous decay of neutral and charged unstable particles of a new type'.

Between July 1950 and March 1951, the chamber on the Pic du Midi produced 10 000 photographs of high-energy showers. Among these were 67 'vees' of which 51 were due to neutral decays and 12 to charged decays. This was decisive proof that the 'vee' particles were genuine. But what was novel was that four of these cases indicated the presence of a neutral particle heavier than the proton. The first 'vees' found were from particles with a mass about half that of the proton; they have since become known as charged and neutral kaons (see pp. 74–75). The neutral particle heavier than the proton is called the lambda (see pp. 76–77). Together, the kaons and the lambda became known as the 'strange' particles because their behaviour was unexpected.

The kaon and lambda were but the first of a whole new family of particles carrying a new property – 'strangeness' – which is akin to electric charge, but which does not exist in ordinary matter. No one had predicted the existence of any of these particles, and their discovery created great excitement. The kaons and the lambda and still heavier relations took some years to understand, as Chapter 5 describes. Researchers were helped in the 1950s by the advent of new powerful particle accelerators that allowed them to simulate the cosmic ray collisions under controlled conditions. Meanwhile, another technique for tracking particles was bringing still more exciting information about the cosmic rays.

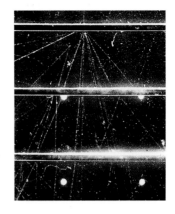

Fig. 4.18 A cosmic ray interacts in a 0.6 cm lead plate across a cloud chamber and produces several particles, including a neutral one that subsequently reveals itself by a 'V' (top left) when it decays into two charged particles. Pictures like this, taken by W. B. Fretter at Berkeley, showed that the 'V' particles were created in high-energy collisions.

Fig. 4.19 The Pic du Midi Observatory, in the French Pyrenees, where the cloud chamber from Manchester University was moved in 1949 to improve its chances of observing 'V' particles.

Powell, Pions, and Emulsions

The cloud chamber had helped immensely in unravelling the content of the cosmic rays at ground level during the 1930s and 1940s. But at the same time, it had become clear that the electrons, positrons, and muons are secondary particles created from very energetic primary radiation as it strikes the atmosphere at high altitudes. Several experiments found that the intensity of cosmic rays varies with the Earth's latitude, with more radiation arriving near the poles than at the equator. This can be explained if the majority of the primary radiation is in the form of charged particles, which will be deflected most by the Earth's magnetic field at the equator. Furthermore, balloon-borne experiments in 1938 revealed an east–west asymmtery in the primary radiation, which implied that the primary cosmic rays must be positively charged. But what kind of particles could they be?

The extremely high energies of the primary cosmic rays at first made it difficult to ascertain their nature. However, in the late 1940s, the development of special photographic emulsions, which could easily be carried aloft by balloons, brought physicists their first beautiful images of the interactions of high-altitude cosmic rays. These emulsions were especially sensitive to high-energy particles; just as intense light darkens photographic plates, so can the passage of charged particles. We can detect the path of a single particle by the line of dark specks that it forms on the developed emulsion. The particle literally takes its own photograph.

Fig. 4.20 Cecil Powell (1903–1969), right, and Giuseppe Occhialini in the physics building at Bristol University in 1947.

Photographic plates had figured in the very earliest work on radioactivity; indeed, it was through the darkening of plates that both X-rays and radioactivity were discovered. The essential feature of photography is that paper, or a glass plate, covered with a thin layer of silver bromide responds to light. (The bromide is usually in the form of an 'emulsion' of crystals mixed with gelatin.) Light affects the silver bromide crystals in such a way that when they are treated chemically – developed – they release some pure silver. The more light landing at a point, the more silver is produced and the darker the image on the developed film or plate. In this way a 'negative' image forms, with dark regions corresponding to places that have received the most light, and which are the brightest on the scene being photographed.

X-rays are an energetic form of visible light, so it is perhaps not surprising that they register on photographs. What may seem more remarkable is that a speck of radium left on a photographic plate should produce an image, as in Fig. 2.9 (p. 21). This happens because the alpha particles from the radium can ionize atoms in the emulsion and, as with light, this darkens the crystals of silver bromide after processing. Becquerel discovered radioactivity through blurred images from uranium salts and, in 1911, M. Reinganum became the first person to record the tracks of individual alpha particles in emulsion.

Despite these early successes, the photographic technique for detecting radiation appeared to suffer from two serious drawbacks. First, the emulsions had to be in thin layers, only fractions of a millimetre thick, so that they could be developed properly. As a result, only particles travelling within the plane of the wafer-thin emulsion left any significant track; particles passing through the emulsion left an almost invisible spot. Secondly, the emulsions available at the turn of the century were sensitive only to slow particles: faster particles passed through so quickly that they did not affect enough light-sensitive grains to form a visible track. To produce denser tracks the emulsions needed to contain more of the active ingredient – silver bromide – but this was technically difficult to achieve.

One person who persevered with the use of emulsions to record the tracks of low-energy particles was Cecil Powell. Powell had been a student at the Cavendish Laboratory under Charles Wilson, the inventor of the cloud chamber, and moved to Bristol University in 1928. In 1935 his team began to build a Cockcroft–Walton accelerator. They initially used a cloud chamber to study the interactions of the particles produced, but Walter Heitler, one of the theorists at Bristol at the time, drew Powell's attention to the work of two Viennese researchers, M. Blau and H. Wambacher. They had been using photographic emulsions to detect cosmic rays; in particular, they had shown that emulsions could be sensitive to protons, not just the heavier alpha particles.

For Powell, the photographic technique had the advantages over the cloud chamber of being simpler and much more accurate in measurements of a particle's range. A set of emulsion-covered plates is sufficient to collect particle tracks; a cloud chamber, on the other hand, is a complex piece of apparatus, needing moving parts so that the chamber can be continually expanded and recompressed. Powell's team tested emulsions by detecting cosmic rays high on mountain tops. Then they turned to detecting the collisions of particles from the accelerator at Bristol.

The technique proved worthwhile. In the latter half of the 1930s, and to a lesser extent during the Second World War, Powell's group studied nuclear collisions not only at Bristol but at a more powerful accelerator at Liverpool University. In some ways they were lucky, for their emulsions had unusually high quality. Moreover, the accelerator provided a copious supply of particles travelling in a well-defined direction, so it was straightforward to arrange the plates in suitable positions for recording tracks. However, after the end of the war, Powell was to extend the technique to study cosmic rays at high altitudes.

Fig. 4.21 An example of Powell's work with emulsions in accelerator beams. This shows the collision of a proton, coming in from the bottom of the picture, with a proton in the emulsion. Notice the 90° angle between the tracks of the two scattered protons (compare Fig. 3.10, p. 41). The shorter proton track is about 0.04 mm long.

In 1945, the newly elected Labour government set up an important scientific committee at the Ministry of Supply in London, chaired by Patrick Blackett. One of the committee's decisions was to encourage nuclear research outside the immediate concerns of national defence. To this end it formed two panels, one to develop accelerators and the other (which included Powell) to investigate special 'nuclear' emulsions, particularly sensitive to energetic subatomic particles. With support from the Ministry of Supply, a research team at Ilford Ltd had by May 1946 produced an emulsion incorporating about eight times as much silver bromide as normal. The improved sensitivity provided a medium that rivalled the cloud chamber in the visual beauty of its images.

These better emulsions were immediately exploited in cosmic ray research in the skilled hands of Powell and Occhialini, who had come to Bristol in 1945 on Blackett's recommendation. Occhialini took some plates to the French Observatory on the Pic du Midi. The results stunned the physicists by revealing a whole new world of nuclear interactions, produced by the primary cosmic radiation entering the atmosphere. 'It was as if, suddenly, we had broken into a walled orchard, where protected trees had flourished and all kinds of exotic fruits had ripened in great profusion,' recalled Powell in later years.

The fruits turned out to be more exotic than the physicists had anticipated, for trapped in the emulsions examined in 1947 was evidence for a new type of particle. This new particle, like the 'mesotron' Anderson had found 10 years previously, had a mass between that of the electron and the proton. But it was slightly heavier than Anderson's particle, and in the tracks revealed by the high-sensitivity emulsions the new particle could be seen to decay after a few tenths of a millimetre into a particle like the 'mesotron'.

Powell and his team had at last really discovered the particle that Yukawa had predicted back in 1935 as the carrier of the strong force – the particle we now call the pion. Anderson's particle, now known as the muon, results from the decay of the pion, but for ten years the pion had remained hidden, and the muon had mistakenly been presumed to be Yukawa's particle. Only with the use of sensitive emulsions at high altitude did the pion's brief life become visible for the first time. Yukawa received the Nobel prize shortly after the pion's discovery, in 1949; Powell was honoured in the following year.

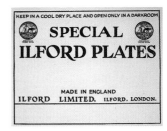

Fig. 4.22 A packet of the new emulsions from Ilford Ltd.

Fig. 4.23 The interaction of a cosmic ray in the emulsion known as Kodak NT4, which was first produced in 1948 in response to the demand for emulsions of greater sensitivity. This was the first emulsion to be completely sensitive to electrons, and it gave recognizable tracks for particles at all velocities – notice how all these tracks are solid lines.

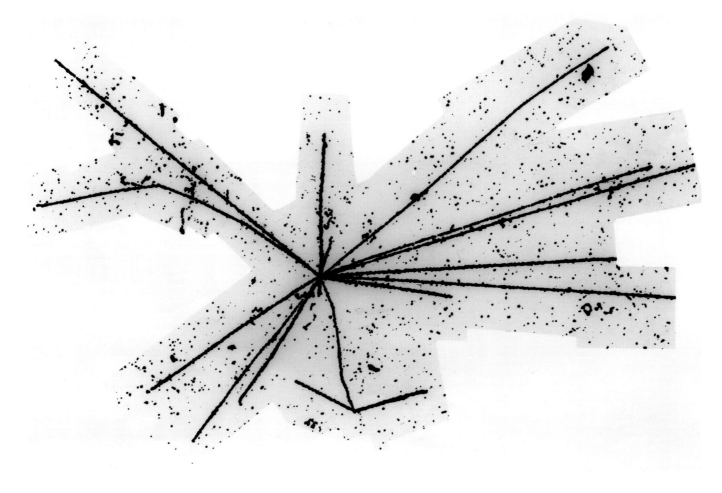

The new emulsions at last enabled physicists to identify the nature of the primary cosmic radiation. Polyethylene balloons, their skins fractions of a millimetre (25 micrometres!) thick, ascended to great altitudes, carrying emulsion. On return to Earth, the emulsions were developed, which revealed the primary rays to be atomic nuclei, moving at nearly the speed of light. Heavy nuclei, with large positive charge, gave thick tracks; lighter nuclei, with smaller net charge, gave finer tracks. It was possible to identify the nuclei of different elements with ease (see Fig. 3.8, p. 40). By 1950, it was clear that the primary rays are mainly protons (86%), with a proportion of helium nuclei (12%), and carbon and oxygen at the level of about 0.5%. Rarer nuclei were identified as more emulsions were exposed, and now many nuclei, up to uranium, have been found.

An important breakthrough in this work was the development of techniques to expose larger amounts of emulsion. Occhialini, while at Bristol, devised a means of successfully processing thick emulsions, up to a millimetre deep. The way to expose a greater depth of emulsion to cosmic rays is to pack many layers together in a stack. This may sound obvious, but recall that the emulsion is held on a backing, generally of glass, to keep it as flat as possible. In the early 1950s, Powell and other researchers discovered how to peel unexposed emulsions away from the glass plates on which they were made – a process known as 'stripping' – so that as many as 100 layers of emulsion can be stacked together. After exposure, the layers of emulsions are carefully separated and returned one by one to the glass plates, so that they can be developed and studied in the usual way.

As with the cloud chamber, the tracks in emulsions provide sufficient clues to allow particles to be identified. But in this case the pictures must be studied through a microscope, for individual tracks are invisible to the naked eye. Moreover, because emulsion is much denser than the gas of a cloud chamber, particles do not travel so far, even at high energies.

One clue to a particle's nature comes from its 'range' – the distance it travels through the emulsion before stopping. The more energetic a particle, the greater its range. In addition, the person looking through the microscope can count the darkened grains; the more dark spots in a given length of track – usually 100 micrometres – the greater the rate of ionization. Heavy ionization – many dark spots – can occur because the particle has a large electric charge; or it can arise because the particle is moving slowly, and is near the end of its range. A third clue comes from the scattering that occurs when the dense material of the emulsion deflects a particle from its straight path. Detailed measurement of the change in angle from one section of track to the next reveals information about a particle's mass.

It is worth pointing out one problem with examining emulsion through a microscope, namely that the depth of focus is typically only 0.5 micrometres, one thousandth the thickness of the material. The darkened grains are only a few tenths of a micrometre across, and so for tracks dipping through the emulsion, the scanner

Fig. 4.24 A balloon to carry emulsions up into the primary cosmic radiation in the high atmosphere awaits launch from Cardington in Bedfordshire, in the 1950s.

Fig. 4.25 Cecil Powell (standing at far right) with his Cosmic Ray Group outside the main entrance of the Physics Department at the University of Bristol in 1949.

can look at only one portion at a time. Most of the pictures of tracks in emulsion are not exactly what one sees through a microscope; they are collages of the view into different layers as the focus is slowly changed.

Examining emulsions, especially those that have been stacked in a sandwich of many layers, can be very time-consuming. In the late 1940s, Powell built up a team of women to help his group in this task; and in publications, he and his fellow physicists were careful to credit the particular person who had found an interesting event. In this way, Powell anticipated the teams of scanners needed for later detectors, such as bubble chambers. He also established the precedent for international teams of scientists from a number of institutions, who collaborated on collecting the data – exposing the emulsions – and then divided the spoils for analysis.

Stacks of emulsion carried by balloons completed the final part of a picture that Hess had begun on his flights in hot-air balloons. Atomic nuclei from outer space fragment on collision with atoms in the upper atmosphere. The fragments consist for the most part of protons, neutrons, and light nuclei, many of which are clearly visible in the photographs. But the nuclear maelstrom also includes pions, which can carry positive or negative electrical charge or can even be uncharged.

The uncharged pions decay rapidly to gamma rays, which produce showers of electrons and positrons as they travel through the atmosphere. The charged pions that are not first absorbed by nuclei in the atmosphere decay in flight, transmuting to muons. These muons traverse the atmosphere with ease and can even penetrate far underground. Although muons are much longer-lived than pions, they also often decay in flight. Powell photographed examples of a pion decaying into a muon, which in turn decayed into an electron (Fig. 4.26). The abrupt changes in direction at each stage result from the simultaneous emission of an unobserved lightweight particle called the neutrino, which is also very penetrating and can even travel right through the Earth.

Thus by the early 1950s, a picture of the whole sequence of processes involved in the cosmic radiation, from the upper atmosphere to below ground, had emerged. It had also revealed rather unexpected things, such as the muon, the 'vees' discovered by Rochester and Butler, and the pion finally identified by Powell and Occhialini. The suspicion arose that a weird world of exotic particle varieties lay undiscovered, and this spurred scientists to build their own particle accelerators so that 'cosmic rays' could be generated with high intensity and to order under controlled conditions. The heyday of cosmic rays was past; it was now to be the turn of the accelerators, as Chapter 6 will describe.

Fig. 4.26 With the development of electron-sensitive emulsions, Powell was able to record the complete decay chain of a charged pion in images such as this one taken in October 1948. The pion comes in to the top of the picture from the left, leaving a strong track. It decays to a muon and an invisible neutrino. The muon proceeds down the page and then itself decays into an electron and a second neutrino. Again the neutrino remains invisible; the electron, however, leaves a faint but clear track. Notice how the muon's track, which is about 0.6 mm long, becomes denser as it slows down before it decays.

5. The Cosmic Rain

When physicists in the late 1920s proved once and for all that cosmic rays are penetrating high-energy particles, they opened up a new means for studying matter. Radioactivity, the phenomenon that had revealed the contents of the once 'indivisible' atom to Rutherford and his contemporaries, had now become a tool used by other scientists, including chemists and biologists. Cosmic rays became the new mystery for physicists to understand.

The cloud chamber was automated in the early 1930s, so that Geiger counters would trigger the expansion of a chamber only when specific patterns of tracks passed through, and it became a valuable tool for studying cosmic rays. Then in the late 1940s, the improved photographic emulsions brought to life an amazing wealth of detail in the tracks of particles captured high on mountain tops, or still higher in balloon-borne experiments. These new techniques revealed for the first time particles unseen in the study of atoms and nuclei at the relatively low energies associated with radioactivity. The new particles did not appear as part of the stable matter of the world about us; rather they proved to be transient objects, formed in the high-energy maelstrom of cosmic ray collisions in the upper atmosphere. That they were discovered at all is due to the ability of cloud chambers and emulsions to provide time exposures of a particle's path, which sometimes recorded the birth and death of a particle on the same image.

Cosmic ray research in the 1930s and 1940s led to the discovery of several new particles and in a sense gave a preview of things to come later in experiments at particle accelerators. Some of the new particles had been predicted by theory. For example, the positive electron, or positron, the first example of an antiparticle, fitted in with the theory of the electron that Dirac had put forward in 1928. The pion also had been predicted, by Yukawa, as the carrier of the strong force. We now know that it was the first example of the group of particles that are classified as 'mesons'. Other new particles were, however, entirely unexpected. The muon was at first confused with the pion, and it was only in the 1950s that physicists came to recognize it as a heavy relative of the electron. Most mysterious of all were the so-called 'strange' particles: the kaon, the lambda, the sigma, and the xi.

A better appreciation of the role of the muon and the strange particles did not come until particle accelerators could mimic the actions of cosmic rays and produce plentiful supplies of the 'extraterrestrials'. With hindsight, we can see that the new particles were the first hints of a deeper level of structure in nature, which is still not fully understood.

The following portraits of particles discovered in the cosmic rays include pictures from these early experiments at accelerators, along with others from more modern machines. These include a number of images from bubble chambers, which superseded the cloud chamber in the mid-1950s. Instead of revealing tracks of ionization as chains of droplets in a vapour, the bubble chamber shows them as strings of bubbles in a superheated liquid. And there are also more modern images, in which tracks form in 'electronic bubble chambers' containing a gas strung with fine wires that detect the ionized trails.

The images show how our knowledge of the particles has increased over the years, as the detectors to study them have increased in sophistication. Though we may not yet understand everything about the particles portrayed in this chapter, they no longer seem exotic. Nowadays physicists make them all routinely at modern high-energy particle accelerators. The particles that were found first in the cosmic radiation have today become useful tools right here on Earth, helping to solve the problems raised by their very existence.

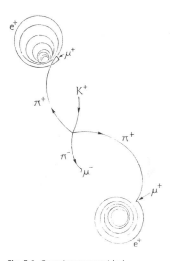

Fig. 5.1 Cosmic rays provided physicists with the first glimpses of new subatomic particles, which were later studied in detail in experiments at particle accelerators. Positrons, muons, pions, and kaons, all figure in this photograph from the 2 m bubble chamber at the CERN laboratory. Near the centre of the image a positive kaon (K$^+$) decays into three charged pions to produce a classic kaon 'signature'. One positive pion (π$^+$) moves up the picture, curling clockwise, while another heads down, also curling clockwise. Both of these pions decay to a positive muon (μ$^+$) and then to a positron (e$^+$); in each case, the muon leaves a short track, while the positron produces a characteristic spiral. The third pion (π$^-$) moves down the picture, curling anticlockwise. It decays to a negative muon (μ$^-$) which collides with the bubble chamber's glass wall before it can decay in its turn. At each decay of the pions and muons, invisible neutrinos are also emitted, their presence revealed only by the abrupt changes in direction of the tracks. (The picture has been coloured to enhance the visibility of the tracks.)

The Positron

All the atoms of matter contain negatively charged electrons and positive protons; the total negative and positive charges are equal, making matter electrically neutral overall. But we can imagine a world in which electrons are positive and protons negative. After all, the definitions of 'positive' and 'negative' are purely arbitrary; what is important for atoms is that electrons and protons have opposite charges and are bound together by the electric force. Nature, however, seems to have made a choice, because all electrons have the same charge, as do all protons. Or do they?

In 1928 a theorist at Cambridge University, Paul Dirac, had been attempting to combine Einstein's theory of special relativity with the equations governing the behaviour of electrons in electromagnetic fields. In so doing, he was led to the remarkable conclusion that particles of the same mass as the electron but of opposite charge must exist. At that time none had been seen and Dirac proposed that there might exist parts of the Universe where positive and negative charges were reversed. Then in 1932, Carl Anderson at Caltech observed a new kind of particle in the cosmic radiation passing through his cloud chamber

(Fig. 4.11, p. 55). The new object was similar in mass to the electron, but it was positively charged. He had discovered the positive electron, or 'positron'. This was the first example of 'antimatter' – an 'antiparticle' with properties opposite to those of the familiar particles.

The French physicists Irène and Frédéric Joliot-Curie later discovered, in 1934, that certain nuclei can spontaneously emit positrons, in a form of radioactivity akin to the emission of electrons. In the familiar form of beta decay, a neutron converts to a proton by emitting an electron (and an antineutrino). However, some nuclei can become more stable by converting a proton to a neutron while emitting a positron (and a neutrino).

Positrons are also created together with electrons when pure energy 'freezes' into matter. One of the most common ways in which this happens is when an energetic gamma-ray photon produces an electron–positron pair. Figure 5.2 is a bubble chamber photograph that has been 'cleaned up' to show only the relevant tracks. Two electron–positron pairs have been produced simultaneously by separate gamma rays, which have entered from the top of the picture. The gamma rays, being neutral, remain unseen, but the electrons and positrons are charged and leave tracks in the detectors. The bottom electron–positron pair is relatively energetic and the two tracks curve only a little in the bubble chamber's magnetic field. The top pair, on the other hand, has less energy and the tracks curl round to form spirals. The reason these particles are less energetic is that the gamma ray that created them spent much of its own energy in knocking an electron (the long track) out of an atom in the bubble chamber's liquid. The image shows clearly the transformation of energy into matter, in accordance with Einstein's equation $E = mc^2$. This is how positrons are formed in the cosmic radiation, from gamma rays with sufficient energy to create the total mass of an electron–positron pair.

Bringing an electron and a positron together reverses the events seen in Fig. 5.2. The two particles mutually convert into energy, in a process called annihilation. In Fig. 5.3 we see a positron annihilating with an electron that was in an atom in the liquid in a bubble chamber. The annihilation produces two photons, which remain unseen until one rematerializes as an electron–positron pair.

Fig. 5.3 In this photograph from the '15 foot' bubble chamber at Fermilab, the highlighted tracks show the birth and death of a positron together with the reappearance of some of its energy in the form of a new positron and an electron. An invisible gamma ray creates an electron–positron pair at bottom right of the image. The electron curls away to the right, the positron to the left. While the electron slows down (curling more), no doubt eventually to form part of an atom, the end of the positron's track must mark its annihilation with an atomic electron to form a pair of gamma rays. The gamma rays continue invisibly in the positron's direction, but one of them soon creates a new electron–positron pair, at top left of the image.

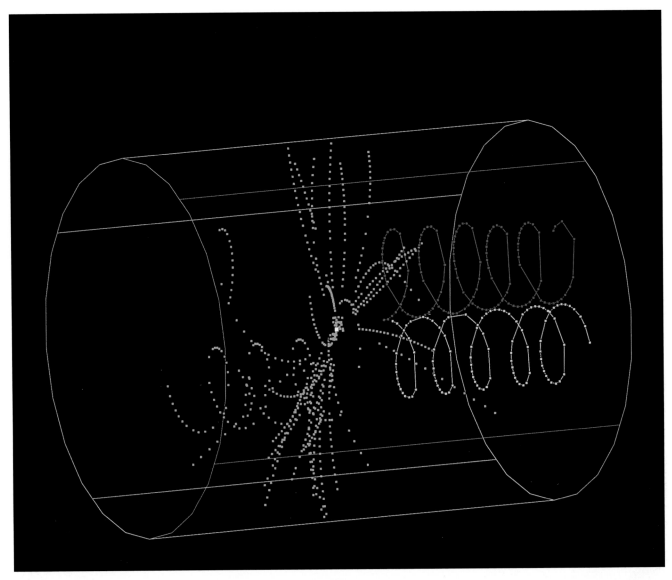

Fig. 5.4 A positron and an electron form perfect matching spirals in this computer reconstruction of particle tracks in the ALEPH experiment at CERN. The cylindrical outline indicates the extent of ALEPH's gas-filled particle tracking chamber (3.6 m in diameter). The dots mark the position of ionization sensed by wires threading the gas, and reveal the tracks of charged particles created in the high-energy annihilation of an electron and positron at the centre of the detector. Joining up the dots reveals the tracks of a low-energy positron (red) and electron (yellow), which appear as if from nowhere, created near the centre of the image by an invisible gamma ray among the products of the original annihilation.

Energy in the Universe is constantly being changed from one form into another, though the total energy remains constant. We are familiar with the conversion of electrical energy into light (as in lamps), and chemical energy into heat (as in fires). Less familiar forms of energy conversion fuel the Sun, which in turn provides the energy for the cycle of life on Earth to continue. Matter is, in a sense, 'frozen energy', and the discovery of the positron revealed a new form of energy transformation – radiant energy converting into matter (electrons) and antimatter (positrons). This may be the way that much of the matter in the Universe formed, freezing out from the radiation of the hot Big Bang with which the Universe began.

Dirac's original theory applied to electrons, and predicted the existence of the positron. Now we know that it applies equally well to the proton, neutron, and many other particles of matter, all of which have antimatter equivalents (see pp. 112–114). This knowledge enables physicists to use electrons and positrons together as tools to create new forms of matter and antimatter. Provided matter exactly balances antimatter, and the rule $E = mc^2$ is obeyed, any variety of particle can emerge; we are not limited to the common forms prevalent on Earth. Figure 5.4 shows the aftermath of the annihilation of an electron and a positron of very high energy. Their total energy materializes as a burst of particles, echoing the formation of matter from high-energy radiation in the early moments of the Universe. But in this image there is also clear evidence that a new electron (yellow) and positron (red) were produced, which spiral away like mirror images of one another.

The Muon

'Who ordered that?' physicist Isidore Rabi once remarked about the muon, the particle we now recognize as being a heavy version of the ubiquitous electron. It seemed unnecessary for nature to have provided more than one variety of the same type of particle. Moreover, the early history of the muon was full of confusion. It was mistaken for an entirely different particle – the pion – which ironically turned out to be the very particle that gives birth to the muon in the first place.

Whereas electrons, protons, and neutrons are the stuff of ordinary matter, muons are typically the stuff of cosmic rays. Atomic nuclei from outer space hit the upper atmosphere and produce a debris of pions. These soon decay to produce showers of negative and positive muons, which rain down on the Earth continuously. But the muon is not a stable particle. Negative muons decay into an electron, a neutrino, and an antineutrino, positive muons into a positron and a neutrino and an antineutrino, and this occurs in 2.2 microseconds when the muons are at rest. However, when muons are in motion, like all particles they experience the subtle slowing of time inherent in Einstein's theory of relativity. A fast-moving, energetic muon has its life prolonged as measured by clocks on Earth.

The majority of cosmic ray muons decay during their flight through the atmosphere, but the most energetic of them survive long enough to penetrate deep underground. In Fig. 5.5 a shower of muons passes through the ZEUS detector, about 25 m below ground at the DESY laboratory in Hamburg. The detector was built to record 'man-made' collisions between beams of electrons and protons (see pp. 154–155), but it is not immune to the

Fig. 5.5 A shower of cosmic rays passes through the ZEUS detector 25 m underground at the DESY laboratory in Hamburg. Some of the muons in the shower are visible through their tracks (green) in the central part of the detector – a cylindrical tracking chamber 1.7 m in diameter. Many more muons also deposit energy in the calorimeters that surround the tracking chamber. These parts of the detector are made from dense material – in this case uranium interlaced with scintillator – and are designed to catch all the energy of many particles. Only the penetrating muons (and the very weakly interacting and hence invisible neutrinos) pass through the whole detector. The deposits of energy show up as bright yellow blocks in this 'end-view' of the detector.

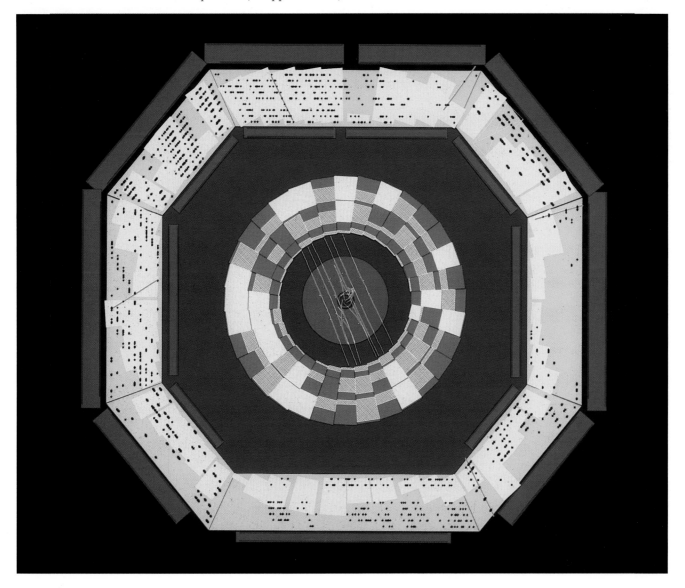

effects of cosmic rays, despite its 25 m covering of rock. Muons can penetrate even deeper than this, as Fig. 5.6 shows. Here a cosmic muon has passed through the Super-Kamiokande detector in Japan (see p. 201), 1000 m underground, and fired its electronics.

It is not only because muons have relatively long lifetimes that they can penetrate so far; it is also because they are heavy, weighing in at some 200 times the electron's mass. When electrons pass through matter, they radiate energy in the form of photons and are rapidly brought to a standstill. The heavier muons, on the other hand, have much less tendency to radiate and slow down.

Despite these differences between muons and electrons, the two particles are very similar in other respects. When Carl Anderson and his colleague Seth Neddermeyer first observed cosmic ray muon tracks in their cloud chamber in the early 1930s, they believed them to be due to ultra-high-energy electrons obeying new laws of physics. But by careful measurement of the tracks, Anderson, Neddermeyer, and others showed in 1936 that the tracks had to be made by a new particle with a mass somewhere between that of the electron and the proton – hence their name for it, the 'mesotron', or middle particle. Figure 5.7 shows one of these mesotrons coming to a halt in Anderson's cloud chamber after passing through a Geiger counter placed across the centre. This is the particle today known as the muon. In Fig. 5.7 the positively charged muon decays to a positron, although the chamber was not sensitive enough to show the latter's track. Analysis of this picture gave one of the first determinations of the mass of the muon – now known to be 210 times that of the electron.

It was this mass that caused the muon to be confused with the pion. Only the year before the discovery of the mesotron, Yukawa had put forward his theory of the strong force, which predicted the existence of a particle weighing about 250 times the mass of the electron. When the mass of Anderson's mesotron was found to be close to this value, an obvious conclusion was that it must be the predicted particle. And so it seemed for a few years, until three young Italians – Marcello Conversi, Ettore Pancini, and Oreste Piccioni – made a surprising discovery in a remarkable experiment that they had begun secretly in Rome during the Second World War.

To begin with, their makeshift laboratory was in a basement near the Vatican City, where they were hiding from the occupying Germans. There they set up their apparatus of Geiger counters, some material to slow down the cosmic ray particles, and some magnetized iron bars, which acted as lenses to concentrate particles of the same electric charge. Their aim was to show that the mesotrons decayed and to measure their lifetime.

Yukawa's theory predicted that the negative mesotrons should be easily captured around the positively charged atomic nuclei and absorbed by the strong force before they could decay. So only the decays of positive mesotrons should be detected. Using iron to slow down the particles, the researchers observed just what they expected. Some of the positive mesotrons were found to stop in the apparatus and subsequently decay later into another particle, but a corresponding number of negative mesotrons simply stopped with no evidence of decay. So up to this point the Italians still believed that the mesotrons were indeed Yukawa's predicted particle. The surprise came when they changed the absorber to a lighter material, carbon, and they began to see negative mesotrons decay. This behaviour completely ruled out the mesotron's identification with Yukawa's particle. Yukawa had invented his particle to explain the strong force, and it should therefore have had a great affinity for nuclear matter: the negative version should have been absorbed by a carbon nucleus just as by an iron nucleus, before the end of its life. Anderson's mesotron was clearly something different.

The complete picture became clear in 1947 when a particle was finally discovered that did fit Yukawa's description completely – the pion. Soon afterwards the mesotron was renamed the muon.

Study of the muon's behaviour has repeatedly affirmed that it is like a heavy electron and is not influenced by the strong force, but physicists have still to solve the mystery of the muon's existence. In the mid-1970s fresh clues came with the discovery of another particle resembling the electron and the muon, the tau (see pp. 162–163), which weighs in at nearly 20 times the muon's mass. Why should there be three kinds of 'electron', and are there heavier ones still undiscovered? These are among the questions that continue to challenge particle physicists into the twenty-first century.

Fig. 5.6 (OPPOSITE) This pattern of light in the Super-Kamiokande detector in Japan reveals a penetrating cosmic ray muon, 1000 m underground. Grey dots mark phototubes on the walls of the cylindrical detector, 41 m tall and 39 m in diameter. The phototubes detect Čerenkov radiation – light produced when a charged particle passes faster than light does through the water filling the detector. The coloured dots indicate phototubes that have sensed light, the purple end of the spectrum indicating the earliest light to arrive. The muon entered through the circular bottom of the cylinder, where the earliest light appears, and left about 120 nanoseconds (120 billionths of a second) later through the side wall near the middle of the image.

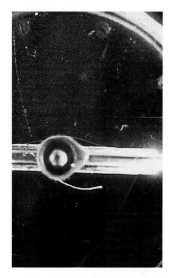

Fig. 5.7 A photograph by Anderson and Neddermeyer of a positive muon coming to rest before it decays in a cloud chamber that was activated by a Geiger counter inside the chamber. (The counter lies horizontally across the centre of the picture; the circular structure is part of the counter.) The incoming muon leaves a faint track at the upper left of the picture. The track curls round and becomes thicker after the muon loses energy in traversing the glass walls and copper cylinder of the Geiger counter. The chamber was not sensitive enough to record the track of the positron produced in the muon's decay. The muon travels 2.9 cm after emerging from the counter.

The Pion

One way to learn about an object is to pummel it. Shake electrons, for example, and they emit electromagnetic radiation; this is how radio waves are produced, when electrons are made to oscillate at particular frequencies. The burst of radiation is released by the disruption of the electric field surrounding the charged electron. What happens if instead we pummel protons? Again we find that in disturbing the tranquillity of a proton at rest, we release a burst of radiation. But this time it is not simply photons of electromagnetic radiation; instead, it consists mainly of particles known as pions. The pions are set free when the nuclear force field associated with the proton is disturbed, and the more energetically we disturb it, the more pions are produced.

Collisions between primary cosmic rays and the upper atmosphere produce positive, negative, and neutral pions in vast numbers. They are unstable, however, and decay rapidly. It is their children and grandchildren that form the bulk of the cosmic radiation near sea level. Particles decay via one or more of the fundamental forces and they always decay into particles lighter than themselves. Since pions are the lightest particles subject to the strong force, they cannot decay into lighter particles under the influence of this force. Pions would therefore be stable particles if they were not also subject to the electromagnetic and weak forces. These step in and cause the pions to decay in the following way.

The positive and negative pions decay within 10^{-8} s to positive and negative muons, which decay in their turn into positrons or electrons. The neutral pion, however, decays very much more rapidly – within 10^{-16} s – to gamma-ray photons, which then spawn the electron–positron pairs in the cosmic radiation. Partly because of the great rapidity of its decay, the neutral pion was not discovered until many years after its charged siblings and it is therefore described separately on pp. 108–109.

In Fig. 5.8 the decay of a positive pion is captured by a device called a streamer chamber, in which tiny luminous streamers form along the trails of ionization in a gas in a high electric field. The successive steps of the decay from pion to muon to positron are clearly visible. At each step the new particle deviates markedly from the path of its parent, indicating that invisible neutrinos are also released in the decay.

The pion had been 'on order' for more than a decade before it was discovered in 1947 in emulsions exposed on the Pic du Midi by Powell's group from Bristol. In 1935, Hideki Yukawa had proposed a new particle, necessary to convey the force between the protons and neutrons of the nucleus. This idea was not entirely new, for Yukawa was building on the early quantum theory of electromagnetic forces developed by Dirac in 1928. In this theory, electrically charged particles – electrons, protons – interact by exchanging bundles of light, or photons.

By analogy, Yukawa argued that protons and neutrons in a nucleus must exchange some particle. He reasoned that because the effect of the strong force is limited to the tiny dimensions of the nucleus, the carrier particle must be heavy – unlike the massless photon. Yukawa calculated that the particle must have a mass between that of the proton and the electron – roughly 15% of the proton's mass. Many physicists, including Yukawa, were at first misled by the discovery of the muon, in 1937, with almost the right mass. Powell's discovery of the pion in 1947 put the record straight.

Within a nucleus, pions form an invisible, evanescent web between protons and neutrons, binding them together. It is only when nuclear particles collide at high energies that we can see the pions liberated. This is how the pions are formed in cosmic rays, and in Fig. 5.9 we see an instance where both the birth and the death of a cosmic ray pion have been captured in emulsion. This picture, taken in 1947, was one of the first observations of a pion and demonstrated its strong affinity for nuclei. It had no sooner been produced at A than it shattered a nucleus at B.

Although the pions have been described as the transmitters of the strong force, they are not classed with the other force-transmitting particles, such as the photon or the W and Z particles. This is because they are now known not to be elementary particles, but composites made up of quarks. The strong force is transmitted by the pions only at the larger, nuclear level; at the deeper level of the quarks, it is transmitted by particles called gluons (see pp. 168–171).

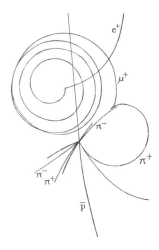

Fig 5.8 (OPPOSITE) The decay of a positive pion is captured in a streamer chamber image of an antiproton annihilating in the neon gas filling the chamber. The antiproton ($\bar{p}$) comes in from the bottom of the picture and interacts to produce a typical starburst of tracks. One of the positive pions (π^+) curls round on the right in the chamber's magnetic field, before decaying to a muon (μ^+) which forms a beautiful spiral. Eventually, the muon decays to a positron (e^+). At each decay the tracks change direction abruptly, indicating the simultaneous emission of an undetected neutrino. The thick tracks not identified in the diagram are nuclear fragments.

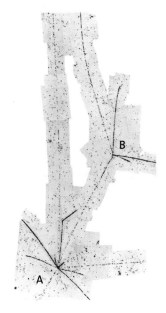

Fig. 5.9 The birth (A) and death (B) of a pion are recorded in this photograph taken by César Lattes, Occhialini, and Powell in 1947. It was one of the first observations of the creation of a pion. The distance between points A and B is about 0.11 mm.

The Kaon

On 15 October 1946, George Rochester and Clifford Butler observed something unusual in their cloud chamber at Manchester University. Two tracks like an upside-down 'V' appeared from a single point beneath a lead plate, as if from nowhere (Fig. 5.10). Their stereo views of the chamber showed that the tracks indeed originated from the same point and did not merely appear coincident from a particular perspective. Nor were the tracks caused by protons knocked out of the gas; the ionization and curvature showed them to be due to much less massive particles. In the following months, Rochester and Butler calculated that the two particles could be the decay products of a neutral particle with a mass some 800 times that of the electron – about half the mass of a proton – unlike anything they had seen before. Seven months after their original discovery, on 23 May 1947, Rochester and Butler found another unusual event, which seemed to be due to the decay of an electrically charged particle with a similar mass to the neutral particle. With these two discoveries, Rochester and Butler had found the first examples of decays of particles we now call kaons.

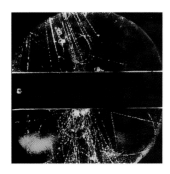

Fig. 5.10 (ABOVE) The original observation of a V particle, recorded in 1946 by Rochester and Butler at Manchester. This image shows a pair of tracks forming a pronounced fork to the right just below the lead plate across the centre of the chamber. This was probably due to a neutral kaon, produced in an interaction in the lead, which decayed into a negative and a positive pion.

Fig. 5.11 (RIGHT) A computer reconstruction of tracks in a modern particle detector echoes the first observation of a neutral kaon. This image shows the outline of an 'end view' of the central part of the cylindrical OPAL detector at CERN. High-energy electrons and positrons travel in opposite directions along the axis of the cylinder, occasionally annihilating to produce a burst of particles that shoot out sideways into the detector. In this instance, the particles include a neutral kaon, which leaves no track before decaying into a characteristic 'V' indicated by the magenta tracks. (The central tracking detector is 3.7 m in diameter.)

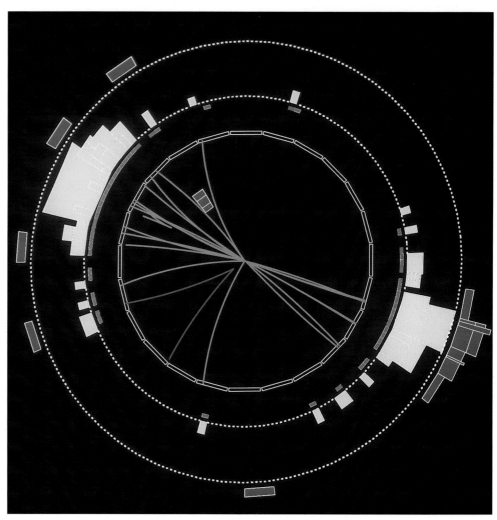

With hindsight we can say that Fig. 5.10 shows the decay of a neutral kaon to two pions (one positive, one negative). The characteristic 'V' can also be seen in Fig. 5.11, which is an image made nearly half a century later in the OPAL experiment at CERN. Cloud chamber or modern electronics: the tools have developed and the imagery too, but the 'signatures' of particular particles remain.

The charged kaon's most common mode of decay, occurring 63% of the time, is to a muon and a neutrino. Less frequently – 21% of the time – a charged kaon will transmute into a charged pion together with a neutral pion, as shown in a beautiful example in Fig. 7.2 (p. 108). However, as Fig. 5.1 (p. 64) shows, the decay of a charged kaon to three charged pions leaves a particularly distinctive signature, which is easy to spot despite the fact that only 5% or so of charged kaons decay this way. The neutral kaon decays most frequently to two charged pions – one positive, one negative – as Figs. 5.10, 5.11, and 5.12 all show, though it too can decay to neutral pions or to combinations of charged pions with muons or electrons and neutrinos.

The many different decays of the kaon posed problems for physicists studying the particles in cosmic rays. From the few events observed in emulsions and cloud chambers it was not entirely clear whether they were dealing with several particles of similar mass, or a single type of particle that could end its life in a variety of ways. Only with the advent of particle accelerators, which produced large numbers of kaons under controlled conditions, did it finally become clear that there is one type of particle and that it can be positive, negative, or neutral.

The kaon was the first of a number of particles found in cosmic rays that were dubbed 'strange'. This name arose because the particles all live for a surprisingly long time – in the case of the kaon, about 10^{-8} s, which is a million billion times longer than expected. The kaon, like the pion, is produced by the strong force; unlike the pion, it should be able to

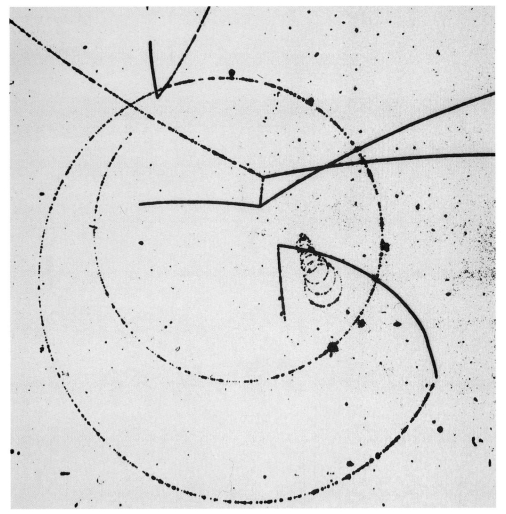

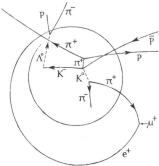

Fig. 5.12 A charged kaon and a neutral kaon appear together on this photograph from the Saclay bubble chamber at CERN. An antiproton ($\bar{p}$) enters from top right and annihilates near the centre of the image with a proton in the bubble chamber liquid, creating a neutral kaon (K^0), a negative kaon (K^-), and a positive pion or pi-plus (π^+).

The pi-plus moves up the picture, leaving a short track before it bounces off a proton in the liquid and is knocked towards the upper left corner, while the proton shoots off right. The negative kaon moves left, leaving a dense track before it interacts with another proton. This interaction produces a neutral lambda particle (see pp. 76–77), which leaves no track but which travels up the picture (dotted track labelled Λ^0) before it decays to produce a V consisting of a proton (p) and a pi-minus (π^-).

The neutral kaon produced in the initial annihilation leaves no track, but moves down and slightly to the right, before it decays to create a characteristic V formed by a pi-minus and a pi-plus. The pi-plus decays to a muon (μ^+), which lives so brief a time as to be indiscernible before it too decays to a positron (e^+), which leaves the broad spiral track. The neutrinos emitted in the pion and muon decays are invisible, but cause the abrupt change in direction of the visible tracks.

decay via the strong force. As we have seen, the pion cannot decay 'strongly' because it is the lightest particle subject to the strong force. The heavier kaon, on the other hand, should be able to decay strongly to produce pions, in a mere 10^{-23} s. Instead, with kaons and certain other particles, the strong force seems to have been cut off by something, and this is what at first seemed strange. To illustrate by how much its effects are postponed, one scientist said, 'It is as if Cleopatra fell off her barge in 40 BC and hasn't hit the water yet.'

The mystery of this postponed death began to be explained in the early 1950s when a whole family of strange particles first became apparent from studies of cosmic rays. The discovery of the kaon was soon followed by that of the lambda, and as the following pages describe, it provided the first clues as to just what strangeness is.

The Lambda

Fig. 5.13 A high-energy proton (yellow) enters from the bottom and collides with a proton at rest in the liquid hydrogen of the '80 inch' (200 cm) bubble chamber at the Brookhaven National Laboratory. The small electron spiral (green) shows that negative particles curl anticlockwise and positive particles clockwise. The collision produces seven negative pions (blue); nine positive particles (red), which include a proton and a positive kaon as well as seven positive pions; and a neutral particle – a lambda. The lambda travels up the picture leaving no track, but betrays its existence when it decays into a proton (yellow) and a negative pion (purple), which curls rapidly to the left. (This picture has been 'cleaned up' to show only the relevant tracks.)

The lambda particle leaves one of the most distinctive signatures in particle track detectors: it writes its own Greek name – Λ. In Fig. 5.13 we see an inverted 'lambda' formed by two tracks emanating from the decay of a neutral particle, produced in a high-energy collision along with 16 charged particles. This decay into a proton and a negative pion is the most common decay of the lambda, occurring 64% of the time, though often the particle vexes physicists trying to track it by decaying into two neutral particles, a neutron and a neutral pion. Like the kaon, the lambda is a 'strange' particle – it lives for 10^{-10} s; but unlike the kaon it is heavier than the proton and neutron. Indeed, it was the first such 'hyperon', or heavy particle, to be found.

The first images of decaying lambda particles came in 1951 from a number of cloud chambers: from Manchester University's device, once it had been raised to the lofty heights of the Pic du Midi; from Anderson's cloud chamber on White Mountain in California; and from Robert Thompson's chamber in Indiana. These images showed clearly tracks of different quality emerging from the 'V' of the neutral decay, but precise identification of the particles was often difficult. The team from Manchester established a neat method of analysing the motions of the particles, and used it to show that their neutral Vs corresponded to two kinds of particle: one about 950 times the mass of the electron, and the other 2250 times more massive than the electron, or some 20% heavier than the proton. The former was the neutral kaon; the latter the lambda particle.

So in the early 1950s, physicists were faced with two varieties of strange particle, which lived far longer than expected. What were they to make of them? Soon after the discoveries, Kazuhito Nishijima and other theorists in Japan, as well as Abraham Pais in the USA, began the process of unravelling the strange code. They proposed that these particles are produced by the strong force in pairs, and that they can be disrupted by the strong force only in pairs. When a pair of strange particles separate from one another, the strong force can no longer act on them as a pair, and their deaths are postponed. Instead, they decay by the much feebler electromagnetic and weak forces that are responsible for the decay of the pion or neutron.

According to the theory of production in pairs, or 'associated production', the lambda is produced along with another strange particle, such as the kaon. Confirmation of this came in 1954 with accelerators that could produce particle beams of high enough energy to create these particles. At the Brookhaven National Laboratory on Long Island, New York, experimenters found that a lambda and a kaon were often produced together. Figure 5.14 is a later bubble chamber photograph from Berkeley, which clearly shows the associated production of a lambda and a kaon, and their subsequent separate deaths.

Associated production was the first step towards solving the puzzle of the strange particles. The next step came in 1954 when American theorist Murray Gell-Mann, and independently Nishijima and T. Nakone in Japan, proposed that 'strangeness' is a new property of matter, akin to electric charge. Just as electric charge is conserved, so is strangeness conserved when the strong force is at work.

A pion and a proton have no strangeness. If they collide, as in Fig. 5.14, and produce a neutral kaon, with strangeness +1, then they must balance the books by producing a particle with strangeness –1, the lambda. This is why strange particles are always produced in pairs. (The allocation of positive strangeness to certain particles and negative strangeness to others is of course arbitrary, just as the allocation of negative electric charge to electrons and positive charge to protons is arbitrary; the fact is that protons and electrons have opposite electric charges, and the positive and neutral kaons have opposite strangeness to the negative kaon and the lambda.)

There is a major difference, however, between strangeness and electric charge. The latter is conserved, as far as we know, under all circumstances. Strangeness, on the other hand, is conserved only in interactions via the strong force. Once created, two strange particles go their separate ways and usually decay via the weak force. The heavier strange particles, the xi and the sigma (see pp. 78–79), can decay to lighter strange particles as long as overall strangeness is conserved. But the two lightest strange particles, the kaon and the lambda, cannot decay into lighter strange particles; instead they decay separately into non-strange particles. And since on average the kaon decays in 10^{-8} s and the lambda in 10^{-10} s, there is a time – however brief – when there is an imbalance of strangeness. Whereas electric charge is conserved always, strangeness leaks away when the weak force acts. Physicists do not yet fully understand why this should be.

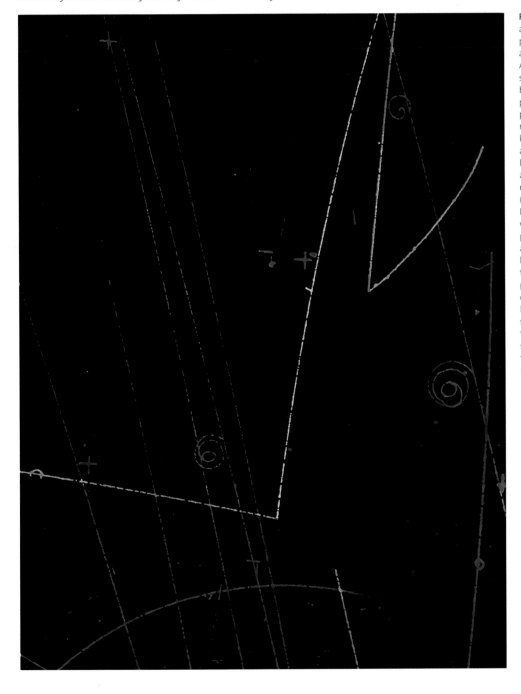

Fig. 5.14 An early photograph of the associated production of two strange particles, taken in a bubble chamber at the Lawrence Berkeley Laboratory. A pi-minus (green) with zero strangeness (S = 0) enters at the bottom right and interacts with a proton (S = 0) in the chamber liquid, producing a lambda (S = −1) and a neutral kaon (S = +1). The neutral kaon and lambda leave no tracks but are revealed when they decay; the lambda decays into a proton (red) and a pi-minus (green), while the neutral kaon decays into a pi-plus (yellow) and a pi-minus (green). Notice how strangeness is conserved when the strange particles are produced, being zero both before and after the initial interaction; this is because their production occurs via the strong force. When each strange particle decays, however, strangeness changes; both the lambda and the kaon decay into particles with zero strangeness. The decays occur via the weak force, and this allows strangeness to change one unit at a time. (The blue tracks are particles not involved in the interaction.)

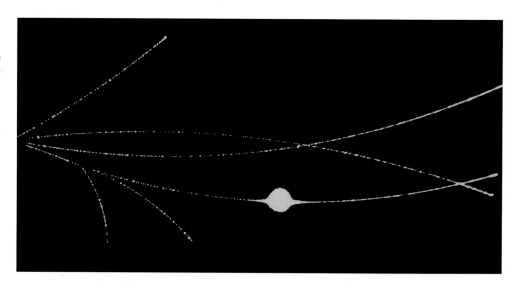

Fig. 5.15 A negative kaon interacts with a proton in the gas in a streamer chamber at the Lawrence Berkeley Laboratory. It produces, from top to bottom, a pi-plus, a pi-minus, a positive kaon, and a negative xi particle. The xi decays to a lambda and a pi-minus, which veers sharply downwards. The invisible lambda decays after a very short distance in a typical 'V', formed by a proton and a pi-minus, which again veers downwards. (Tracks not involved in the interaction have been removed from this photograph.)

The Xi and the Sigma

Two more strange particles – the negative xi or cascade particle, and the sigma – were discovered shortly after the lambda, and they helped to confirm the picture of strangeness that theorists were developing. In 1952, the cosmic ray group at Manchester chanced upon a startling image recorded by their cloud chamber on the Pic du Midi. By good fortune a particle never seen before had entered the chamber and decayed within it. The new particle decayed to a neutral lambda and a pi-minus, the lambda recognizable through its decay to a proton and another pi-minus. This proved to be the first time that a particle had been seen to descend to a proton in a sequence, or cascade, of decay steps. This simile led the physicists of the time to name it the cascade, today summarized by the Greek letter 'xi', or Ξ.

With the xi-minus, we begin to see strangeness as a property carried by particles in discrete amounts like electric charge. The xi has two units of negative strangeness and descends to a proton by shedding them one at a time. First it sheds one unit by decaying to the singly strange lambda, and then the lambda sheds its own unit when it decays to a proton and a pion.

This is not, of course, the whole story, since strangeness *is* conserved when particles are produced by the strong force. But by going back in time, to the production of a xi, we can see how the strangeness books are balanced at the time of creation. Figure 5.15 is a streamer chamber photograph in which a negative kaon, carrying one unit of negative strangeness, enters from the left and strikes a proton in the chamber's gas. The collision gives rise to two new strange particles, a xi and a positive kaon. The xi has strangeness −2 and the positive kaon strangeness +1, giving the same total of −1 strangeness brought into the interaction by the original negative kaon. At the next stage, the doubly strange xi loses one unit of negative strangeness when it decays into a singly strange lambda and a pion. The lambda decays in its turn into a proton and a pion, shedding its strangeness of −1. The positive kaon escapes the image before it too decays, and strangeness finally 'leaks away'.

In 1953, the year after the discovery of the xi, a group of Italian physicists identified a new strange particle in emulsion exposed to cosmic rays, and a similar object was also observed in a cloud chamber by a team from Caltech. The particle was positively charged, decayed to a proton, and analysis of the tracks showed that it was 30% heavier than the proton; as a result it became known at first as the 'superproton'. Later in the same year, a negatively charged version of the particle was found in accelerator experiments, and in 1956 a neutral version was identified in a bubble chamber at Brookhaven's Cosmotron accelerator. The three particles are today known as the positive, negative, and neutral sigma particles, after the 's' of superproton, and each carries one unit of negative strangeness.

Figure 5.16 shows the separate production and decay of both a positive and a negative sigma in a bubble chamber exposed to a beam of negative kaons. In the lower half of the photograph, one of the kaons collides with a proton in the bubble chamber liquid to produce a positive sigma (the short track) and a negative pion. This positive sigma takes a different decay path from that first noted by the Italians in 1953, transmuting to a positive pion and an invisible neutron. In the upper half of the photograph, another kaon interacts with a proton and produces a negative sigma, together with a negative pion and two positive pions. This sigma decays to a negative pion and another invisible neutron.

The theory of associated production and the concept of strangeness developed by Gell-Mann, Nishijima, and Nakone served to explain the observed behaviour of the strange particles discovered between 1947 and the end of the 1950s. It led to the prediction of the existence of the neutral sigma, and of the neutral xi, whose discovery is described in a separate portrait (see pp. 110–111). But how and why strangeness occurs in the first place remained a mystery.

The first steps towards its solution came in the following decade, when Gell-Mann and the Israeli physicist Yuval Ne'eman developed the classification of particles that became known as the Eightfold Way, and used it to predict successfully the existence of a particle with three units of strangeness, the omega-minus (see pp. 118–119). Shortly afterwards Gell-Mann went further and proposed the existence of a new level of elementary particles, the quarks, one of which is the strange quark. These theoretical developments were made possible only by the use of increasingly powerful particle accelerators and sophisticated new techniques for detecting the subatomic debris produced in high-energy collisions.

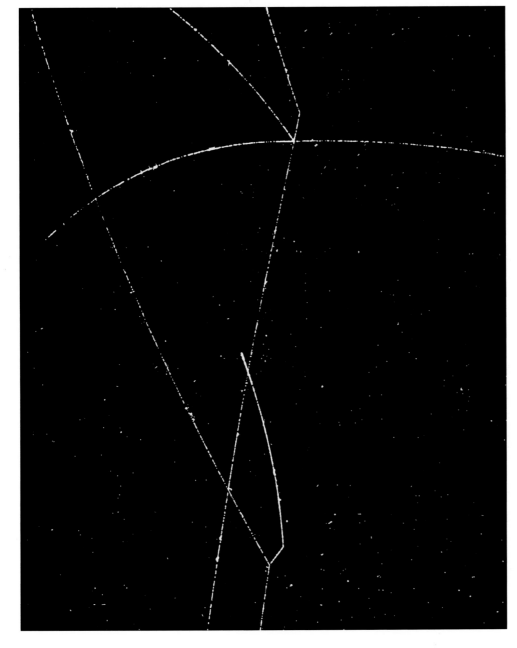

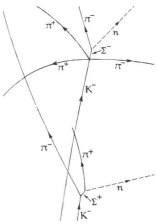

Fig. 5.16 This picture from the '72 inch' (183 cm) bubble chamber at the Lawrence Berkeley Laboratory shows the characteristic kinked tracks due to the decays of negative and positive sigma particles. The particles are produced in the interactions of negative kaons (K⁻). At the top of the picture, a sigma-minus (Σ^-) decays to a pi-minus (π^-) and a neutron (n), which leaves no track. Below it a sigma-plus (Σ^+) decays to a pi-plus (π^+) and a neutron, which again remains invisible. (This picture has been 'cleaned up', with all but the relevant tracks removed.)

6. The Challenge of the Big Machines

Fifty kilometres west of Chicago, sprawling over 2700 hectares, lies a site that incorporates a mid-western village to house visitors and their families, its own herd of American buffalo, and a number of fascinating examples of modern architecture. A typical American theme park? No, this is the Fermi National Accelerator Laboratory, better known as Fermilab – a prime example of a modern particle physics laboratory with a machine that currently produces the highest-energy particle beams in the world.

Fermilab's centrepiece is the Tevatron, a ring 6.3 km in circumference, etched on the Illinois prairie. The machine itself lies buried in a tunnel, but it is marked out above ground by a service road. The tunnel contains 1000 electromagnets, designed to steer protons on a circular course until they reach energies as high as 1 tera electronvolt (TeV) – a million million electronvolts (eV), where 1 eV is the energy gained from acceleration through 1 volt.

These are no ordinary electromagnets, however; they are superconducting magnets – cooled to a mere 4.7 degrees above the absolute zero of temperature, or −268.3 C. Electromagnets usually consist of coils of copper wire, which create a magnetic field when a current flows through the wire. The higher the current, the stronger the field – but only up to a point. The magnetic field 'saturates' at a value of about 2 teslas – roughly 100 000 times the strength of Earth's magnetic field. Increasing the current beyond this point heats up the magnet rather than strengthening the field. However, the higher the energy of a particle beam, the stronger the field needed to steer it round a ring of a given size. The Tevatron's ring originally contained copper-coiled magnets, which could steer protons up to a maximum energy of 500 GeV (0.5 TeV). To double this energy to 1 TeV required stronger fields, and a different technology for building magnets.

The Tevatron magnets have coils formed from superconducting cables of niobium-titanium alloy. A superconductor is a material in which electric currents can flow with practically no resistance, provided the temperature is extremely low – typically a few degrees above absolute zero. Superconducting magnets have two major advantages over conventional electromagnets. First, they produce higher magnetic fields, and secondly, because the current meets so little resistance, they can achieve these stronger fields with less electrical power.

Originally, when the Tevatron started up in 1985, its superconducting magnets shared the tunnel with the ordinary electromagnets of the Main Ring – the earlier 500 GeV machine – which served to provide the protons for the Tevatron. But in 1997, the Main Ring delivered protons for the last time, and when the Tevatron started up again three years later, its protons came from a new machine, the Main Injector.

An accelerator like the Tevatron boosts protons almost to the speed of light – 300 000 kilometres per second. But this cannot be done in a single step. The protons, which are initially almost stationary, must be accelerated through different stages, rather like going up through the gears when accelerating a car from rest. First gear at the Tevatron is a Cockcroft–Walton generator, a machine that is a direct descendant of the device built in the Cavendish Laboratory in 1932 (Fig. 2.30, p. 33), but which looks like a science fiction fantasy. Here protons are accelerated to an energy of 750 keV (0.00075 GeV), or about 4% of the speed of light. The protons originate as the nuclei of hydrogen atoms in a simple cylinder of compressed hydrogen gas. Before the first stage of acceleration, the gas is

Fig. 6.1 An aerial view shows the site of the Fermi National Accelerator Laboratory (Fermilab), near Chicago, with its two large accelerator rings marked out by service roads above the underground machines. The oval ring at bottom left is the new Main Injector, with a circumference of 3 km, built to feed high-intensity beams of protons and antiprotons into the Tevatron, which occupies the original 6 km circumference ring to its right. The straight lines heading diagonally from the Tevatron ring towards the top right corner of the picture are service roads running alongside the beam lines that feed particles from Fermilab's accelerators to its fixed-target experiments.

Fig. 6.2 The Cockcroft–Walton generator at Fermilab provides the first stage in the acceleration of the protons. Electrons are added to hydrogen atoms in the cube-shaped structure (an ion source) to the right of the picture. The resulting negative ions – each consisting of one proton and two electrons – are accelerated to 750 keV. They then pass, in a pipe, through the wall behind the source on their way to the second stage, the linear accelerator (linac). Negative ions, rather than protons, are used in the first stages of acceleration because they are easier to inject into the later stages and enable the machine physicists ultimately to get more protons into the Tevatron. The dome-topped structure contains components of the 'Cockcroft ladder' that builds up a high direct voltage (750 kilovolts) from the alternating mains. The dark columns are supports with smoothly curving 'corona rings' (the shiny collars) which help to prevent unwanted discharges. The technician is grounding any remaining electricity as a safety precaution after the generator has been used.

ionized, splitting the hydrogen atoms apart to yield electrons, protons, and negative hydrogen ions – hydrogen atoms to which an extra electron has stuck. It is these negative ions – each a proton with two electrons – rather than bare protons, that are used in the first stages of acceleration because they are easier to inject into the later stages and enable the machine physicists to get more protons to their final destination in the Tevatron.

From the Cockcroft–Walton generator, the protons (still in the guise of hydrogen ions) move on to second gear in the 150 m long linear accelerator or 'linac'. This consists of a series of copper cylinders in which electric fields accelerate the protons to 400 MeV (0.4 GeV) or about 70% of the speed of light. Here, as in most particle accelerators today, including the Tevatron itself, the electric fields are set up by radio waves in hollow copper vessels called 'cavities'. Radio waves, like all forms of electromagnetic radiation, are coupled vibrating electric and magnetic fields. When pumped in to a copper cavity of the correct size and shape they will form a 'standing wave', rather like a sound wave in an organ pipe but varying between regions of positive and negative electric field. Provided that the field is in the correct direction, protons entering such a cavity will absorb energy from the radio waves and will therefore be accelerated.

After the linac, the hydrogen ions move into third gear in the Booster, a small accelerator ring only 150 m in diameter. It is here that the ions lose their two electrons as they fly through a thin carbon foil, only 1.5 microns (0.0015 mm) thick, so that bare protons emerge into the machine. Electromagnets around the Booster steer the protons round and round the ring. On each circuit, the protons receive a small accelerating 'kick' from electric fields. After 20 000 circuits, taking only 1/30 of a second altogether, the protons have an energy of 8 GeV and are travelling at 99% of the speed of light. Now they are ready for injection into the ring of magnets that forms the Main Injector, the fourth gear at Fermilab.

The Main Injector lies adjacent to the Tevatron, like a smaller sibling with a circumference of 3 km. This is a new machine, which started up in 1999, with the task of providing the Tevatron with far more protons at a time than the original Main Ring did. It can accelerate 30 thousand billion (3×10^{13}) protons at a time, up to an energy of 150 GeV – but it does more than feed the Tevatron.

One of the Main Injector's tasks is to direct protons at 120 GeV onto targets to create secondary beams of particles for experiments – in particular, neutrinos and kaons. This requires carefully extracting protons for up to a second at a time, while the beams continue to race round 100 000 times a second. The extracted protons strike special targets of carbon

Fig. 6.3 (LEFT) A view inside part of the 150 m long linear accelerator or 'linac' at Fermilab, which takes the protons (as negative ions) from 750 keV to 400 MeV (400 000 keV). It consists of a series of hollow copper cylinders ('tanks') in which oscillating electric fields are set up by radio waves. A series of 'drift tubes' runs along the centre. The tubes are spaced so that the particles pass between them when the field is in the direction to accelerate them, but are shielded within the tubes when the field is in the direction to slow them down.

Fig. 6.4 (RIGHT) An electromagnet being manufactured for Fermilab's Main Injector, which takes 8 GeV protons (and antiprotons) and accelerates them to 150 GeV. This is part of a dipole – two-pole – magnet, to bend the path of the particles round the Main Injector ring. The brown part contains the coil of copper wire through which the electric current will pass to create the magnetic field, with a north pole to one side of the coil and a south pole to the other. The final magnet will consist of two of these coils, located above and below the path of the particles, with a well-defined magnetic field between them to guide the particles around the ring.

or beryllium to produce showers of pions and kaons. The pions are allowed to decay to produce a neutrino beam, while the kaons can be separated out to form a kaon beam, each beam serving its own experimental area.

The Main Injector also directs 120 GeV protons onto a special nickel target to produce antiprotons, as many as 200 billion (2×10^{11}) in an hour. The antiprotons are accumulated in the small magnet rings of the 8 GeV Antiproton Source before being sent back into the Main Injector for acceleration to 150 GeV.

The Tevatron is Fermilab's fifth gear – the final stage in acceleration of both protons and antiprotons, taking them up to 1000 GeV, at 99.99995% the speed of light. Antiprotons, the antimatter versions of protons, have negative rather than positive electric charge, and this means that they can travel round the Tevatron's ring of superconducting magnets at the same time and at the same velocity as the protons, but in the opposite direction. Once the particles are at 1000 GeV, or 1 TeV, the two beams are allowed to collide head on – and the Tevatron has reached its final goal.

The Tevatron is a wonder of the modern world, its operation dependent on split-second timing and the reliable functioning of thousands of individual components, each of which can cause a breakdown if it fails. Every aspect of the system is monitored by banks of

Fig. 6.5 For 14 years the big ring at Fermilab contained two accelerators, one on top of the other in the same 3 m wide tunnel, as seen in this photograph taken in 1998. The 6 m long electromagnets in the upper ring – red and blue – guided protons and antiprotons as they were accelerated to 150 GeV; the yellow and red superconducting magnets in the lower ring form the Tevatron, and took the particles on the last part of their journey to 1000 GeV (1 TeV). In 1997–99, the upper ring of red and blue magnets was removed and its job was taken over by the completely separate Main Injector. The Tevatron, now occupying the tunnel in splendid solitude, remains fully operational.

powerful computers. In the main control room, the accelerator physicists can call up colour displays to show the status of thousands of parameters in the complex of machines.

Similarly, each of the experiments incorporates microprocessors to control the simpler aspects of the apparatus, as well as larger computers to take overall charge of the operation of the experiment. The researchers are not quite redundant, however! They must be on hand night and day while the experiment is running to see that nothing untoward happens. Like a continual industrial process, the average experiment at a modern particle accelerator requires a team of experts and technicians to work shifts and keep a 24-hour watch.

Such complexity is very different from the experiments that Rutherford performed nearly a century ago. It is even a long way from the cloud chamber experiments that first revealed new particles in the cosmic radiation. Yet the forerunners of the Tevatron and the other modern giant particle accelerators were invented in the early 1930s, just as cosmic ray research was entering its heyday.

The Whirling Device

In November 1927, in a presidential address to the Royal Society, Rutherford wished for 'a copious supply of atoms and electrons which have an individual energy far transcending that of the alpha and beta particles from radioactive bodies'. His words inspired physicists and engineers both in America and in Britain. In Rutherford's own Cavendish Laboratory, Cockroft and Walton built a machine that produced in 1932 the first nuclear disintegrations from artificially accelerated particles (see p. 32). But the invention that was to lead directly to today's giant accelerators was a different type of machine, the cyclotron. It was the inspiration of one man, Ernest Orlando Lawrence, who arrived at Berkeley in 1928 to be associate professor of physics.

The 27 year old Lawrence had originally intended to continue his researches on photoelectricity, but in 1929 he came across the doctoral thesis of Rolf Wideröe, a Norwegian engineer working in Germany. Wideröe had put into practice an idea for accelerating particles that had been suggested five years earlier by a Swedish physicist, Gustaf Ising. Lawrence immediately saw a way to improve Wideröe's device still further, and thereupon changed the course of not only his own future but also that of particle physics.

Ising and Wideröe had considered accelerating particles to high energy through a series of small pushes from relatively low accelerating voltages. In Wideröe's design, the particles travel through a series of separate metal cylinders in an evacuated tube. Within the

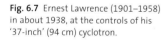

Fig. 6.7 Ernest Lawrence (1901–1958) in about 1938, at the controls of his '37-inch' (94 cm) cyclotron.

cylinders there is no electric field and the particles simply coast along. But across the gaps between the cylinders Wideröe set up electric fields by means of alternating voltages, which switch between positive and negative values. He matched the frequency of the alternating voltage with the length of the cylinders, so that the particles would always feel a kick, not a brake, as they emerged into a gap. In this way, the particles could be accelerated every time they crossed between one cylinder and the next. This is the basis of the operation of the modern linear accelerators, or linacs, used in the preliminary stages of acceleration at today's big rings.

Lawrence's inspired idea was to use a magnetic field to bend the particles into a circular orbit. Then they could pass across the same accelerating gap many times, rather than travel through a succession of gaps as Wideröe's scheme required. Lawrence saw that if the particles are accelerated on each circuit, they must spiral outwards as they increase in energy and become more resistant to the bending influence of the magnetic field. But he also realized that as the radius of the orbit increases, so does the particle's speed, with the result that the time taken for each circuit remains constant. Despite their spiralling orbit, the particles can still cross a gap at equal intervals of time and remain in step with an alternating accelerating voltage.

The principle underlying Lawrence's 'whirling device' was to place two hollow semicircular metal cavities, or 'Ds', between the circular north and south poles of an electromagnet. A gap separates the two Ds and an electric field across the gap accelerates the particles as they cross it on the first half of their circuit. On the second half of the circuit, the particles cross the gap again, but in the opposite direction; so if the particles are to be accelerated again, the electric field must have changed direction. To accelerate the particles continuously, the electric field in the gap must switch back and forth. All Lawrence needed to do was to match the frequency at which the electric field switched with the time taken for the particles to complete the circuit. Then particles issuing from a source at the centre of the whirling device would spiral out to the edge and emerge with greatly increased energy.

Lawrence announced the successful operation of his first 'cyclotron', as the whirling device had become officially known, to the American Physical Society in January 1931. Together with his research student Stanley Livingston, he had built a machine that accelerated protons to an energy of 80 keV. As a result, Lawrence received a grant of $500 from the National Research Council for work on a larger, useful machine.

A year later, Lawrence and Livingston, together with a new research student, David Sloan, successfully operated a cyclotron with a diameter of 28 cm – the '11 inch' – and reached the magical figure of 1 MeV. But in their zeal to improve the design of their accelerator, the team at Berkeley had neglected to exploit its applications. Instead, Cockroft and Walton observed the first artificially induced nuclear transmutations at Rutherford's Cavendish Laboratory.

Fig. 6.8 Rolf Wideröe (1902–1996), around 1920.

Fig. 6.9 (LEFT) Lawrence's first successful cyclotron, built in 1930, was only 13 cm in diameter and accelerated protons to 80 keV.

Fig. 6.10 (RIGHT) The basic components of a cyclotron are an electric field to accelerate the particles and a magnetic field to curve them round on a circular path. In practice, the magnetic field is supplied by two electromagnetic 'pole pieces'. These generate a vertical north–south field through the path of the particles, which are contained in a horizontal plane. The electric field is provided across a gap between two hollow D-shaped metal vacuum chambers. Particles from a radioactive source at the centre are accelerated when they cross the gap between the Ds. By applying an electric field oscillating at radio frequencies, the direction of the field can be changed so that it is in the correct direction to accelerate the particles each time they cross the gap. The particles curl round in the cyclotron's magnetic field, but as they increase in energy, they curl less, so that they spiral outwards until they emerge from the machine.

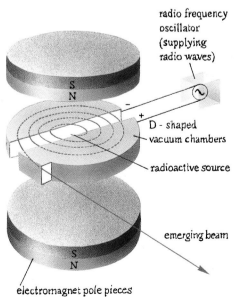

radio frequency oscillator (supplying radio waves)

D - shaped vacuum chambers

radioactive source

emerging beam

electromagnet pole pieces

Berkeley did however become the Mecca for accelerator builders, and spawned cyclotrons at other laboratories not only in the USA but across the world. Places such as Cornell, Princeton, Chicago and Michigan, Liverpool and Birmingham, Paris, Stockholm, and Copenhagen were soon to have cyclotrons of their own. Nor were these machines used only to do physics. Work at Berkeley had shown the importance of making radioactive isotopes for use in medicine, biology, and chemistry. The cyclotron gradually became a tool of the new research field of 'nuclear science'.

By 1939, the diameter of the largest cyclotron at the Radiation Laboratory at Berkeley had risen to 1.5 m – the '60 inch'. During the early 1930s, while the USA was sunk in a deep economic depression, Lawrence had, amazingly enough, created 'big science' – research on an industrial scale that involved the collaboration of scientists and engineers, many technicians and support staff, and large sums of money. Lawrence's talents lay not only in daring to reach for goals that were increasingly challenging, but also in a remarkable ability to raise the funds to support enterprises that were increasingly costly.

In November 1939, Lawrence was awarded the Nobel prize for physics for his invention and development of the cyclotron. Five months later he had a promise of $1.4 million from the Rockefeller Foundation to build a giant 100 MeV cyclotron – the '184 inch' – based on an enormous magnet with poles 4.6 m (184 inches) in diameter. Lawrence wanted to produce the supposed carrier of the strong force, later called the pion, and he believed that bombarding nuclei with alpha particles accelerated in his proposed machine would do the trick. With double the charge of protons, alphas would be accelerated to double the energy, or 200 MeV, and Lawrence calculated that 150 MeV alphas would be energetic enough to release the pions from the clutches of the strong force. But the Second World War intervened, and although Lawrence got his 4.6 m magnet, it was as a 'mechanism of warfare' which was used in a method he had devised to separate the fissile isotope uranium-235 from the much more common uranium-238.

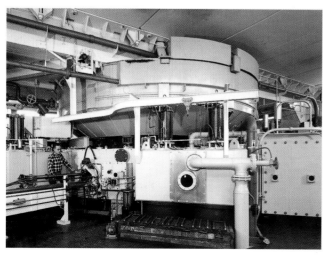

The pause for war had one fortuitous consequence. The original design for the '184 inch' would probably never have produced the desired beam of 150 MeV alpha particles. One effect of Einstein's theory of special relativity is that as objects approach the speed of light they become increasingly heavy. The cyclotron works on the principle that the particles always take the same time to complete a circuit, but this ceases to be true when special relativity applies. The heavier a particle becomes, the longer it takes to complete a circuit; eventually it will arrive too late at the gap between the cyclotron's Ds to catch the alternating voltage during the accelerating part of its cycle.

In the smaller cyclotrons made before the war, this effect was insignificant. But for protons of about 25 MeV, at about one fifth the velocity of light, the increase in mass of about 2% is enough to begin to make itself felt. This is really the

practical limit for a proton cyclotron. At 100 MeV, and approaching half the speed of light, protons are over 10% heavier than they are at rest. But Lawrence, in characteristic spirit, was not to be deterred by relativity. In 1939, he had hoped to beat the increase in mass by using the brute force of a very high accelerating voltage, taking protons to 100 MeV within a few turns. By the end of the war a more subtle technique had come to light, and one that could go far beyond the limit of 25 MeV.

Ed McMillan, 'conscripted' during the war from Berkeley to work on the atomic bomb at Los Alamos, and Vladimir Veksler in the Soviet Union, independently thought of the same idea to enable the cyclic accelerator to break free from the constraints of relativity. They proposed adjusting the frequency of the applied voltage so that it remains in step with the particles as they take longer to circulate.

A machine operating at variable frequency could no longer accelerate a continuous stream of particles, as the cyclotron had done. Changing the frequency to keep in time with higher-energy particles would mean that any particles still at lower energies would become out of step. Instead the 'synchronized' cyclotron, or synchrocyclotron, would take particles from the source a bunch at a time, and accelerate these bunches out to the edge of the magnet. The frequency of the accelerating voltage would meanwhile decrease to compensate for the particles' increasing mass. The final energy of the particles is then limited only by the strength and size of the magnet.

When McMillan returned to Berkeley after the war, his idea for varying the cyclotron frequency was applied to the design of the '184 inch'. The great magnet was relieved of its uranium enrichment duties and could at last be incorporated in a particle accelerator. At the beginning of November 1946, the new synchrocyclotron produced its first beam – deuterons with an energy of 195 MeV. But before the physicists at Berkeley began to search for pions, they were overtaken by events in cosmic ray research. Powell and his colleagues found the charged pion early in 1947. However, as a 'consolation prize', Berkeley was rewarded with the discovery of the neutral pion two years later.

Fig. 6.14 Members of Lawrence's 'Rad Lab' relax at a party in Berkeley in 1939. Lawrence, fork in hand, is seated at the head of the table to the left; McMillan is between the two women with polka dot dresses at the same table.

Fig. 6.15 Edwin McMillan (1907–1991) explains the principle of phase stability.

Man-made Cosmic Rays

Lawrence's 4.6 m synchrocyclotron complemented cosmic ray studies by producing copious supplies of pions to order. But even as it did so, in 1947, the cosmic ray physicists found the first of a series of exotic new particles. The kaon and its fellow strange particles were significantly heavier than the pion; some were even heavier than the proton!

Lawrence's machine was not powerful enough to produce these heavy particles. It was limited by the strength of the magnetic field and the diameter of the magnet's poles: once the accelerated particles reached a certain energy their orbits could no longer be contained between the poles. As so often in the history of accelerators, the cry went up for 'more energy'. But Lawrence's 4.6 m magnet was as large as it was practical to make, so how could higher energies be reached?

The solution was to alter not only the frequency of the accelerating voltage to match the increasing energy of the particles, but also to increase the magnetic field. If the magnetic field is strengthened continuously as the circling particles gain energy, they can be kept on more or less the same orbit instead of spiralling outwards. Moreover, the enormous single magnet of the cyclotron can be replaced by a doughnut-like ring of smaller magnets, each with a profile like a 'C'. The particles travel through a circular evacuated pipe held in the embrace of the magnets; they are accelerated during each circuit by an alternating voltage of varying frequency, which is applied at one or more places around the ring; and they are held on their circular course through the pipe by the steadily increasing strength of the magnetic field. Such a machine is called a synchrotron, and it is still the basis of large modern accelerators, such as the Tevatron at Fermilab.

When McMillan returned to Berkeley after the war, he set about building a prototype electron synchrotron. It was easier for technical reasons to begin with an electron machine rather than a proton device. But although electron synchrotrons were to play an important role during the next 30 years, it was proton synchrotrons that became the order of the day, both at Berkeley and at other laboratories around the world.

Fig. 6.16 The Cosmotron at the Brookhaven National Laboratory was the first proton synchrotron to come into operation, in 1952, accelerating protons to an energy of 3 GeV. The magnet ring was divided into four sections (the nearest is clearly visible here) each consisting of 72 steel blocks, about 2.5 m x 2.5 m, with an aperture of 15 cm x 35 cm for the beam to pass through. The machine ceased operation in 1966.

In 1947, the US Atomic Energy Commission approved the building of proton synchrotrons at two competing sites – Berkeley on the West Coast and the Brookhaven National Laboratory on Long Island, New York. The machine at Brookhaven was designed to reach 3 GeV, so that its beam of protons would produce pions in profusion after colliding with a suitable target. Berkeley's preliminary goal was to find the antimatter counterpart of the proton, the negatively charged antiproton. The antielectron, or positron, had been discovered in cosmic rays by Carl Anderson in 1932. Detecting the antiproton would provide the missing link in establishing that the laws of physics are symmetrical between matter and antimatter. Theory suggested that an energy of just over 6 GeV would be necessary to produce antiprotons from the collisions of protons with a target, so Berkeley aimed for this higher energy.

The 3 GeV machine at Brookhaven, the Cosmotron, became the first proton synchrotron to operate, in 1952, and it led the field for two years. Early experiments there complemented well the work done on strange particles with cosmic rays. It discovered the negatively charged partner of the positive sigma particle found in the cosmic radiation. And, more importantly, it provided the first concrete evidence that the 'vees' formed by the decays of two kinds of strange particle – the kaon and the lambda – always emerge together. This did much to strengthen the theory of associated production, which had predicted that strange particles are always produced in pairs.

Meanwhile, in California, the Bevatron was nearing completion at Berkeley. By November 1954, it was delivering 10^{10} protons per pulse at 6.2 GeV, and in 1955 a number of teams began the hunt for the antiproton. There were already faint indications that such an object might have been found in cosmic ray experiments in Europe; Berkeley did not want to be eclipsed by the cosmic radiation yet again.

The first antiproton searches at Berkeley used the tools of the cosmic ray physicists – emulsions and cloud chambers. But because the antiprotons were rare, the photographs revealed no signs of the anticipated nuclear starburst that would result from a proton–antiproton annihilation. The collisions between the accelerated protons and

Fig. 6.17 From 1954 to 1993, the Bevatron at the Lawrence Berkeley Laboratory accelerated protons up to an energy of 6 GeV. The Bevatron's ring of magnets weighed 10 000 tonnes – five times greater than the magnet ring in the Cosmotron.

protons at rest in a target would produce only one antiproton for every 50 000 pions. What the physicists needed were techniques that would automatically sift out the occasional antiprotons from the large 'background' of pions before the information about the particles was recorded.

Two teams, led by Edward Lofgren and Emilio Segrè, planned to seek out antiprotons in this way. They designed a series of detectors to determine the momentum and velocity of the particles created in the collisions. If you know a particle's momentum and velocity, you can calculate its mass; and if you find a particle with the same mass as the proton, but with negative instead of positive charge, you can be fairly certain that you have found one of the very rare antiprotons.

The hunt began with the selection of negatively charged particles from the debris produced by collisions of protons with a target inside the Bevatron's magnet ring. This was the easy part. The Bevatron's magnetic field bends positive and negative particles in opposite directions, so a beam of negative particles was selected by suitably aligning a hole in the accelerator's shielding with the internal target. A more delicate problem was to pick out from this beam the particles with the same mass as the proton, while ignoring the light pions and the slightly heavier kaons.

The first step was to use a magnetic field to spread the particles out according to their momentum, much as a prism spreads out visible light according to wavelength. A magnet bends particles with high momentum less than those of lower momentum, and so a suitably placed slit or collimator will allow through a narrow beam of particles, all of which have more or less the same momentum. Now 'all' that was needed was to measure the velocity of each particle, and thereby calculate its mass.

Segrè, together with Owen Chamberlain, Clyde Wiegand, and Tom Ypsilantis, chose to use two ways of pinpointing the velocity so as to be doubly sure that they had indeed captured an antiproton. One method used two scintillation 'counters', which produced a flash of light each time a charged particle passed through. Modern plastic scintillators are the descendants of the scintillating materials Rutherford used in his scattering experiments. But whereas Rutherford and his colleagues had to use their own eyes to see and count the flashes, by the 1950s electronic components made the process automatic. Each tiny burst of light is converted to a pulse of electricity, which is then amplified to produce a signal suitable for feeding into coincidence counting circuitry of the kind Bruno Rossi had invented in the 1930s. In this way, two or more scintillation counters can reveal the flight path of a particle as it produces flashes in each counter.

Segrè and his colleagues set up two scintillation counters 12 m apart. At the particular momentum the physicists had selected, antiprotons would reach the second counter

Fig. 6.18 Members of the team who discovered the antiproton surround Edward Lofgren, leader of the other team of antiproton hunters at Berkeley. Left to right: Emilio Segrè, Clyde Wiegand, Lofgren, Owen Chamberlain, and Thomas Ypsilantis.

eleven thousandths of a microsecond later than the faster, lighter pions. By using lengths of cable to delay the signal from the first counter, the physicists could make it coincide in time with the second signal, but only if the signals came from antiprotons. When a pion passed between the counters, it would make the journey too quickly, and the signals would no longer coincide. In this way, coincident signals were used to reveal the passage of the elusive antiprotons.

The second method the team used to measure the velocity of their particles was based on the Čerenkov effect, named after the Russian physicist Pavel Čerenkov, who discovered the phenomenon in 1934. When a particle moves extremely swiftly through a material, it can create a kind of shock wave of visible light, known as Čerenkov radiation. The crucial factor is that the particle must be moving faster through the material than ordinary light does in the same substance. The Čerenkov radiation emerges at an angle to the particle's path, and the greater the particle's velocity, the larger this angle becomes. By choosing the right material, you can create a window that reveals only particles above a certain velocity.

Segrè's team in fact used two Čerenkov counters, one containing an organic liquid ($C_8F_{16}O$), the other built from fused quartz. The liquid Čerenkov counter produced a signal for any particle moving faster than an antiproton, and so was able to identify the pions; the quartz counter, specially designed by Wiegand and Chamberlain, revealed only particles with a velocity close to that expected for the more sluggish antiprotons.

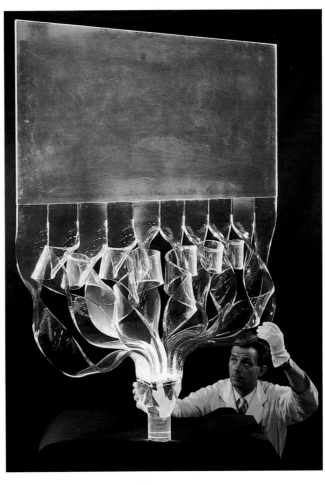

By early August 1955, Segrè and his colleagues had set up their apparatus at the Bevatron, and on September 21 they obtained their first evidence for antiprotons. Barely a month later, the team was confident enough to send a paper announcing the discovery to the *Physical Review*. They had sifted out 100 antiprotons from a background of 5 million pions. The Bevatron had done its stuff, and Segrè and Chamberlain shared the Nobel Prize in 1959. (See Fig. 7.6, p. 112, for the first image of an antiproton.)

Fig. 6.19 A large rectangular sheet of plastic scintillator is prepared for an experiment at CERN. The curly structures are acrylic 'light guides' which collect light emitted in the scintillator and concentrate it in the circular 'pipe' at the bottom. This pipe will fit against the surface of a photomultiplier tube, which converts the light into small electrical signals. The whole structure will be carefully wrapped, first in reflecting foil and then with layers of black plastic, to make it completely light tight. Notice how the light guides are designed to be all of the same length, so that the light from different parts of the scintillator arrives at the phototube at nominally the same time.

Fig. 6.20 Čerenkov radiation is responsible for the blue glow in the water surrounding the core of a nuclear reactor. Energetic charged particles from the reactor travel through the water faster than light does and as a result emit the Čerenkov radiation.

Fig. 6.21 Donald Glaser (b. 1926) inspects a xenon bubble chamber at the Lawrence Berkeley Laboratory in the early 1960s. Xenon is useful because it forms a dense liquid in which gamma rays readily become 'visible' by converting to electron–positron pairs.

Fig. 6.22 The track of a cosmic ray passes through Donald Glaser's first bubble chamber, a small glass phial holding a mere 3 cl of diethyl ether.

The discovery of the antiproton confirmed the promise of the Bevatron. With twice as much energy as the Cosmotron, Berkeley's new accelerator could discover new heavy particles, as well as investigate the behaviour of known lighter particles at high energies. But the high energies also made it more difficult to detect the particles. Energetic particles could hurtle all too easily through a cloud chamber without decaying or interacting with the atoms in the chamber's thin gas. For example, to record the whole life of a strange particle, from production to decay, at the Bevatron's energies would have required a cloud chamber 100 m long! In addition, cloud chambers are slow devices. The cycle of recompression after an expansion can take up to a minute; the Bevatron, on the other hand, delivered pulses of protons every two seconds.

What was needed was a detector that would capture the long tracks of high-energy particles and operate quickly. Gases were much too tenuous for the job. Liquids, on the other hand, were a more promising alternative, because their much greater density means they contain far more nuclei with which the high-energy particles can interact. But how do you make particle tracks visible in a liquid? The cloud chamber had depended on the production of liquid droplets in a gas, but liquid droplets in a liquid are like the proverbial black cat in a coal heap. The problem was to turn the black cats white.

The solution came not from Berkeley, but from a young physicist at the University of Michigan – Donald Glaser. Glaser had done his PhD at Caltech, where he had come under the influence of Carl Anderson in the late 1940s, just when the strange particles were causing consternation in the physics community. In 1949, Glaser moved to Michigan to begin teaching and research, and three years later he had the brilliant idea of how to make visible the tracks of particles passing through a liquid. He had worked out how to make the cats in the coal heap white.

A homely example of the effect Glaser wanted to harness is the action of opening a bottle of beer. The fall in pressure as you release the bottle's cap causes bubbles to rise through the liquid. Glaser's idea was to hold a liquid under pressure and very close to its boiling point. If you lower the pressure in these circumstances, the liquid begins to boil – an effect familiar to mountaineers, who can brew up a cup of tea on a mountain top at a lower temperature than is possible at sea level. But if you lower the pressure very suddenly, the liquid will remain liquid even though it is now above its boiling point. This state is known as 'superheated liquid' and because it is unstable, it can be maintained only so long as no disturbance occurs in the liquid.

Glaser realized that charged particles shooting through a superheated liquid will create a disturbance and trigger the boiling process as they ionize the atoms of the liquid along their paths. For a fraction of a second, a trail of bubbles will form where a particle has passed, and this trail can be photographed. But you must act quickly, or the whole liquid will begin to boil violently. Glaser therefore planned to release the pressure and then immediately restore it. Particles entering the liquid during the critical moments of low pressure would leave trails that could be photographed. The immediate restoration of pressure would mean that the liquid was once again just below boiling point, and the whole process could be repeated.

In the autumn of 1952, Glaser began experiments to discover if his 'bubble chamber' would work. After thoroughly considering possible liquids, he chose to use diethyl ether. With a small glass vessel holding just 3 centilitres of the liquid, he successfully photographed the tracks of cosmic rays. But he faced an uphill battle in developing his invention. He was refused support by the US Atomic Energy Commission and the National Science Foundation. They said his scheme was too speculative. And his first paper on the subject was rejected on the grounds that it used the word 'bubblet', which was not in the dictionary. But his luck changed in 1953, when a chance meeting brought the bubble chamber to fruition.

Glaser's first talk on his idea was to be given on the last day of the American Physical Society's meeting in Washington DC in April 1953. Among the participants at the meeting was Luis Alvarez, a distinguished physicist with a long record of important discoveries. He was involved with the Bevatron, which was still under construction at the time, and was concerned about the problem of how to detect the high-energy particles that the machine would produce.

On the first day of the meeting, Alvarez was sitting at lunch with colleagues from his wartime days at Los Alamos. On his left was a young man who had not experienced those times and was missing out on the reminiscences. Alvarez started talking to him about physics and the current ideas. The young man was Glaser, who complained to Alvarez that his ten-minute talk had been allocated to the final slot on the Saturday, by which time most people would be on their way home. Alvarez admitted that he too would be unable to attend the talk for that very reason. Sheepishly, he asked Glaser what he was going to report. Glaser explained how he had invented the bubble chamber and built a small version 2 cm in diameter. Alvarez was impressed; he realized immediately that this was the breakthrough he had been looking for.

That night Alvarez told his colleagues from Berkeley what he had learnt and suggested that it might be possible to build a big chamber filled with liquid hydrogen. This makes an ideal target for nuclear collisions because hydrogen is the simplest form of matter. Alvarez's colleagues were won over as dramatically as he had been. They all agreed that this was the way to proceed, and on their return to California they set about designing a large hydrogen-filled bubble chamber.

The idea of using hydrogen instead of ether made the work more difficult. Hydrogen becomes a liquid only when cooled to a chilly 20 degrees above absolute zero, or −253 C! But by the end of the year, only eight months after Alvarez had talked to Glaser, one of the group at Berkeley, John Wood, had observed tracks in a hydrogen-filled bubble chamber. It was only 3.5 cm in diameter, but it proved that the idea worked. Moreover, Wood made the important discovery that he could obtain clear images of tracks despite accidental boiling of the hydrogen induced by flaws in the chamber's walls.

Glaser, and everyone else, had originally thought that ultra-smooth walls were needed, and had therefore concentrated on glass chambers. Now that it was clear that smoothness was not so crucial, Alvarez's team turned to a construction based on metal walls with glass windows. They first built a chamber 6 cm in diameter and then a 10 cm chamber, which was ready for testing on the Bevatron in November 1954. The team then designed a 'big' chamber, 25 cm in diameter, which began regular work at the Bevatron in 1955. But Alvarez was already thinking much bigger.

In early 1955, before the 25 cm chamber was even complete, he proposed building a 75 cm chamber. Like Topsy, this grew in Alvarez's imagination until he eventually settled for a monster, 50 cm wide, nearly 40 cm deep, and 180 cm long. The '72 inch' would hold 17 cubic litres of liquid hydrogen and the window would contain 800 square centimetres of glass – which of course had to be thick enough to withstand the pressures inside. Even Lawrence, director of the laboratory at Berkeley and daring pioneer of the cyclotrons, was amazed at Alvarez's audacity. 'I don't believe in your machine,' he told Alvarez, 'but I do believe in you, and I will help you to obtain the money.'

The monster was not cheap. By the time it was completed in 1959, the '72 inch' had cost over $2 million. It was a far cry from Glaser's first tiny chamber. It occupied its own

Fig. 6.25 The '80 inch' (200 cm) liquid hydrogen bubble chamber at the Brookhaven National Laboratory in 1965. The stainless-steel chamber is almost totally obscured by the surrounding magnet coil, the huge steel magnet yoke, and equipment to expand the chamber and to keep the liquid hydrogen cool. Together the assembly weighed some 450 tonnes and stood about 7.2 m high, but it could be moved up and down, from side to side, and even rotated on a turntable, according to the desires of the experimenters. The hydraulic ram to move the apparatus sideways is visible at the lower left. The man on the platform just above the ram is removing one of three automatic cameras that photographed tracks in the chamber. The particle beam entered the chamber through the vertical rectangular 'window', seen to the right of centre. The chamber took 250 man-years to design and build, between 1959 and 1963, and cost in the region of $6 million.

building, complete with crane, compressors, and a magnet drawing 3 MW of power. Detecting the smallest fragments of matter had become big business.

As in the case of the cyclotron, other laboratories followed Berkeley's lead and built bubble chambers of various sizes and filled with a variety of liquids. One that was to gain fame in the 1960s was the '80 inch' (200 cm) hydrogen bubble chamber at the Brookhaven National Laboratory. This was fed particles from the successor to the Cosmotron, an accelerator known as the Alternating Gradient Synchrotron or AGS.

The operation of a bubble chamber is always intimately tied to the operating cycle of the accelerator that feeds it. In the case of the '80 inch', the expansion of the chamber began some 15 milliseconds before the burst of particles from the AGS was due. The expansion was accomplished by the withdrawal of a large piston, 90 cm in diameter and 80 cm high. The withdrawal of the piston through just 1 cm reduced the pressure inside the chamber from over 5 atmospheres to 2 atmospheres.

The particles entered the chamber when the piston was fully withdrawn, the pressure at its minimum, and the liquid superheated. Then, about one millisecond later, an arc light flashed on for a fraction of a millisecond. The flash illuminated the trails of bubbles formed by charged particles, and exposed the film in three or four cameras viewing the chamber. The delay between minimum pressure and the flash allowed the bubbles to grow to a diameter of about 10 millionths of a metre, large enough to show up on the photographs. Meanwhile, the piston moved back in towards the chamber, increasing the pressure again, and the film in the cameras was automatically wound on to the next frame. It then took about a second for the chamber to 'recover' and be ready for the next expansion.

One of the most famous discoveries of the '80 inch' occurred in 1964 when researchers found one photograph – out of a batch of some 80 000 – with the tell-tale pattern of tracks that betrayed the birth and death of the omega-minus particle (see Fig. 7.12 and pp. 118–119). In a typical bubble chamber experiment, over a million photographs may be taken, occupying hundreds of reels of film. How do the researchers cope with all this information, and find rare events like the production of the omega-minus?

The first task in examining a bubble chamber photograph is to spot the interactions that look interesting and to identify the particles that have produced the tracks. Sometimes certain tracks are instantly recognizable, like the tight spirals formed by low-energy electrons, but generally only careful measurement of the tracks gives the correct identification. The techniques used are basically the same as in the interpretation of cloud chamber or emulsion pictures.

For instance, the curvature of a track in a magnetic field reveals a particle's charge and its momentum. But this is not usually sufficient to label a track correctly because two particles of different mass and energy can have the same momentum. Often the

Fig. 6.26 A scanner at Fermilab works on a photograph taken at the laboratory's 4.6 m (15 ft) bubble chamber. A spray of particles appears as if from nowhere, produced by the violent interaction of a neutrino which, being neutral, leaves no track. The photograph is projected onto a table to enable the scanner to make accurate measurements of the positions, lengths, angles, and curvature of the particle tracks. From these measurements physicists can identify the various types of particles involved in the event.

only way is to assign identities to the different tracks, and then to add up the energy and momentum of all the particles emerging from an interaction. If they do not balance the known values before the interaction, the assumed identities must be wrong, and others must be tested, until finally a consistent picture is found.

Identifying particles through such trial-and-error calculations is the kind of repetitive job at which computers excel. The first machine devised at Berkeley in the late 1950s to scan bubble chamber photographs was nicknamed the Franckenstein after its creator Jack Franck. The device projected an image of a stereo pair of photographs. A human operator then set a photosensitive 'scanner' at the beginning of each track. The scanner would follow the tracks, bright against the dark background, and under the operator's control would punch out information about the trajectories on cards that a computer could read. The computer used this data to reconstruct the tracks in its 'brain' and compare them with preprogrammed patterns that the physicists thought might occur. Programming the computer to do this took two years. Even so, Franckenstein could analyse only 100 or so interactions a day, while the Berkeley '72 inch' was photographing thousands of interesting interactions in the same time. By the late 1960s, however, the machines had improved to the point where more than 100 photographs could be analysed per hour.

Strong Focusing

The bubble chamber, invented in 1952, was to be the workhorse of particle physics for the best part of 30 years, displaying the tracks of particles at ever increasing energies. But the early synchrotrons were soon to be superseded, and even as the Bevatron and the Cosmotron spewed forth their first protons, the ideas for a new breed of synchrotron were already on paper.

The basic concept of the synchrotron had reduced the huge magnet with circular poles, which the synchrocyclotron required, to a magnetic ring formed from smaller sectors. Even in the best-behaved synchrotron beams, however, the particles cannot all crowd exactly on to the ideal orbit. There is an initial spread to the beam as it is injected into the ring; collisions with residual air molecules in the vacuum chamber can deflect the paths slightly; and the particles will have slightly differing energies. To keep a hold on as many particles as possible, the poles of the C-shaped magnetic sectors in the first synchrotrons were gently shaped to provide a weak magnetic focusing, so that stray particles returned to the ideal orbit. The result was that the particles oscillated about the perfect path, weaving from outside the main orbit to within it, as they were nudged first this way and then that by the magnetic forces. But the focusing was weak; to keep hold of as many particles as possible, the 'racetrack' they whirled around had to be broad.

In the Bevatron, for example, the vacuum chamber through which the beam travels was 30 cm high and 120 cm wide; this formed the 'filling' for a magnetic doughnut weighing 10 000 tonnes. So the Bevatron gave nearly ten times the energy of the 4.6 m synchrocyclotron for two and a half times the weight – but 10 000 tonnes of iron was still a large amount. An even larger machine, which started up at the Dubna Laboratory in Moscow in 1957, accelerates protons to 10 GeV, but its racetrack is 40 cm by 150 cm, and it weighs a colossal 36 000 tonnes. At 10 GeV the weak-focusing synchrotron was in danger of becoming a dinosaur; how could it be saved?

Stanley Livingston and his colleagues at Brookhaven, Ernest Courant and Hartland Snyder, already had the rescue package for the synchrotron in 1952. They proposed a method for focusing the particle beams strongly, so that they swung less far from the ideal orbit. Unbeknown to the physicists at Brookhaven, a Greek engineer, Nicholas Christofilos, had patented the same idea in 1950; he later joined the rival establishment at Berkeley. The principle behind strong focusing is to shape the magnet pole faces so that they guide a deviant particle quickly back towards the middle of the vacuum chamber; such a particle will naturally swing across to the other side of the ideal orbit, but it will be swiftly directed back again, and criss-cross the vacuum chamber many times on its way round the machine. But there is one catch; a magnet shaped to focus the beam in the horizontal plane tends to defocus the beam vertically. To avoid losing as many particles in one direction as they were hoping to save in the other, Livingston and his colleagues, and Christofilos, realized that they needed to alternate two shapes of magnet.

Fig. 6.27 The 10 GeV proton synchrotron at the Dubna Laboratory in Moscow, the biggest to rely on the principle of 'weak focusing'. The magnet weighs a total of 36 000 tonnes, and must accommodate a beam up to 1.5 m wide.

Fig. 6.28 CERN's proton synchrotron – the 'PS' – accelerated its first protons to 24 GeV in 1959. The machine was later upgraded to run up to 28 GeV and today continues to form a vital part of a complex of interconnected machines that has accelerated beams of protons, antiprotons, electrons, positrons, and heavy ions.

The first would focus horizontally, but defocus vertically; the second would have the opposite effect, focusing vertically, but defocusing horizontally. With this combination, the net result would be a tightly controlled beam that could be kept in a smaller vacuum chamber, and which could be guided by smaller magnets with smaller jaws. The concept of strong focusing was first used in an electron synchrotron at Cornell University, in Ithaca, New York. There, Robert Wilson, who had worked with Lawrence at Berkeley, built a 1.5 GeV machine which started up in 1954.

Meanwhile in Europe, a number of nations had come together under the auspices of UNESCO with the idea of rebuilding the shattered remnants of European unity in some project that no country could afford on its own. Particle physics, with its huge accelerators, was a natural choice. This was the origin of CERN, the 'Conseil Européen pour la Recherche Nucléaire'; and as early as May 1951, a board of consultants chosen personally by Pierre Auger, UNESCO's Director of Natural Science, suggested building not only a modest synchrocyclotron, but also an accelerator that would be the biggest and best in the world.

In 1952, when a 'Provisional CERN' was officially set up, the task of exploring the possibilities for the big machine was given to one of four special study groups. The still untested potential of strong focusing was not lost on the experts involved; indeed, the European plans for a giant accelerator had spurred Livingston and his colleagues to devise a means of strong focusing. So a team began to design a strong-focusing synchrotron that could reach up to 25 GeV, four times higher than the Bevatron.

By the end of September 1954, CERN officially came into being, with a permanent Convention ratified at first by nine European countries, and in the following five months by three more. Now it had become the European Organization for Nuclear Research, but the acronym CERN stuck, and has stayed with it ever since. A site for the laboratory had been chosen on the outskirts of Geneva, and the synchrotron designers had already moved there in the previous October. By then the team had proposals for the new machine fit for the rest of the world to approve at an international conference that was also attended by representatives from Brookhaven. They too had plans for a 25 GeV machine, which the US Atomic Energy Commission approved shortly afterwards. The race was on.

CERN crossed the finishing line first. Its proton synchrotron – the 'PS' – accelerated protons to 24 GeV on 24 November 1959. The machine had been completed on schedule, less than seven years after the Convention was signed, and to cost – around £10 million. John Adams, who had led the accelerator team to its triumph, celebrated by opening a

Fig. 6.29 John Adams (1920–1984), on the day after CERN's new proton accelerator had reached 24 GeV, beating the record of 10 GeV held by the laboratory at Dubna. In his left hand he holds a picture of a monitor display confirming the energy; in his right hand he holds the (empty!) vodka bottle given by Dubna's director to be drunk once the record was broken.

bottle of vodka given to him by Vladimir Nikitin from the Dubna Laboratory in Moscow. Until that night Dubna had held the energy record with its 10 GeV machine. The following day the bottle was sent back to Dubna, empty of vodka, but containing a photograph of the instrument screen that proved that 24 GeV had been reached by the PS. The 10 GeV record was well beaten.

The PS contrasted completely with the Synchrophasotron – the name the Soviets used for their synchrotron at the Dubna Laboratory. With 100 strong-focusing magnet sectors arranged around a ring of 100 m average radius, the total weight of iron in the PS is 3200 tonnes – less than one tenth the amount in the weak-focusing 10 GeV machine at Dubna. Moreover, in the PS, the vacuum 'tank' of the weak-focusing machines has shrunk to an elliptical 'pipe' 14.5 cm across and 7 cm high.

Spark Chambers

CERN was first in the race to 25 GeV, but Brookhaven was not far behind, and in 1960 the AGS (Alternating Gradient Synchrotron) began operation. On 29 July, CERN's record was broken, when the beam in the AGS reached 30 GeV. By the following December the physicists at Brookhaven had begun experiments. The American researchers had a tradition of designing and operating experiments at Berkeley, especially on the Bevatron, as well as on the Cosmotron at Brookhaven. For the Europeans, working on such a grand scale was new, and after the brilliant success of the new accelerator at CERN it took some time for them to develop large experiments. During the 1960s, Brookhaven claimed several notable 'firsts', but the work at CERN was equally important, especially in the development of new kinds of detector to explore the territory that the 30 GeV machines had opened up. In particular, new detectors emerged to challenge the supremacy of the bubble chamber.

A bubble chamber can provide a complete picture of an interaction, but it has some limitations. It is sensitive only when its contents are in the superheated state, after the rapid expansion. Particles must enter the chamber in this crucial period of a few milliseconds, before the pressure is reapplied to 'freeze' the bubble growth. But how do you tell which incoming particles will produce interesting reactions? The question echoes the earlier difficulties with cosmic rays in cloud chambers. In that case, the problem of deciding when to take pictures could be solved because the cloud chamber has a 'memory'. Its expansion can be triggered after particles have passed through, using a signal from external counters that indicate that something interesting might have happened.

A bubble chamber cannot be triggered in this way; the expansion must occur before the particles arrive. And because the whole cycle of expansion and recompression takes about 1 second, the collection of rare events can take a long time. To study large numbers of rare interactions requires a more selective technique. In the 1960s, the spark chamber proved the ideal compromise.

Like the technique of coincidence counting, which was so vital in the discovery of the antiproton, the spark chamber was spawned from work on cosmic rays. Marcello Conversi – one of the Italians who had helped to identify the muon during the Second World War – had invented 'flash tubes' in the mid-1950s, and these became widely used to study cosmic ray showers. Flash tubes are sealed glass tubes filled with neon, which are arranged in layers between metal plates, rather like rafts. Charged particles ionize the gas in the tubes, and if a high voltage is applied to the plates, luminous discharges – sparks – occur in the tubes that the particles traverse. A photograph of the array of tubes end-on reveals the tracks of the particles. Around the same time, researchers in the UK and Japan independently developed a different way of putting the sparks to use. They did away with the glass tubes and applied the high voltage across plates with gas between them. They had invented the device that became known as the spark chamber.

The basic spark chamber consists of parallel sheets of metal separated by a few millimetres and immersed in an inert gas such as neon. When a charged particle passes through the chamber it leaves an ionized trail in the gas, just as in a cloud chamber. Once the particle has passed through, you apply a high voltage to alternate plates in the spark chamber. Under the stress of the electric field, sparks form along the ionized trails. The process is like lightning in an electric storm. The trails of sparks can be photographed, or

Fig. 6.30 Fred Ashton at work on an array of flash tubes at Durham University. The glass tubes, seen end on, are a little more than 1.5 cm in diameter and are filled mainly with neon. They are stacked between metal plates, across which a high voltage is applied when a charged particle has passed through. The particle ionizes the gas in the tubes it crosses and they 'flash' under the influence of the electric field between the plates, revealing the path of the particle.

their positions can even be recorded by timing the arrival of the accompanying crackles at electronic microphones. Either way, a picture of particle tracks for subsequent computer analysis can be built up.

The beauty of the spark chamber is that like the cloud chamber it has a 'memory' and can be triggered. Scintillation counters outside the chamber, which respond quickly, can be used to pinpoint charged particles passing through the chamber. Provided all this happens within a tenth of a microsecond, the ions in the spark chamber's gaps will still be there, and the high-voltage pulse will reveal the tracks. Any longer, and the ions will have been swept away by a low-voltage 'clearing' field that mops up unwanted ions.

A still better version of the spark chamber was invented in the 1960s by Frank Krienen at CERN. His idea was to subdivide the plates of the spark chamber into sheets of parallel wires, a millimetre or so apart. As before, when a charged particle travels through the spark chamber's filling of inert gas it leaves an ionized trail; a high voltage applied to alternate planes of wires provokes sparks to form along the trail. But the pulse of current associated with each spark is sensed only by the wire or two nearest to the spark. So by recording which wires sensed the sparks you have a reasonably accurate (to within a millimetre) idea of where the particle has passed. Notice how there is no longer any need for one stage in the data analysis – the film scanning necessary to convert visual information into numbers. The wire spark chamber produces information ready for a computer to digest with little further processing.

Fig. 6.31 Spark chambers in operation at CERN in 1969. Sparks fly along the ionized trails of charged particles in layers of gas sandwiched between metal plates. A high voltage applied across adjacent plates in the spark chambers makes the gas flash where it has been ionized.

Fig. 6.32 Frank Krienen with one of his wire spark chambers in 1963.

Fig. 6.33 (LEFT) Yuval Ne'eman (b. 1925) in 1966.

Fig. 6.34 (RIGHT) Murray Gell-Mann (b. 1929) in 1956.

Wire spark chambers became popular in the late 1960s, and several ways of recording the information from the wires were developed. As well as bypassing the need for film-scanning, the wire chambers offered the additional advantage of a faster response. This is because for electronic recording, the sparks do not have to grow as large as they do if they are to be photographed, and this in turn means that the chamber 'recovers' more quickly – in other words, the ions from one set of sparks can be mopped away more rapidly in preparation for the next trigger pulse. Wire spark chambers can be operated up to 1000 times per second – 1000 times faster than most bubble chambers.

The wire spark chamber fitted in particularly well with the computer techniques for recording data that were developed in the 1960s. Signals from many detectors – scintillation counters, Čerenkov counters, wire chambers – could be fed into a small on-line computer. The computer would not only record the data on magnetic tape for further analysis off-line, but could also feed back information to the physicists while the experiment was in progress. Sets of chambers with wires running in three different directions provided enough information to build up a three-dimensional picture of the particle tracks. And the computer could calculate the energy and momentum of the particles and check their identification, exactly as in bubble-chamber analysis.

Experiments based on spark chambers, scintillation counters, and Čerenkov counters proved a useful complement in the 1960s to the bubble chambers at CERN and Brookhaven. The spark chambers allowed the rapid collection of data on specific interactions; bubble chambers, on the other hand, gave a far more complete picture of events, including the point of interaction or 'vertex'. The 'electronic' and 'visual' detectors were complementary, and together they proved a happy hunting ground for the seekers of previously unknown particles. First in cosmic rays, then at the accelerators, physicists were discovering an ever growing 'zoo' of particles. This apparent complexity of nature at the subatomic level was compounded by the discovery in the 1950s of the 'resonances' (see pp. 115–117), extremely short-lived energetic states of the common proton and heavier particles.

The confusion began to be resolved in 1962. Theorists Murray Gell-Mann and Yuval Ne'eman had realized independently that the known particles, including the resonances, could be fitted symmetrically into a series of 'families'. Gell-Mann called this beautiful symmetry the 'Eightfold Way' after the Buddha's 'Eightfold path to truth'. At a meeting at CERN in 1962, Gell-Mann proposed the existence of a new particle that would make one of the Eightfold Way's 'families' complete. He called it the omega, for obvious reasons, and the race was on to discover it. In February 1964, a team at Brookhaven studying the interactions of kaons in the '80 inch' bubble chamber found the vital evidence to clinch the theory – the decay of a particle that could only be the predicted omega (Fig. 7.12, p. 118). Within weeks, similar evidence at CERN confirmed Brookhaven's discovery.

The Supersynchrotrons

With the discovery of the omega particle, the 'Eightfold Way' symmetry proposed by Gell-Mann and Ne'eman seemed on firm ground. But it did not tackle the fundamental problem of the early 1960s. Why is nature so complex? Why is there this great diversity of particles? Gell-Mann's answer was that the observed particles are not in fact fundamental objects, but are built from more basic building blocks, which he called 'quarks'.

The idea of quarks took some time to catch on, the more so because there was little evidence that quarks could be knocked out of protons, say, in the same way that protons can be knocked out of nuclei, or electrons out of atoms. But the search for simplicity, and the possibility of discovering 'free' quarks at higher energies, drove particle physicists in the early 1960s to consider still bigger machines, which would reach out to energies far beyond those of Brookhaven's AGS and CERN's PS.

In 1967, the Soviet Union again became world leader, with a proton synchrotron that could reach 70 GeV. This machine, at the Institute for High Energy Physics at Serpukhov, outside Moscow, was for five years the world's biggest particle accelerator. Physicists from Europe and the USA were keen to join in and investigate the new energy region; détente was the order of the day, at least in particle physics. Then the big machine at Fermilab started up in 1972, accelerating protons to 200 GeV. By 1976 it had successfully reached 500 GeV, but the route had by no means been easy.

Although the machine's design originated at the Lawrence Berkeley Laboratory, it was in reality the baby of Robert Wilson, one of the pioneers of electron synchrotrons at Cornell University. In 1967, Wilson became head of the project to build the new machine at a site known as Coon Hollow, to the west of Chicago. He faced the unenviable task of being allocated $250 million for its construction, rather than the $350 million the designers at Berkeley had said it would need.

Wilson decided not only to accept the financial challenge, but to exceed it by building a machine that would reach 500 GeV – two and a half times the original design energy! He has described this decision as 'close to bravado', but it was a calculated attempt to attract the right sort of people to the project. Wilson's 'band of stout-hearted men' put together a design for a machine with a ring exactly 2 km in diameter. Even with this large circumference, in order to reach 500 GeV the team still needed to design electromagnets that would produce magnetic fields nearly 20% higher than had been achieved before. To help do this, they used separate magnets to bend and to focus the beam – a new concept in synchrotrons.

In an attempt to speed up sluggish funding, Wilson announced in 1969 that the machine would be ready a year early, in July 1971. He would have been correct but for two factors. First, the 6.3 km long beam pipe turned out to have obstructions, sufficient to prevent protons from making a complete circuit of the ring. The desperate machine-builders even tried employing a ferret, named Felicia, to help in pulling magnets on wires through the pipe in an attempt to clear it! But the most devastating problem concerned the magnets. These were installed in a tunnel completed during a frozen Illinois winter. In summer, the tunnel became dripping wet, and when powered up, nearly half the magnets destroyed themselves in showers of sparks.

But by March 1972, the machine was accelerating protons to 200 GeV. It still proved a difficult beast to handle. To begin with it produced a proton beam only 50% of the time. Despite these drawbacks, many groups from the association of universities that runs Fermilab, and from outside the USA, successfully performed experiments in an exciting new energy region. And in May 1976, the machine reached 500 GeV – Wilson's dream had been fulfilled.

Fig. 6.35 Robert Wilson (1914–2000) in sodbusting pose in 1969 at the ground-breaking for Fermilab's Main Ring. As founding director of Fermilab, he led the construction of the world-beating accelerator, which started at an energy of 200 GeV in 1972 and reached 500 GeV in 1976.

Fig. 6.36 The tunnel of the Super Proton Synchrotron (the SPS), in the weeks before the machine accelerated its first protons in 1976. The SPS became the world's first proton–antiproton collider in 1981, and remains in use to this day as part of CERN's network of accelerators.

Meanwhile, Europe had also joined in the bid for higher energies. After some bitter wrangling among the member states, CERN had decided to build a machine that would take protons to 400 GeV. The Super Proton Synchrotron (SPS), as it became known, took six years to build, and it delivered its first beams almost as Fermilab's accelerator was finally reaching 500 GeV. The two machines delivered proton beams at more than 10 times the energy of their predecessors, but almost as soon as they had begun plans were afoot to make the ultimate proton collision energies far higher.

As early as 1973, Wilson had begun a programme to develop superconducting magnets at Fermilab, with the aim of providing the stronger magnet fields necessary to guide higher-energy particle beams. And at a meeting in November 1978 led by Leon Lederman, who was about to take over from Wilson as the laboratory's director, a scheme was conceived to double the energy of Fermilab's protons. The idea was to build a ring of superconducting magnets in the same tunnel as the 500 GeV synchrotron, or 'Main Ring', which would be 'demoted' to the role of injector to the new synchrotron. The superconducting magnets would be designed to fit neatly beneath the conventional electromagnets of the existing machine. No project had used superconducting technology on this scale before, and Fermilab had to create its own assembly line to build, test, and modify the new magnets. The laboratory had also to build what was at the time the world's largest liquid-helium plant, to provide the coolant needed to keep the superconducting magnets at their operating temperature of 4.7 degrees above absolute zero. By June 1982, Fermilab was ready to turn off the Main Ring to make way for installation of the new ring beneath the old. A year later, on 3 July 1983, Lederman and his colleagues held their breath as the first protons circulated round the Tevatron – named for its ultimate top energy of 1 TeV (1000 GeV).

By 1983, the SPS had also been modified to reach much higher collisions energies – in this case by converting the machine to a proton–antiproton collider, as Chapter 8 describes

(see pp. 143–145). A similar conversion was also agreed in the original plans for the Tevatron, and by 1987 it too began operation with beams of protons and antiprotons colliding head on. Both the SPS and the Tevatron continue working to this day, though the Main Ring at Fermilab has now been replaced by a new machine, the Main Injector, as described earlier in this chapter (see p. 82).

The SPS and the Tevatron became the foremost proton synchrotrons of the mid-1980s. But what of electron machines? Near Stanford in California, not far south from Berkeley, lies the machine that was for 20 years the world's most powerful electron accelerator. This is not a circular synchrotron; it is a 3 km long linear accelerator, the longest linac in the world.

Why a linear accelerator? Electron synchrotrons work perfectly well apart from one fundamental problem: high-energy electrons radiate away energy when they travel on a circular path. The radiation – known as synchrotron radiation – is greater the tighter the radius of the orbit and the higher the energy of the particle. Protons also emit synchrotron radiation, but because they are 2000 times as massive as electrons, they can reach much higher energies before the amount of energy lost becomes significant. But even at only a few GeV, electrons circulating in a synchrotron radiate a great deal of energy. And this lost energy must be paid for by pumping in more energy through the radio waves in the accelerating cavities.

It was for these reasons that physicists at Stanford decided to build a huge linear machine, at the Stanford Linear Accelerator Center, or SLAC. The origins of SLAC go back as far as 1934, when William Hansen of Stanford University began to consider how to build a linear electron accelerator. This machine would be similar to Wideröe's pioneering device, but it would need a powerful source of high-frequency radio waves to accelerate the high-velocity lightweight electrons. (Recall that the frequency must match the time for the particles to travel from one accelerating gap to the next.)

Fig. 6.37 The 3 km (2 mile) long linear accelerator at the Stanford Linear Accelerator Center (SLAC). The electrons start off from an electron 'gun' where they are released from a heated filament, at the end of the machine near the bottom left of the picture. The electrons in effect surf along radio waves set up in a chain of 100 000 cylindrical copper 'cavities', about 12 cm in diameter. The machine is aligned to 0.5 mm along its complete length and is situated in a tunnel 8 m below ground. The surface buildings that mark out the line of the linac contain the klystrons (see Fig. 6.40, p. 104), which provide the radio waves.

Fig. 6.38 (LEFT) William Hansen (1909–1949) with a section of his first electron linear accelerator, which operated at Stanford University in 1947. It was 3.6 m long and could accelerate electrons to an energy of 6 MeV.

Fig. 6.39 (RIGHT) Russell Varian (1898–1959), left, and his brother Sigurd (1901–1961). In 1937 at Stanford University, they together developed a powerful new source of radio waves – the klystron.

Hansen was soon joined by Russell and Sigurd Varian, who had been working in their own private laboratory on a means for generating and detecting radio waves of centimetre wavelength. Sigurd had been an airline pilot and he was keen to develop better navigational aids for aircraft; Russell, a graduate of Stanford, had been working in radio and television research and had the right kind of expertise to invent a suitable device.

The arrival of the brothers at Stanford, unpaid except for $100 for materials, proved a turning point. While the Varians had no ostensible interest in accelerating electrons, together with Hansen they developed the 'klystron' – a powerful source of radio waves that has become standard equipment in electron and proton accelerators, as well as for other applications including the transmission of television signals. Russell Varian worked out the design of the klystron in 1937; and with the combination of Hansen's experience and Sigurd's skill, the first device was operating in August the same year.

After the Second World War, Hansen's thoughts turned once again to building a linear electron accelerator, in which the particles were fed energy by the new powerful source of radio waves – the klystron. He developed a series of machines, each more powerful than the previous one, and by 1953 Stanford could boast an electron accelerator 63.6 m long, which reached an energy of 600 MeV.

Fig. 6.40 The power to accelerate the electrons along the 3 km linear accelerator at SLAC comes from devices called klystrons. These powerful sources of radio waves (in fact, microwaves) are installed in the 'klystron gallery' that runs along the surface, 8 m above the accelerator itself. The klystrons – the structures with the red cylinders – are about 12 m (40 ft) apart.

It was at about this time that the idea of 'the Monster' began to grow in the minds of several researchers at Stanford, including Wolfgang Panofsky. At a meeting at Panofsky's home on 10 April 1956, the first unofficial ideas about the Monster were recorded: length, 3 km (or more precisely, 2 miles); energy, 15 GeV or more. By 1959, President Eisenhower had recommended federal funding for the machine as a national facility. Thus SLAC was born, and in January 1967 'the Monster' reached its design energy of 20 GeV.

The machine at SLAC was later improved to reach an energy of 30 GeV, and ultimately achieved a maximum energy of 50 GeV. But some of its most intriguing results came in the first few years of operation. It was then that the first clear evidence emerged that protons and neutrons are built from still smaller objects. SLAC's high-energy electrons, fired from one end of the accelerator, proved the ideal tool to probe protons.

In 40 years, from the first accelerators of Lawrence and the other pioneers, ideas of the basic constituents of matter had changed enormously. In 1932 – the year the neutron was discovered – the picture had been of four basic particles: proton, neutron, electron, and the then hypothetical neutrino. Assuming Dirac was right – and the experiments at Berkeley in the mid-1950s showed that he was – then each of these 'elementary' particles should have its complementary antiparticle, thus bringing the grand total to eight. By 1973, the picture had at first become increasingly complex – with the discovery of many particles, especially at the Bevatron – and had then swung back again to simplicity. In 1973, the basic constituents of matter seemed to be the electron, the muon, two neutrinos (see pp. 120–123), and three varieties of quark – and of course their antiparticles. But as Chapter 8 describes, the course of physics was to change again the following year, in 1974.

Fig. 6.41 End Station A at SLAC is one of the experimental areas where the electron beam finally emerges from the linear accelerator. The beam enters through the pipe coming in from the left and collides with the 'target' (surrounded by concrete blocks). The electrons scatter from the target's nuclei through a variety of angles and three 'spectrometers' record the results. One is the tall grey cylinder behind the target area; the second is the structure incorporating the large yellow 'container', with the third partially hidden behind it. The spectrometers contain banks of different detectors to track the scattered particles, and magnets to measure their momentum. They can be rotated to different positions around the target along the rails that are visible. The scale of these instruments, which provided the first direct evidence for quarks, is given by the man standing at the foot of the yellow 'container'.

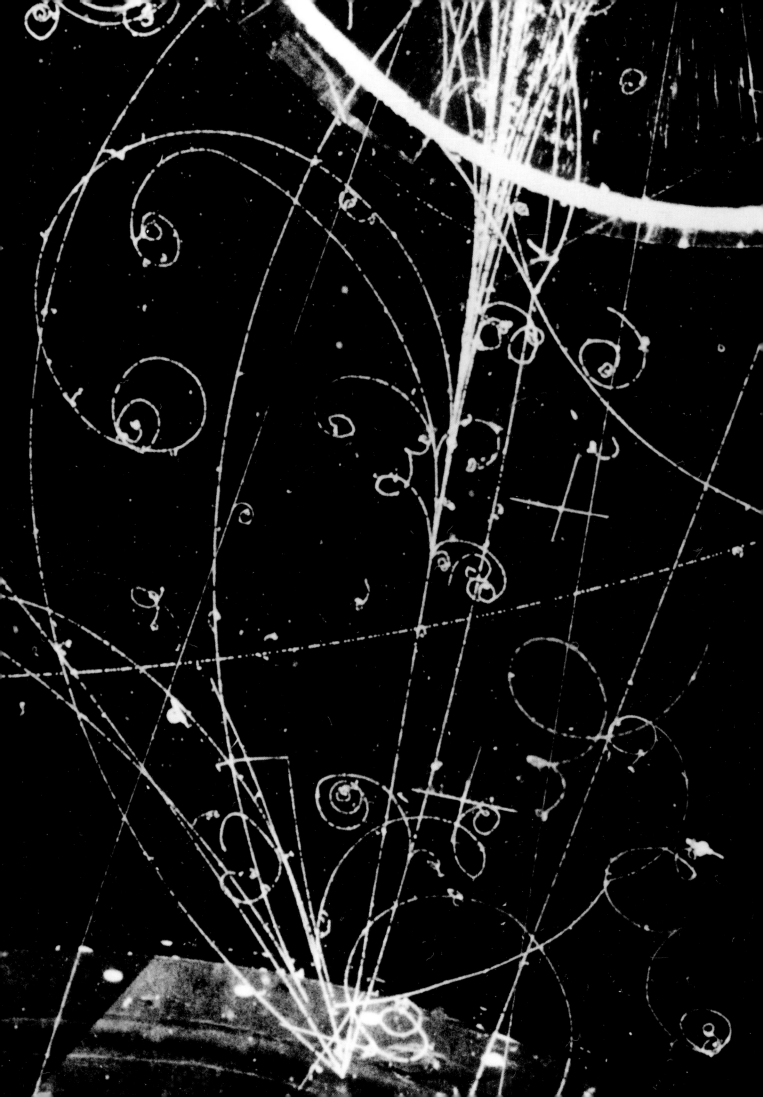

7. The Particle Explosion

The year 1952 was the beginning of a new era in particle physics. It saw the invention of a new type of detector – the bubble chamber – which was to dominate discoveries for the best part of 30 years; and it witnessed the first of a new breed of accelerator – the synchrotron – designed with the express purpose of creating man-made versions of the particles found in cosmic rays. The subject, which had first been a branch of nuclear physics, then a branch of cosmic ray research, had now graduated to become a fertile field of discovery in its own right. By the early 1960s, particle physicists were almost falling over each other in their efforts to find new particles.

Experiments at accelerators enabled the physicists to fill in gaps in the patterns of particles that were beginning to emerge. The first particle to be discovered at an accelerator – the neutral pion – completed the pion family of three. Similarly, the neutral xi, when at last discovered in a bubble chamber, provided a partner for the negative xi, which had been found in cosmic rays. With increasing amounts of energy at their disposal, experimenters confirmed Dirac's theory of antimatter, finding antiparticles for each of the known particles. The antiproton, the antineutron, the antilambda, and so on, followed in quick succession.

Though the bubble chamber provides pictures that in some cases can be read almost as easily as a book, or which are readily adaptable to become works of art, it is not always the best detector to use. In many cases, particularly with the rarer, shorter-lived particles, electronic techniques based on particle 'counters' have proved invaluable. In the following pages, we see how well bubble chamber and counter experiments complemented each other during the 1950s and 1960s, rather as cloud chamber and emulsion techniques had done previously. Particles such as the pi-zero and the antiproton first succumbed to counters. And counter techniques were vital not only in demonstrating the existence of neutrinos, but also in showing that there were two different types of neutrino, one associated with the electron and the other with the muon; a pattern was also emerging among the leptons.

Having already revealed an amazing wealth of particles, in the 1960s the two techniques together brought the startling evidence that the protons and neutrons are not the last word regarding the structure of matter. First came signs of the 'resonances' – very short-lived states that carry all the hallmarks of being complex vibrating structures. Their discovery came with the ability to analyse automatically hundreds of thousands of bubble chamber pictures. It was in fitting the resonances into the pattern of particles that Gell-Mann and Ne'eman produced evidence for a new symmetry of nature. The discovery of the omega-minus in 1964 vindicated these ideas and led to the suggestion that protons, pions, strange particles, and so on are all built from smaller entities – the quarks.

Then, at the end of the 1960s, electronic counter experiments at the Stanford Linear Accelerator Center (SLAC) began to show how high-energy electrons could penetrate the proton and pinpoint its granular nature. Sixty years after Rutherford had prised his way into the atom with alpha particles, the physicists at Stanford echoed his experiments but at a new, deeper level.

The effects that betray the presence of the quarks within the proton and other related particles are subtle in the extreme and stretch our concept of 'seeing' further than with any of the particles encountered so far in this book. Yet the effects are quite as real as with the other particles; it is just that we must look harder to perceive nature's deepest secrets.

Fig. 7.1 The bubble chamber came to dominate images of subatomic particles in the 1960s, providing beautiful swirling pictures with their own aesthetic appeal in addition to their scientific merits. This photograph is from Fermilab's 4.6 m (15 ft) bubble chamber, filled with liquid hydrogen and subjected to a beam of neutrinos. The neutrinos are electrically neutral and so do not produce tracks themselves, but one of them has interacted with a proton in the bubble chamber's liquid, producing the spray of charged particle tracks that shoots up from the bottom of the picture.

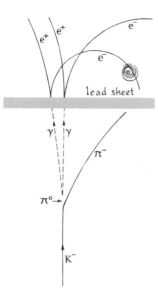

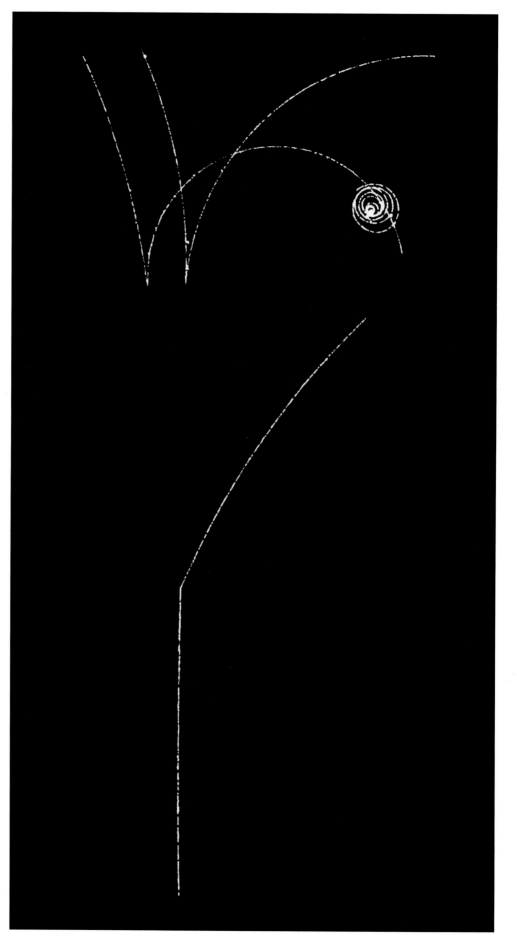

Fig. 7.2 A negative kaon (K⁻) decays in a bubble chamber at the Lawrence Berkeley Laboratory, producing a negative pion (π⁻) and a neutral pion (π⁰). The neutral pion immediately decays into two gamma rays (γ) whose paths are marked by the dotted lines in the diagram. The gamma rays strike a lead sheet in the chamber and each turns into an electron (e⁻) and a positron (e⁺). The bubble chamber's magnetic field curls the negative particles clockwise, the positive ones anticlockwise. The tight spiral towards the end of the track of the lower electron is another electron, which has been knocked out of an atom in the bubble chamber liquid; other extraneous tracks have been removed.

The Neutral Pion

The neutral pion, or pi-zero, was the first unstable subatomic particle to be discovered with the aid of an accelerator. It is the neutral partner to the positive and negative pions described on pp. 72–73, which were first observed in the interactions of cosmic rays. The pi-zero is produced just as readily as its charged siblings in cosmic ray collisions, but its lack of electric charge means that it behaves differently and is much more difficult to detect.

Charged pions live for 10^{-8} s before decaying into other particles, and as a result they leave relatively long tracks in cloud and bubble chambers. The pi-zero, on the other hand, lives for a mere 10^{-16} s, a hundred million times more briefly. And because it is electrically neutral, it does not leave any track in a cloud or bubble chamber. This invisibility is compounded by the fact that 99% of the time the pi-zero decays into two very energetic photons, or gamma rays, which are also neutral. So how do you detect a short-lived particle that leaves no track and which decays into two particles that also leave no tracks?

The answer is to use some special tricks to make the elusive particle reveal itself. All detectors of pi-zeros incorporate a dense material such as lead, which will coerce the invisible gamma-ray photons into producing electron–positron pairs. This effect is seen in Fig. 7.2, a photograph taken in a bubble chamber at the Lawrence Berkeley Laboratory. A negative kaon entering at the bottom produces a visible negative pion, which curves away to the right, and an invisible pi-zero. The pi-zero immediately decays into two gamma-ray photons which shoot, equally invisibly, towards the top of the picture. But where they meet a sheet of lead inserted into the bubble chamber, they turn into two visible electron–positron pairs.

Lead has proved to be an ideal material for 'converting' gamma-ray photons into electron–positron pairs. Today, many experiments incorporate stacks of lead-doped glass blocks – just like the 'lead crystal' of fine glassware – to detect pi-zeros. The lead encourages the photons to convert to electrons and positrons, which in turn radiate Čerenkov light in the glass. A phototube on the end of a block detects the light and hence registers the arrival of a pi-zero. Lead will also induce electrons and positrons to radiate photons, which can in turn produce more pairs, and so on, creating 'a shower' of charged, visible particles. This is seen in Fig. 7.3, where a cosmic ray entering a cloud chamber initiates a sequence of avalanches of electrons and positrons in successive lead sheets placed across the chamber.

It was the quantity of gamma rays in the cosmic radiation that first led theorists to propose the existence of the pi-zero. In 1948, Robert Oppenheimer at Berkeley and two of his students, H.W. Lewis and S.A. Wouthuysen, published a paper suggesting that cosmic gamma rays originate from the decay of neutral pions, but no one could be certain.

In 1949, R. Bjorkland and colleagues searched for the particle at Berkeley's new '184 inch' synchrocyclotron, and they employed electronic methods. When the protons accelerated in the cyclotron struck metal targets inside the machine, they produced a copious supply of pions. If neutral pions existed, they should have been produced too but should have immediately decayed into two gamma-ray photons. Two strategically placed holes in the concrete wall shielding the cyclotron allowed any gamma rays to escape and collide with foils of tantalum which, like lead, induced the production of electron–positron pairs.

The electrons and positrons emerging from the tantalum foil were deflected in opposite directions by a magnetic field and detected in coincidence by devices known as proportional counters. The degree to which the magnetic field deflected the particles, together with the amount of ionization they produced in the gas-filled tubes of the proportional counters, enabled the experimenters to determine the particles' energies. This in turn revealed the energy of the gamma rays that had produced the electrons and positrons. The results fitted well with the energies expected if the gamma rays did indeed come from neutral pions decaying in flight. No other explanation could account for the measured distribution of energies; the first evidence for the neutral pion had been found.

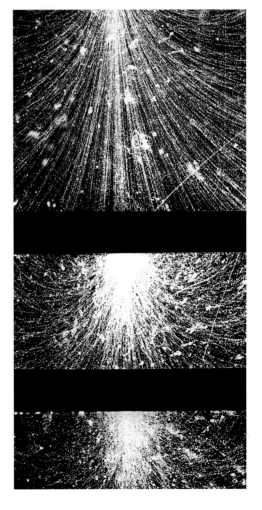

Fig. 7.3 Pi-zeros produced in high-energy cosmic ray interactions in the upper atmosphere decay swiftly to gamma rays and can generate 'showers' of electrons and positrons. Here, such a shower has been produced in a cloud chamber subjected to a magnetic field. A high-energy cosmic ray initiates a cascade of electron–positron pairs at the top of the chamber, which curve in opposite directions in the magnetic field. The electrons and positrons produce further gamma rays, through annihilation and radiation, and when these interact with two successive sheets of lead the process repeats, regenerating the shower each time.

Fig. 7.4 An extremely rare event showing the production and decay of a xi-zero. A negative kaon (pale purple) enters a bubble chamber at Berkeley from the bottom of the image and strikes a proton to produce a neutral kaon and a xi-zero, which leave no tracks, together with a negative pion (green) and a positive pion (pale blue). The pions move away to the right, curling left and right respectively in the chamber's magnetic field. The neutral kaon moves towards the left, leaving a barely visible gap, for it decays almost immediately into its own pair of negative and positive pions (green and pale blue respectively). The xi-zero travels up the picture, in almost exactly the same direction as the incoming negative kaon, until it decays into a neutral lambda and a pi-zero, both unseen. The lambda, only marginally deflected, continues up the picture and then creates a characteristic 'V' as it decays in its turn into a proton (dark purple) and a negative pion (green, bending left). The pi-zero decays almost immediately into a single gamma ray photon (not marked because it leaves the bubble chamber undetected) and an electron–positron pair (red and yellow respectively). This pair points back to the exact spot where the xi-zero decayed; without the electron and positron this point would be unknown. (Extraneous tracks have been removed from this image.)

The Neutral Cascade

The neutral 'cascade' particle, or xi-zero, is, like the pi-zero, difficult to detect because it decays into two neutral particles – one of which is the pi-zero! Figure 7.4, which has been coloured to aid identification, shows a particularly good example of a xi-zero observed in a bubble chamber at the Lawrence Berkeley Laboratory.

A negative kaon (pale purple) enters at the bottom and produces a total of four particles: a xi-zero, a K-zero, a pi-plus (pale blue), and a pi-minus (green). The K-zero decays almost immediately into two pions, revealing itself only by the tiny gap between its birthplace and the point where it decays. The xi-zero, on the other hand, travels invisibly for some distance before decaying into a lambda-zero and a pi-zero. Being neutral, these particles also leave no tracks, but they show themselves by their own decays. The lambda turns into a proton (dark purple) and a pi-minus, while the pi-zero experiences one of its rare (one in a hundred times) decays into an electron–positron pair (red and yellow) and a single gamma-ray photon. The electron and positron curve more in the bubble chamber's magnetic field than the other, more massive particles.

The xi-zero was not discovered until 1959, although its existence had been predicted earlier. It was required by the 'strangeness' scheme of Gell-Mann and Nishijima, first put forward in 1953, which provided a simple rule that ties together particles with very similar masses. In this scheme, the proton and neutron are charged and neutral partners with masses just under 1 GeV, but with no units of the property Gell-Mann called 'strangeness'. Similarly, the negative xi discovered in cosmic rays, with a mass a little over 1.3 GeV, was expected to have a neutral partner of more or less the same mass – the xi-zero. The two xi particles each carry two units of negative strangeness – we say they have a strangeness of –2. This strangeness is divested one unit at a time when they decay. So the xi-zero decays into a pi-zero with no strangeness and a neutral lambda with strangeness –1. The neutral lambda in turn loses its unit of strangeness when it decays into a proton and a pion, neither of which is strange.

The neutral xi remained undiscovered until, late in 1958, Luis Alvarez and his group at Berkeley began a concerted effort to find the elusive particle in the interactions of negative kaons in their 38 cm (15 inch) bubble chamber, in the months before the famous '72-inch' chamber began work. The negative kaon, or K-minus, has a strangeness of –1. So to create a xi-zero, with strangeness of –2, while keeping the total strangeness the same, requires that a particle of strangeness +1 must be produced together with the xi-zero.

 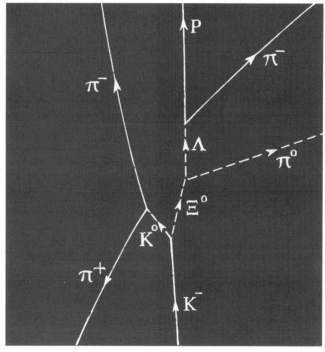

Out of thousands of pictures, Alvarez's team found one example (Fig. 7.5) in which the visible tracks provided enough information to suggest that a xi-zero had been produced. By measuring the angles and momenta of the appropriate tracks, they concluded first that the two pions at the left came from a K-zero; this was the vital particle with strangeness +1. They then showed that the 'vee' formed at the top by a proton and a pi-minus came from a neutral lambda. This 'vee', however, did not point directly back to the origin of the K-zero; instead another neutral particle must have been produced, and it was this particle's decay that the 'vee' of the lambda pointed towards. Calculations based on the masses, angles, and momenta of the particles involved showed that this neutral particle must have approximately the same mass as the xi-minus. It bore all the hallmarks of the xi-zero.

(Note that a neutral particle that creates a 'vee' does not necessarily 'point' along a line bisecting the 'vee'; it depends on the masses and momenta of the two particles in the 'vee'. In the case of the lambda's 'vee', shown in the diagram accompanying Fig. 7.5, the relatively massive proton travels onwards in almost exactly the same direction as the neutral lambda, whereas the relatively lightweight pion is deflected sharply to the right.)

The discovery of the xi-zero was a *tour de force* that demonstrated the power of both the liquid hydrogen bubble chamber and the analysis methods developed in the early days at Berkeley. Balancing the angles and energies of the charged particles revealed where the unseen neutrals had passed. The discovery also dramatically confirmed the predictive power of the strangeness scheme, which could now be used as a firm basis for ideas of a more fundamental nature.

Fig. 7.5 Although dramatically different in appearance, this xi-zero discovery picture, from the 38 cm (15 inch) hydrogen bubble chamber at Berkeley, almost reproduces the pattern of decays shown in Fig. 7.4. A negative kaon (K⁻) with a faint and patchy track strikes a proton and produces a neutral kaon (K⁰) and a neutral xi (Ξ⁰). The neutral kaon decays into a pair of positive and negative pions (π⁺, π⁻). The xi-zero, as shown by the dotted lines in the diagram, decays into a lambda (Λ) and a pi-zero (π⁰). The lambda turns into a proton (p) and a pi-minus; the pi-zero presumably decays almost instantly into two gamma rays, which travel undetected in the direction indicated for the pi-zero.

Antimatter

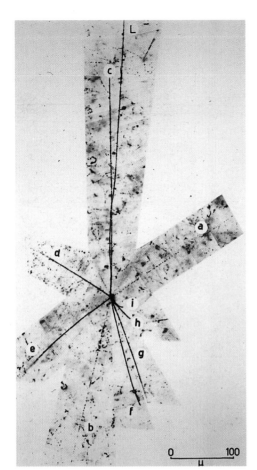

Fig. 7.6 The first image of an antiproton annihilation 'star', found in emulsion exposed to antiprotons at the Bevatron in 1955. The antiproton enters the picture at the top (the track marked L) and travels about 430 micrometres before ending its life in an explosive act of mutual destruction with a proton. Nine charged particles emerge from the point of annihilation and move outwards, their tracks forming a characteristic star-like pattern. The tracks marked *a* and *b* are probably pions, the others probably protons.

Fig. 7.7 (OPPOSITE) An antiproton (pale blue) strikes a proton in a bubble chamber at Berkeley. In the resulting annihilation, the energy released rematerializes as four positive pions (red) and four negative pions (green). In the bubble chamber's magnetic field, the negative pions and the negative antiproton curve in a clockwise direction, the positive particles anticlockwise. The two lower pions have less energy than the others and therefore curve more and leave thicker tracks. The one on the left travels only a short distance and stops when it is captured by a proton. The one on the right ends by decaying into a muon (yellow) and an invisible neutrino. Tracks not involved in the interaction, including the characteristic curlicues of low-energy electrons knocked from atoms, are coloured dark blue.

Our Earth, the Solar System it inhabits, and indeed our whole Galaxy and the millions of other galaxies in the Universe, all seem to consist of matter built from electrons, protons, and neutrons. But in 1932, in his experiments with cosmic rays, Carl Anderson discovered the 'antielectron', or positron – a particle just like the electron except that it carries a positive rather than negative electric charge. The positron was in fact already required by equations in a theory written by Paul Dirac in 1928, and its discovery implied that other antiparticles should exist, for the equations applied equally well to protons, neutrons, and many other subatomic particles discovered since the early 1930s.

For each and every variety of matter there should exist a corresponding 'antimatter' – opposite in properties such as electric charge and strangeness, but with identical mass. Thus we can conceive of antiatoms in which positrons orbit antinuclei built from antiprotons and antineutrons. Dirac's theory also predicted that matter and antimatter are doomed never to coexist. When a particle meets its antiparticle, the two annihilate – a catastrophic process in which the mass of the two objects is converted instantly to energy according to the equasion $E = mc^2$. This energy can 'evaporate' as photons, or rematerialize as new particles and antiparticles that rush away from the point of their creation.

Today, we still have no firm evidence that large-scale clumps of antimatter, built from antiatoms, exist anywhere in our Universe. But physicists can readily make antiprotons, antineutrons, and other antiparticles in high-energy collisions at accelerators, and they can manipulate them to probe the mysteries of the subatomic world. Yet it was more than 20 years after Anderson's discovery of the positron that experiments proved the existence of the antiproton, and several more years before physicists could feel certain that for every particle of matter there exists an appropriate antiparticle.

The antiproton was the first of several antiparticles to be discovered at the Lawrence Berkeley Laboratory. Its well-defined attributes – the same mass as the proton, but with negative charge – made it a suitable subject for study with electronic counting techniques, as described on p. 90. In this way, Emilio Segrè and his colleagues found the first signals of antiprotons in 1955. At the same time they sought visual confirmation of their discovery. Protons accelerated in the Bevatron produced many particles when they hit a target. The negatively charged ones were filtered off and focused by magnetic fields and used to bombard stacks of emulsion. Occasionally a rare antiproton among the other negative particles should meet a proton in the emulsion, annihilate, and produce a distinctive starburst of particle tracks.

The exposed emulsions were studied both in Berkeley and by Eduardo Amaldi's group at Rome University. The Italians found the first proton–antiproton annihilation 'star' (Fig. 7.6), shortly after the discovery of the antiproton. Later, Segrè's team reinforced the discovery by finding a star in which the total energy of all the particles produced in the annihilation clearly added up to more than the energy of the incoming antiproton. Because energy cannot be created from nothing, this proved that the star did not result simply from the decay of the incoming particle; it had to result from the mutual annihilation of two particles – proton and antiproton.

Figure 7.7 shows an antiproton star captured later at Berkeley with the `72 inch' liquid hydrogen bubble chamber. The image has been colour-coded to make it easier to identify the different particles. The incoming antiproton annihilates with a proton and produces eight pions – four positive (red) and four negative (green) with almost perfect symmetry.

The discovery of the antiproton opened the way for the search for its counterpart, the antineutron. When a proton and an antiproton have a near miss, they escape destruction but may neutralize each other's charge. The proton turns into a neutron and the antiproton turns into an antineutron. The antineutron is living in a hostile world of matter, and it is only a question of time before it annihilates with a neutron or a proton, producing a distinctive burst of energy.

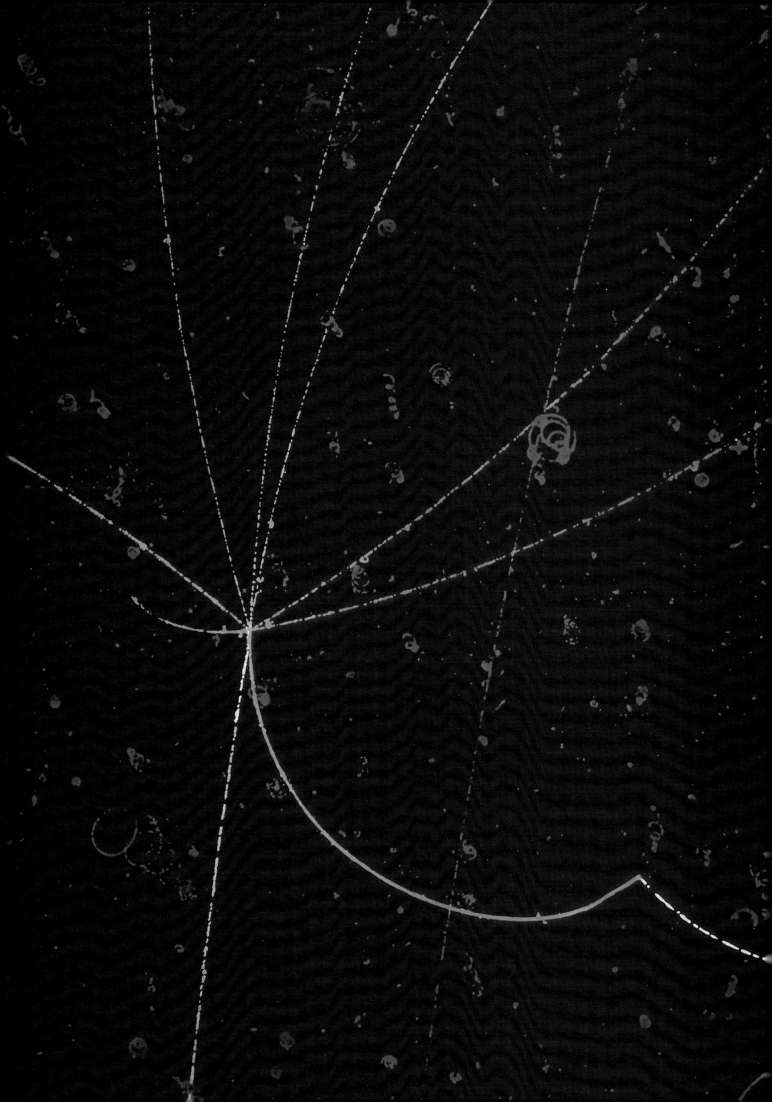

Fig. 7.8 In 1995 a total of 23 000 events correctly 'triggered' the apparatus of experiment PS210 at CERN, designed to detect the production of atoms of antihydrogen. A careful analysis reduced these 23 000 triggers to 11 events that satisfied the criteria for being due to antiprotons and positrons – the component particles of antihydrogen. These graphs show 'time-of-flight' (in nanoseconds) along the horizontal axis, while the vertical axis marks the deflection in a magnetic field compared with that expected for antiprotons (and should therefore be zero). The results of all 23 000 triggers are shown in the top graph, but only 94 of these – shown in the middle graph – had recorded energies expected for a positron and an antiproton. These are further reduced to the 11 events of the lower plot, by demanding that the energies of the two gamma rays detected are consistent with electron–positron annihilation. This is the signature of the first created atoms of antimatter.

Bruce Cork and colleagues at Berkeley decided to use this process of 'charge exchange' to hunt for antineutrons. They used a tank of liquid scintillator to detect annihilations of antineutrons produced by antiprotons generated from the Bevatron. An annihilation in the scintillator created a burst of charged particles and this in turn produced a large, characteristic pulse of light, which was detected by photomultiplier tubes. In 1956, Cork's team found 114 antineutron annihilations in this way.

By the early 1960s, antimatter equivalents of most of the particles then known had been discovered, and today antiparticles – mainly positrons and antiprotons – are used routinely as tools of science, technology, and medicine. In high-energy physics, the largest and most successful modern accelerators are the colliders, in many of which beams of electrons and positrons, or protons and antiprotons, are accelerated in opposite directions inside the same beam pipe and brought to head-on collision at selected points.

Despite our familiarity with antiparticles, the question of the existence of antimatter in bulk somewhere in the Universe remains a vexed issue. Surely matter and antimatter must have been produced in equal amounts in the energetic outpouring of the Big Bang. So where has all the antimatter gone? Why can we find no evidence for antigalaxies containing antistars, which might be orbited by antiplanets populated by antielephants and anticockroaches? Finding the explanation for this mystery is one of the major challenges at the beginning of the twenty-first century.

As a first step into the antiuniverse, atoms of its simplest element – antihydrogen – were made for the first time at CERN in 1995. Only eleven were made, and almost immediately lost in the detectors that proved their very existence. Making the necessary antiprotons and antielectrons (positrons) was relatively easy for CERN, but making the different antiparticles join together to form an antiatom required some extra ingenuity. This is because antiprotons are made 'on the move' when high-energy protons collide with a target, but to form an antiatom, an antiproton and a positron must have little relative motion. To overcome this, the physicists at CERN used the antiprotons to make positrons when they encountered a target of xenon gas. There was then a small chance an antiproton would pick up a positron – rather as a moving bus can gather up a person running alongside.

The few antiatoms created in this way split apart again as soon they entered the first section of the detection system. In each case, the positron was immediately annihilated to create the trademark pair of gamma rays, while the antiproton survived long enough to be deflected by a magnetic field and have its journey between scintillation counters timed to confirm its identity. The top graph in Fig. 7.8 shows 23 000 events 'triggered' in the apparatus during 1995. They are arranged along the horizontal axis according to their journey time or 'time of flight' (in nanoseconds) and on the vertical axis according to how much their deflection by the magnetic field differed from that expected for antiprotons (so 0 represents the expected deflection). In the middle graph this number has been reduced to 94, by requiring that the energies recorded were equivalent to those expected for a positron and antiproton. Finally, after demanding that the energies of the two gamma rays detected are each close to 0.511 MeV (that is, half the 1.022 MeV released when an electron and positron annihilate) only the eleven events of the bottom graph remain. These, in essence, are the fingerprints of the first known atoms of antihydrogen. It may not seem much, but it paved the way for a project that began at CERN in 2000, designed to make 100 atoms of antihydrogen a second and store them in a magnetic bottle.

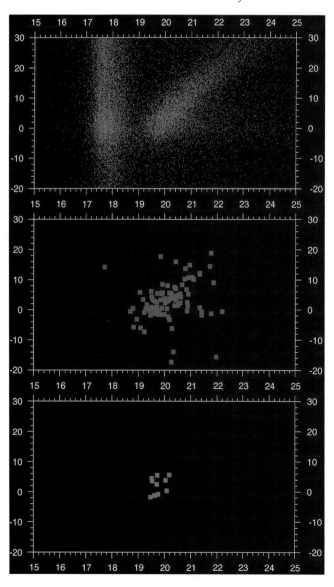

The Resonances

Many of the particles that we have met so far live brief but visible lives. A particle moving near to the speed of light travels a few millimetres in a lifetime of 10^{-11} s, and its track can be seen clearly in enlargements of emulsion and bubble chamber photographs. But there are many particles whose lives are millions of times shorter. They do not produce visible tracks in detectors; instead, physicists infer their existence from the longer-lived particles into which they decay.

These extremely short-lived particles are generally known as 'resonances'. Their lifetimes are of the order of 10^{-23} s. A span of 10^{-23} s fits into a millionth of a second as does a millionth of a second into three thousand years! It is no wonder the resonances leave no visible tracks: even at the speed of light they travel barely further than their own diameter. Something that dies before moving from its birth site, even if travelling at the speed of light: can such a bizarre thing be said to have existed at all? When the first resonance was discovered in the early 1950s, many physicists were initially reluctant to accept its reality for this very reason.

The first intimation of resonances came from the work of Enrico Fermi and his group at the University of Chicago. They wanted to study the way that pions interact with protons because they believed that this was the best strategy for understanding the nuclear forces. In 1951, a new synchrocyclotron started operation in Chicago and the following year Fermi and his team used it to produce pions at six different energies, which they then fired at the protons in a target of liquid hydrogen.

Many pions passed straight through the empty space in the hydrogen atoms, while others bounced off or were absorbed. In front of the target, two small pieces of scintillator, 2.5 cm across, recorded the number of incoming pions, while two larger pieces behind recorded how many of them had traversed the hydrogen. The researchers found that as they increased the energy of the pions, fewer particles reached scintillators behind the target. This effect occurred with both positively and negatively charged pions, although it was more noticeable with the positive ones. It seemed that they had perhaps caught the first glimpse of a phenomenon that at a larger scale is familiar to physicists and engineers – resonance.

It is resonance that enables an opera singer's voice to break a wine glass. The singer hits a note whose frequency exactly matches the natural frequency of the glass; the glass vibrates in response to the sound waves and shatters. Resonance also occurs at the atomic level, as when the sodium atoms in a street light absorb electrical energy and in turn radiate it as light. These resonant systems have the property of absorbing energy in the same characteristic way: if you plot the energy absorbed on a graph, it will rise smoothly to a peak as the frequency (or wavelength) changes, then fall again.

In the experiment at Chicago, Fermi and his colleagues altered the energy of the pions, which quantum theory tells us is the same as altering the wavelength or frequency associated with them. The curve that they plotted as they fired increasingly energetic pions at the hydrogen target resembled a resonance curve; but the team could not confirm this idea because the Chicago cyclotron did not produce sufficiently energetic pions to show whether the curve reached a peak and then fell away. But in 1953, a team at the Brookhaven National Laboratory repeated the experiment with higher-energy pions produced by the newly commissioned Cosmotron, and revealed a definite peak in the pions' absorption (Fig. 7.9). Somehow pions of a certain energy could excite protons into a resonant state. This state was so well-defined that it was given a name – the 'delta'.

How do we know that the lifetime of the delta resonance is as short as 10^{-23} s? The information comes from the width of the resonance 'spike' in the graph of the energy distribution. Quantum theory relates the lifetime of the resonance and the width of the spike in such a way that resonances with brief lives have relatively broad widths, while longer-lived states have narrower, sharper spikes. The widths of the delta and many other resonances are very large, implying that their lifetimes are indeed unimaginably brief.

For a long time the delta resonance was an isolated and unexplained phenomenon. Resonances had been seen before only in complex structures whose component parts can absorb energy in changing from one configuration to another. This is what happens when electrons absorb energy in a sodium atom: they are displaced for a brief period to a higher energy level. But in the 1950s the proton was regarded as a single, indivisible entity. No one dared take the discovery of the delta as a serious hint that the proton might itself consist of

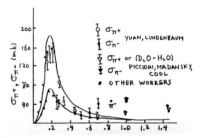

Fig. 7.9 Early data on pion–proton scattering show a typical 'resonance' peak in this sketch by Luke Yuan for the 4th Annual Rochester conference on High Energy Nuclear Physics in December 1954. Yuan, from the Brookhaven National Laboratory, and his colleague Sam Lindenbaum provided some of the first evidence to support the theory that pions can 'excite' short-lived resonances of the basic proton and neutron.

more elementary particles.

But we now know that the proton does indeed contain smaller constituents and that the delta was but the first of a whole range of resonances. A modern image of resonances, Fig. 7.10, shows what happens to a proton excited by an electron beam at the Jefferson Laboratory in Virginia. The shapes from black through red to green show the results as the electrons are scattered through ever larger angles. In each case three or four distinct bumps – resonances – can be seen like a range of mountain peaks. The way that the shape of the range changes with increasing scattering angles is one of the clues that shows the proton to be made of yet smaller particles – quarks (see pp. 124–127). This modern electronic imagery is the culmination of some 40 years of hunting resonances.

It was not until the early 1960s, nearly a decade after Fermi's work, that other examples of particle resonances were discovered. At the Lawrence Berkeley Laboratory, physicists had accumulated millions of bubble chamber photographs, which they set about analysing with the aid of computers. They could measure the energy of the various particles in each interaction, add the energies together in different ways, and plot the results as a graph showing how often a particular total energy occurred in each type of interaction. This kind of graph in effect records the energy distribution, or spectrum, of the particles produced in the interaction, and a resonance shows up as a spike or 'emission line' in the spectrum.

Two young students, Stan Wojcicki and Bill Graziano, found evidence for resonances of the lambda in bubble chamber film at Berkeley in 1960. They analysed pictures where a kaon had struck a proton to produce a lambda and two pions. They expected that the energy of each of the particles would be distributed smoothly, since particles emerging from an interaction usually share the available energy between them in a democratic way, now one being more energetic, now another. Instead, and quite unexpectedly, their graph of the kinetic energies of the lambda and the pions displayed a distinctive spike. They concluded that this 'emission line' must be due to the extremely brief presence of a resonating lambda, which then gives rise to the longer-lived 'normal' lambda. The Y-star (Y*), as it became known, was the first of a long list of resonances discovered at Berkeley, for it turns out that not only protons and lambdas, but also pions and kaons have resonant states.

One way of revealing resonances is the diagram invented in the early 1950s by Richard Dalitz. The basic idea is this. When a resonance decays its decay products can move off in

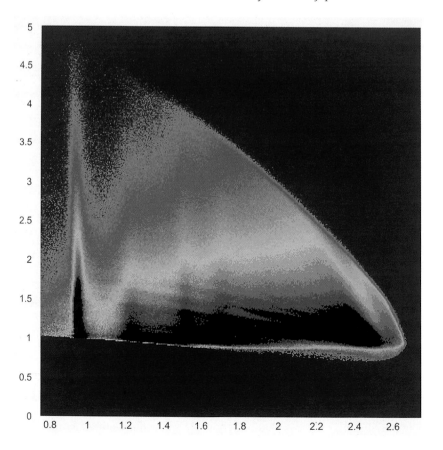

Fig. 7.10 Dark peaks reveal resonances in data from an experiment at the Jefferson Laboratory, Virginia, where electrons are scattered by protons. In this image the colours give a measure of the number of events at each point, with black corresponding to the most events, shading through orange to blue. The horizontal axis shows the energy (in GeV) equivalent to the mass of the resonance of the proton excited by the electron. So data at just less than 1 GeV show where the electron has scattered 'elastically' from the proton without changing it in any way, as in a billiard-ball collision. Higher values correspond to proton resonances such as the delta at about 1.2 GeV and one near 1.5 GeV and so on. At the highest energies, towards the right, there are many resonances, which overlap to smear out the ripples in the data. The vertical axis shows the momentum transferred from the electron to the object it scatters from and corresponds in effect to increasing scattering angles. So at larger angles, the number of events decreases, indicated by the general change in colour, and the effect of the resonances becomes gradually smeared out.

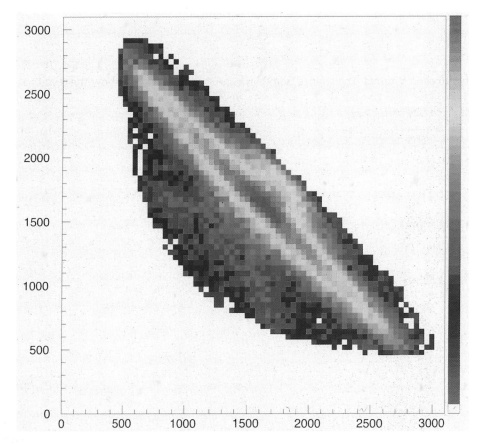

Fig. 7.11 Each point in this diagram corresponds to an event detected by the Crystal Barrel experiment at CERN, which studied low-energy proton–antiproton annihilations at LEAR. The resulting plot – known as a Dalitz plot after Richard Dalitz, who first analysed data in this way – reveals resonances rather like a contour map shows the presence of mountains. The events plotted here all contain a pi-minus, a pi-zero, and an eta meson; the horizontal axis is related to the mass recoiling against the pi-zero in each event, the vertical axis to the recoil against the pi-minus. The coloured scale corresponds to the number of events at each point, with the purple end of the spectrum representing 1, while green shades are equivalent to 10 to 20, and red represents 200. The bright red diagonal band – like a mountain ridge – is due to the resonance known as the rho-minus, which has decayed into the pi-zero and pi-minus particles in many of the events. The small bulge around the centre of the ridge, and the green 'spurs' to the left-hand side, are due to a more massive resonance, known as '$a_2(1320)$' after its mass of 1320 MeV, which has decayed to a pion and an eta.

any direction and with any speed provided the total energy and momentum add up to that of the original resonance. In Dalitz's technique, events are plotted as points on a diagram where the position of each event is determined by the speed of the particles and the directions in which they emerge. Then you look at the resulting 'Dalitz plot' to see if there is a concentration in some region, which is what would happen if a resonance, with some specific mass, had briefly given rise to the particles.

The Dalitz plot in Fig. 7.11 is from the Crystal Barrel detector at CERN's LEAR (Low Energy Antiproton Ring), in which slow antiprotons annihilated with a target of protons and neutrons. The annihilations produced new particles, such as pions and their heavier relatives known as 'etas' (η). As in the previous example, these particles were often not the direct products of the annihilations but were instead the decay debris from short-lived resonances as Fig. 7.11 shows.

The plot is coloured to show the number of events at each point. Red represents the most, yellow fewer, and the blue end of the rainbow represents the least. The result is like a contour map where red is the high ground and blue represents the lowlands. The broad ridge in red reveals the 'rho' resonance, which decays to two pions. (The rho is an excited version of the pion, just as the delta is an excited version of the nucleon.) If you drew a cross-section through the ridge, you would obtain a bump similar in shape to Fig. 7.9. Running to the left of the red ridge, and slightly below, are smaller 'spurs' that show up in green. These are due to a resonance known as the 'a_2', which decays to a pion and an eta.

Careful analysis of the shape of the hills often shows that there are several individual resonances that have combined and overlapped to give the total bump. The cataloguing of these ephemeral resonances has revealed that they are all due to a common deeper structure: resonances exist because the proton, the pion, the kaon, and so on, are built from smaller particles called 'quarks'. In much the same way that the constituent electrons rearrange themselves to form excited (resonating) states of atoms, so do the constituent quarks give rise to resonating states of the particles that are built from them.

The Crystal Barrel and similar experiments have accumulated many millions of events and discovered several resonances. In addition to resonances made from quarks, there appear also to be some that are made from gluons – the quantum bundles of the strong force field that normally binds the quarks to one another. These are known as 'glueballs' (see p. 171).

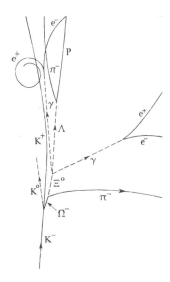

Fig. 7.12 This historic picture from the 200 cm (80 inch) hydrogen bubble chamber at Brookhaven shows the first observation of the omega-minus. A negative kaon (K⁻) collides with a proton to produce three particles: an omega-minus (Ω⁻), a positive kaon (K⁺), and an unseen neutral kaon (K⁰), represented by a dotted line in the diagram. The omega-minus travels a short distance (2.5 cm) and then decays, emitting a pi-minus (π⁻) that veers sharply to the right, and a neutral xi (Ξ⁰) which decays into three more neutral particles – a lambda (Λ) and two gamma ray photons (γ). These neutrals, also marked by dotted lines in the diagram, finally reveal themselves by decaying into visible 'V's: the gamma rays into electron–positron pairs (e⁻, e⁺), the lambda into a proton (p) and a pi-minus.

The Omega-minus

Figure 7.12 shows one of the most famous pictures in particle research, a physicist's Mona Lisa. One of a set of 80 000 photographs from the '80 inch' bubble chamber at the Brookhaven National Laboratory on Long Island, it was the first picture to show the production and decay of an omega-minus particle, and it caused tremendous excitement when it was announced in February 1964. Here was the final piece in a jigsaw that had been accumulating over the previous few years.

The proliferating quantity of subatomic particles, including the resonances, encouraged the theorists to try to find some order in the confusion. In 1960–61, Murray Gell-Mann of

Caltech and Yuval Ne'eman, a member of Israeli Defence who had been granted a leave of absence to study physics in London, independently proposed a method for classifying all the particles then known. This method became known as the Eightfold Way, as suggested by Gell-Mann. What the Periodic Table of the elements had done for atoms and chemistry, the Eightfold Way did for particles and high-energy physics. In the Eightfold Way, the particles are classified into 'families' according to such characteristics as their electric charge and their strangeness. Figure 7.13 shows two such families, one with eight members (an 'octet') and one with ten members (a 'decuplet'). Each particle has a particular position in its family, according to the quantity of electric charge and strangeness the particle has.

A central clue to the deeper meaning of these patterns is that certain properties are conserved when particles interact or decay. A proton and an antiproton, for instance, have electric charge of +1 and –1 respectively, giving a net charge of zero. When they collide and annihilate, they can create four negative pions (–4) and four positive pions (+4), as in Fig. 7.7, (p. 113) since the net charge remains zero. But they cannot create three negative and four positive pions: that would give a net charge of +1 and violate the law of conservation of electric charge. Strangeness is another property that is conserved when strange particles are produced by the strong force, but it can be divested by fixed amounts when they decay.

These properties of charge and strangeness, together with a particle's intrinsic rate of spin, completely define that particle in the Eightfold Way. Each particle fits into a position in one of the many families of particles.

In 1962, the Eightfold Way was still very new and poorly understood by all but a handful of theorists. Some critics thought its clever symmetries more accidental than fundamental, especially since it served only to classify known particles. In July 1962, Gell-Mann and Ne'eman attended an international conference at CERN and both were in the audience when a group from the University of Los Angeles announced the discovery of two new resonances, a negative and a neutral xi-star (Ξ^{-*} and Ξ^{0*}). Both Gell-Mann and Ne'eman realized that the two xi-stars would almost complete a new decuplet in the Eightfold Way.

This family is the inverted pyramid shown in Fig. 7.13. It contains four resonances with no strangeness (the deltas), three sigma-star resonances with one unit of strangeness, and the two newly discovered xi-star resonances with two units of strangeness. Only the tenth, triply strange member of the family was missing.

The next day's session, 10 July, included a review of the known strange particles. At the end of the talk the chairman called for comments from the floor. Both Ne'eman and Gell-Mann raised their hands. The chairman called for Gell-Mann, who was at the time the acknowledged leader in theoretical physics. Gell-Mann then strode to the blackboard and announced his prediction of a triply strange particle to complete the new decuplet.

Gell-Mann called it the omega-minus: minus because it should have negative charge, and omega – the last letter in the Greek alphabet – because it would complete the decuplet. Moreover, by extrapolating from the masses of the nine resonances already in the family, Gell-Mann could predict its mass. Heavier still than the two xi-stars, it should weigh in at 1680 MeV.

The challenge was on for the experimenters, and teams at Brookhaven and CERN set about scouring thousands of bubble chamber pictures. In February 1964, the Brookhaven team found the first 'gold-plated' example of an omega-minus – the picture shown in Fig. 7.12. Calculations gave a mass of around 1686 MeV. CERN found a similarly clear example a few weeks later.

At last, after years of mounting confusion, a viable classification system existed for many subatomic particles. The Eightfold Way clearly worked, but why it worked remained a mystery. Why did the particles fit so neatly into their various families? What principle underlay this subatomic ordering? The answer, as we shall see, is quarks.

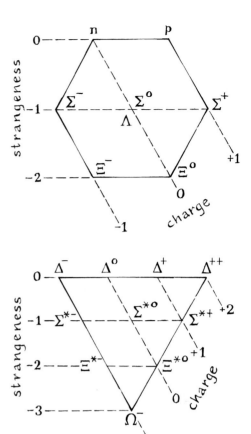

Fig. 7.13 These diagrams show two 'families' of particles – an octet and a decuplet – described by the 'Eightfold Way', the classification scheme proposed by Gell-Mann and Ne'eman in 1962. The members of each family are arranged according to their qualities of strangeness (vertical axis) and electric charge (diagonal axis).

The octet (top) consists of the familiar neutron (n) and proton (p), the trio of sigma particles (Σ), the lambda (Λ), and the two xi particles (Ξ). The central location in the octet is shared between the neutral sigma and the lambda, which is also neutral. (These two particles have the same values of charge and strangeness; in fact, they contain the same quarks, but in two slightly different arrangements.)

The decuplet (bottom) does not suffer from such complexities. Its ten members consist of four delta resonances (Δ), three sigma-star resonances (Σ^*), two xi-star resonances (Ξ^*), and the triply strange omega-minus (Ω^-) – the missing particle that Gell-Mann and Ne'eman predicted. (Note that one of the deltas has a double ration of positive charge.)

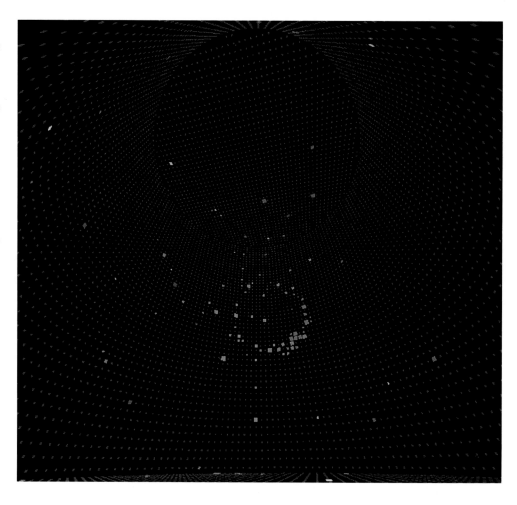

Fig. 7.14 A computer display of the interior of the Super-Kamiokande detector shows a ring of light created by the interaction of a neutrino from the Sun. The detector is a huge cylindrical tank, 41 m high and 39 m in diameter, filled with 50 million litres (12 million gallons) of ultrapure water. Its walls are lined with more than 11 000 light-sensitive phototubes indicated here by the grey dots. The neutrino has knocked an electron into motion, giving it enough energy to move faster than light does through the water. As a result, the electron emits a cone of Čerenkov radiation (see p. 91), which forms a ring when it falls on the walls of the detector and is picked up by the phototubes.

Neutrinos

The neutrino is one of the most pervasive forms of matter in the Universe, yet it is also one of the most elusive. It has no electric charge, very little mass, and it can travel as easily through the Earth as a bullet through a bank of fog. As you read this sentence, billions of neutrinos are hurtling through your eyeballs at almost the speed of light, but unseen. There are an estimated 100 to 1000 neutrinos in every cubic centimetre of space.

An intense 'wind' of neutrinos, emanating from the nuclear processes in the Sun, plays continually upon the Earth. In addition there are lesser breezes of neutrinos and antineutrinos emitted from the collapse of stars and other catastrophic processes in our Galaxy and beyond. The number of neutrinos far exceeds the number of cosmic rays (see p. 49), but unlike the cosmic rays the neutrinos fly past unnoticed. Because the Earth is so transparent to neutrinos, as many of them rain up through our beds by night as rain down on our heads by day!

The neutrinos have no direct effect on us, but theorists have come to believe that they play a crucial role in the processes that formed and continue to shape our Universe. They are regarded today as one of the truly elementary particles of matter.

Although the neutrino interacts extremely rarely with other forms of matter, experimenters have devised ways of making the particle reveal itself. These methods rely on a straightforward 'brute force' technique: direct enough neutrinos at a large enough target and you are bound to detect a few interactions.

Solar neutrinos arrive at our planet at a rate of about 60 billion per square centimetre per second – just enough for a few to be captured in detectors that are far underground, away from the 'noise' of cosmic rays. The neutrinos are emitted when protons turn to neutrons in the heart of the Sun in the chain of reactions that turns four hydrogen nuclei (4 protons) into a single nucleus of helium (2 protons and 2 neutrons) with a release of energy that eventually reaches us as sunlight. Astrophysicists can calculate how many neutrinos should accompany the sunlight. But a celebrated experiment in the Homestake gold mine in South Dakota, run by Ray Davis from the Brookhaven National Laboratory, for 30 years detected only one third of the neutrinos expected.

Since the 1980s, Davis's experiment, which uses the chlorine in a huge tank of dry-cleaning fluid to detect the solar neutrinos, has been joined by other detectors. These have employed a variety of means to snare the neutrinos and discover why such a large proportion seems to go missing. Figure 7.14 shows the fragile signature of a solar neutrino captured in the Super-Kamiokande detector (see p. 201). This is the biggest solar neutrino detector so far – a huge tank of pure water buried deep in a mine in the Japanese 'Alps'. Like other detectors before it, Super-Kamiokande finds too few solar neutrinos, but the puzzle is now being solved by this and other new detectors, as Chapters 10 and 11 describe.

In other experiments today, particle physicists produce beams of neutrinos to order, and use them to explore the elementary constituents of matter. Though they are still elusive, neutrinos have been partially harnessed. Yet it is less than 50 years since their existence was first proven. In the 1920s, physicists studying the radioactive beta decay of atomic nuclei were faced with a conundrum. In the beta decay process, an atomic nucleus was supposed to emit an electron, while at the same time changing into a nucleus with one more unit of positive charge. (In fact, a neutron in the original nucleus turns into a proton, but in the 1920s the neutron was still unknown.) The new nucleus and the emitted electron should always have shared the available energy in the same way. But experiments showed a quite different pattern: the electron emerged with a range of energies. This apparently violated the fundamental law of conservation of energy.

The Austrian physicist Wolfgang Pauli took what was an extremely bold step at that time by suggesting in 1930 that a hitherto unsuspected particle was responsible. The theory was simple: if three objects – the nucleus, the electron, and a mystery particle – shared out the energy of beta decay, then the energy released in the decay could be shared between them in any number of ways. And that would explain why the electrons produced in beta decay do not always have the same energy.

Pauli's proposed particle had to be quite bizarre compared with those already known. It had to be neutral, have little or no mass, but it had to spin about its axis like an electron or a proton. Enrico Fermi named it the neutrino – Italian for 'little neutral one' – and he gave it respectability by incorporating it into his theory of beta decay in 1933.

The neutrino remained a hypothesis until, early in the 1950s, two physicists at the Los Alamos National Laboratory in New Mexico were inspired by the idea of 'doing the hardest physics experiment they could think of'. Clyde Cowan and Fred Reines decided to show that neutrinos could do something, however rarely, and therefore had physical reality.

They thought at first that atomic bombs could provide a suitably copious supply of neutrinos, or rather antineutrinos, emitted in the decays of neutrons released in the explosions. In the end they realized that their proposed experiment should work equally well with antineutrinos produced in the more controlled conditions of a nuclear reactor.

Cowan and Reines chose to look for the process of 'inverse beta decay', in which a proton captures an antineutrino and converts to a neutron, at the same time emitting a positron. Work with a prototype detector encouraged them to build a full-scale apparatus at the Savannah River reactor in South Carolina. They called it Project Poltergeist because of their quarry's apparent undetectability.

To show that inverse beta decay had indeed occurred, Poltergeist was designed to detect two separate bursts of gamma rays emitted as a result of the process. The first burst of gamma rays would come when the positron annihilated with an electron. The second burst would be produced when the neutron was captured by a cadmium nucleus in 'target' tanks containing cadmium chloride (Fig. 7.15). The timing of the two bursts was the crucial element. The positron would annihilate almost instantaneously, but the energetic neutron would need to be slowed by successive collisions before it could be captured by a cadmium nucleus. Cowan and Reines calculated that if inverse beta decay was really occurring, the interval between the two gamma-ray bursts should be in the region of 5 microseconds (five millionths of a second). In the summer of 1956, Poltergeist triumphantly recorded gamma-ray bursts separated by 5.5 microseconds, as Fig. 7.16 shows, and on 14 June Cowan and Reines sent Pauli a telegram to say that the neutrino had finally been found.

Neutrons are not the only particles that can give rise to neutrinos and antineutrinos.

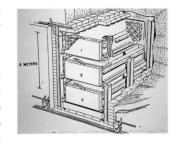

Fig. 7.15 The 10 tonne antineutrino detector built by Cowan and Reines contained three tanks of liquid scintillator (1, 2, 3) in the form of a 'double-decker' sandwich. The filling between the decks consisted of two smaller tanks of water (A, B) in which cadmium chloride was dissolved. The idea was that an antineutrino would react with a proton in the water to produce a neutron and a positron. The positron would annihilate into gamma rays almost immediately; the neutron would slow down and be captured by a cadmium nucleus, giving off more gamma rays several microseconds later.

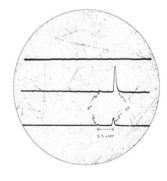

Fig. 7.16 An oscilloscope displays three horizontal lines – traces showing the signal from each of the three tanks of scintillator in Fig. 7.15. An antineutrino interacted in the tank of water B, located between the lower two scintillator tanks in the detector. Each of the lower oscilloscope traces shows a small 'blip' (arrowed) due to the burst of gamma rays from the positron annihilation. They are followed 5.5 microseconds later by a larger pulse due to the gamma rays emitted after the capture of the neutron by a cadmium nucleus.

Pions and muons, for instance, also do so. But the case of the muon presented physicists in the 1950s with some puzzles. Neutrinos and antineutrinos should automatically annihilate each other, but it was clear that the neutrino and antineutrino produced together in the decay of a muon did not annihilate. This led theorists to consider that there might be two types of neutrino. Because some neutrinos always seemed to be emitted together with an electron or positron (as in neutron decay), while other neutrinos were emitted with a positive or negative muon (as in pion decay), the two types came to be known as the electron-neutrino and the muon-neutrino. A muon would thus decay to an electron and an antielectron-neutrino, together with a muon-neutrino.

This idea of two neutrinos became established when the advent of the 30 GeV proton synchrotrons at CERN and Brookhaven made possible the creation of beams of neutrinos, as proposed in 1959 by Bruno Pontecorvo at Dubna in the former Soviet Union and Melvin Schwartz at Columbia University in New York. The trick is to use a combination of electric and magnetic fields to select pions produced when the accelerator's protons strike a target. The pions are then allowed to decay, producing a beam of muons and muon-neutrinos. After an appropriate distance – a few tens of metres at the energies of the CERN and Brookhaven accelerators – a massive wall of solid iron, many metres thick, filters out the muons and most other particles remaining in the beam. Only the extremely penetrating neutrinos pass through and enter a detector in which a tiny fraction of them will interact. A set-up of this kind was used in the famous two-neutrino experiment at Brookhaven in 1962, when Schwartz and several colleagues observed that neutrinos produced in association with muons gave rise to muons, never to electrons.

Figure 7.17 shows the contrasting signals of electron- and muon-neutrinos that have interacted in a more modern detector, called NOMAD for Neutrino Oscillation MAgnetic Detector. The upper image shows the footprint of an electron which has deposited most of its energy in one layer of the detector, yielding the large signal indicated by the long spike. In the lower image, by contrast, a penetrating particle has been produced, which leaves a faint track as it shoots through the detector. This is the distinctive mark of a muon.

Fig. 7.17 Electron-neutrinos can be distinguished from muon-neutrinos through their distinctive 'footprints' in particle detectors, as in these computer reconstructions of tracks in the NOMAD experiment at CERN. In both cases the neutrino has entered the detector from the left and interacted in a series of drift chambers (marked by the green lines) and other detectors which lie within the coil of an electromagnet. In the upper interaction, a charged particle leaves a track in the drift chambers that leads to a large amount of energy deposited in the electromagnetic calorimeter (outlined in red), as indicated by the thick yellow lines to the right. This particle is most likely an electron, indicating an interaction in which an electron-neutrino has given rise to an electron. In the lower example, a charged particle track in the drift chambers lines up with further track segments in the muon chambers, which lie to the right beyond the magnet iron. This indicates the interaction of a muon-neutrino.

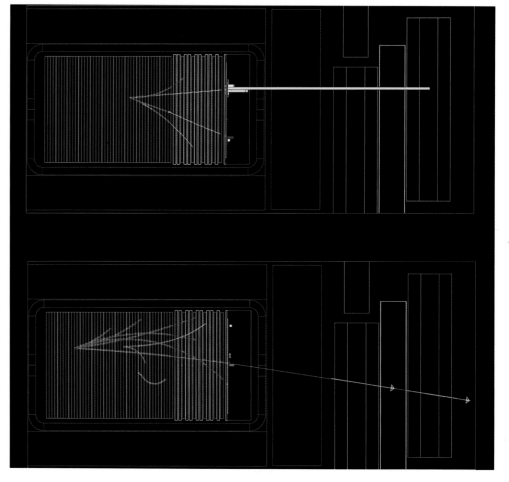

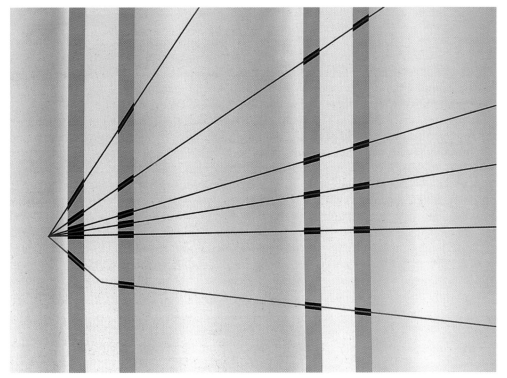

Fig. 7.18 This reconstruction shows one of the four events that revealed the tau-neutrino for the first time in the DONUT experiment at Fermilab. It illustrates part of the 90 cm long 'sandwich' of iron (variable light brown), emulsion (dark brown bands), and plastic (yellow) used to detect the interactions of a beam of tau-neutrinos entering from the left. The iron plates are 1 mm thick, the layers of emulsion 0.1 mm. The plastic is the substrate for the emulsion. A tau-neutrino has interacted in the iron plate at left to produce several charged particles. They leave short segments of track (highlighted with black) in the emulsion layers, allowing the paths of the particles to be reconstructed. The bottom track (red) runs less than a millimetre before it kinks, revealing that the particle has decayed into another particle (green track). This is the signature of the short-lived tau particle (see pp. 162–163) and it is the proof that the event was initiated by a tau-neutrino.

Today, particle physicists recognize the existence of three types of neutrino: the electron-neutrino, the muon-neutrino, and the tau-neutrino. Discovered in 1975 at SLAC, the tau (see pp. 162–163) is a heavier version of the muon, just as the muon is a heavier version of the electron. Together, the electron, the muon, the tau, and the three neutrinos form the family of 'leptons' – particles that do not feel the strong force. As far as we can tell, the leptons, like the quarks described in the next section, are fundamental, structureless particles. After the discovery of the tau many experiments found circumstantial evidence for the tau-neutrino's existence, but direct observation proved elusive. The NOMAD detector at CERN was designed, in part, to reveal tau-neutrinos, but it was an experiment at Fermilab, called DONUT (Direct Observation of NU-Tau), that finally achieved this goal in June 2000. An intense beam of neutrinos hit a target of iron interleaved with emulsion – similar to that used 30 years previously to study cosmic rays (see p. 60). Only one in every million million of the tau-neutrinos in the beam hit an iron nucleus in the target, turning into a visible tau, as in Fig. 7.18, where the tau has left a one millimetre long trail in the emulsion. The experiment initially recorded six million potential candidates for exploration, from which a precious four tau-neutrino events were found. The tau-neutrino is truly 'one in a million'.

Despite the elusiveness of neutrinos, physicists now produce and manipulate them at will. At CERN and Fermilab, high-energy neutrino beams have been used to probe protons and neutrons, and reveal the quarks deep inside. Today beams are being designed for experiments to study the neutrinos themselves and discover whether they can change from one type to another (see Chapter 10). And it turns out that neutrinos are also ideal for studying one of nature's most spectacular cosmic phenomena – the supernova.

When the fuel runs out at the centre of a star and the nuclear reactions cease, the star collapses catastrophically, releasing energy in the dramatic display we call a supernova – a 'super new star'. But the light emitted is only 1% of the enormous energy thrown out. The remainder is in the form of invisible neutrinos and antineutrinos. A supernova exploded 170 000 years ago in the nearby galaxy known as the Large Magellanic Cloud. Light and neutrinos flooded across space. When they started their journeys, stone age was state of the art; by the time they reached Earth modern science was the rule. On 23 February 1987, within a few seconds at 07.35 GMT, 300 million neutrinos traversed an enormous tank of ultra-pure water located 600 m down in a salt mine under the bed of Lake Erie in Ohio. The science of neutrino astronomy came of age at that moment as, for the first time, neutrinos from a distant supernova were detected (Fig. 7.19).

Fig. 7.19 A circle of yellow 'crosses' on this computer display reveals the detection of a neutrino emitted by a supernova 170 000 light years from Earth. Supernova SN1987a appeared in the Large Magellanic Cloud on 23 February 1987, flooding the Earth with neutrinos and antineutrinos as the star Sanduleak −69 202 collapsed catastrophically. Eight of these elusive particles interacted in the IMB detector, a tank of 10 000 tonnes of ultrapure water located 600 m below ground in a salt mine in Ohio. In this example, the antineutrino has interacted with a proton in the water to create a neutron and a positron. The charged positron moves through the water faster than light does, and creates a cone of Čerenkov light about its path (see p. 91). When the light strikes the wall of the detector, it is picked up by phototubes that line the detector, each 'cross' in the circle representing the light detected by one phototube.

Quarks

Fig. 7.20 In this display of a proton--antiproton collision in the UA1 detector at CERN, two pairs of back-to-back high-energy jets are seen. The tracks are colour-coded according to momentum: lower momenta are the red and yellow end of the spectrum, higher momenta are blue and purple. The proton and antiproton, containing quarks and antiquarks respectively, came in from left and right The low-momentum tracks going out sideways are due to particles that materialized from quarks and antiquarks which made glancing collisions. The two high-momentum jets shooting out to top and bottom are from a quark and an antiquark that met head on and rebounded violently at 90° to their original directions. (See Fig. 7.21 for an explanation of the event at the level of the quarks and antiquarks.)

Up and down, charm and strange, top (or truth) and bottom (or beauty) – these are the rather whimsical names given by physicists to the six known quarks which, together with the six leptons, are now believed to be the fundamental constituents of matter. But these departures from the usually dry standard of scientific phraseology mask what is perhaps the most important development of modern physics.

In Chapter 2, we described how physicists established a new layer of reality when they discovered that the stuff everything is made of consists of atomic nuclei orbited by electrons. A few years later, a more fundamental layer was established when it was shown that the nuclei of atoms are not elementary but consist, in their turn, of protons and neutrons. Quarks are what protons and neutrons are made of, as are many other particles found in cosmic rays and created in accelerators. They are an even more fundamental layer of matter and one that the evidence to date suggests is truly elementary.

Quarks occur in clusters, either in pairs or triplets. The proton and neutron, for instance, are both clusters of three quarks; the proton is made of two up quarks and a down quark, and the neutron from two down quarks and an up. No one has yet seen a single or 'naked' quark, and most physicists believe quarks cannot exist as individual free particles. The

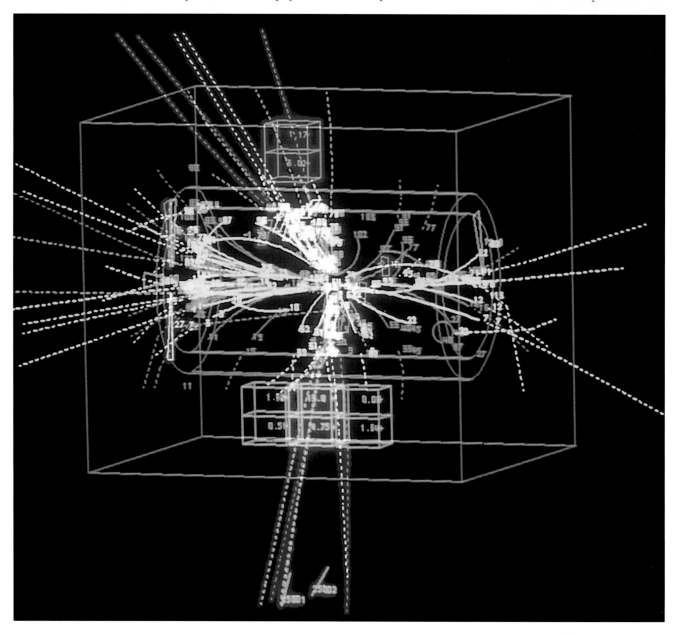

strong force, which rules the quarks, appears to bind them so tightly together that they cannot be prised apart. When a proton or antiproton is driven to high energy in an accelerator and smashed into another proton or antiproton we do not see individual quarks and antiquarks emerging. Instead, within an instant – some 10^{-23} s – additional quarks and antiquarks are created out of the energy of the collision, and these re-form in new pairs and triplets to produce sprays of particles, as in Fig. 7.20.

This image shows a computer display of a head-on proton–antiproton collision recorded at CERN. A quark and an antiquark within the original colliding particles have annihilated and converted into pure energy, which has immediately rematerialized as a new quark and antiquark. The quark and the antiquark fly off in opposite directions, but they cannot exist as individual entities: instead they catalyse the conversion of energy into mass and generate additional quarks and antiquarks to form clusters of detectable particles such as pions, kaons, protons, and their antiparticles.

As we have seen in this chapter, there were hints in the 1950s and early 1960s that protons and pions and many other particles might not be as elementary as they seemed. One clue was the existence of the excited states of subatomic particles known as resonances. Physicists knew that atoms could be excited because they have an underlying substructure, and analogy suggested that the same should be true of excited protons and pions. Another clue was the neat manner in which the Eightfold Way classified particles into families. Surely these patterns reflected a deeper level of structure.

The man who 'invented' quarks and thereby helped to solve the conundrum was Murray Gell-Mann, the theorist from Caltech who had developed the Eightfold Way. In 1964, Gell-Mann proposed the existence of three quarks – up, down, and strange – which were all he needed at that time to explain all the particles other than leptons.

Why 'quarks'? The story goes that Gell-Mann liked the sound of the word and only later discovered it in a line in James Joyce's *Finnegans Wake* – 'three quarks for Muster Mark'. Gell-Mann was not the only person thinking along these lines. The same idea was put forward, also in 1964, by George Zweig, a fellow theorist from Caltech who was visiting CERN at the time. But the hypothesis of quarks was slow to catch on. There was no firm evidence for them. And they were also decidedly bizarre objects by the standards of the day. In particular, they would carry 1/3 or 2/3 of the basic unit of electric charge, which was unheard of. All other particles carried charge in full units – 0, 1, 2, ... Thirds were beyond the pale.

The first solid evidence for quarks came towards the end of the 1960s from studies at the Stanford Linear Accelerator Center (SLAC) with the three enormous spectrometers located in the giant End Station A experimental hall (Fig. 6.41, p. 105). The experiments were similar in principle to the work of Geiger and Marsden in 1908, in which they bombarded atoms with alpha particles and discovered that atomic nuclei exist. In the modern analogue, electrons accelerated to high energy in SLAC's linear accelerator were fired at protons. The very high-energy electrons, which at that time could be produced only in SLAC's 3 km long linear accelerator, were a vital part of the experiment. This is because an electron in motion behaves like a wave, and the wavelength depends on the electron's energy. High-energy electrons travel in short waves, enabling them to penetrate something even as small as a proton; lower-energy electrons, by contrast, travel in longer waves and deflect off the proton as a whole.

If the proton was just a singular, elementary particle, the lightweight electrons fired at it should bounce off with almost the same energy as they arrived with; very little energy would go into making the massive proton recoil. But if the proton was a composite particle, consisting of quarks, the result would be very different. The quarks inside a proton would not be still, but would be in constant motion; in fact, the proton could be described as a vibrant cluster of quarks. So the electron might encounter a very energetic quark or merely a quiescent one.

In the experiments at SLAC, electrons would bounce from the quarks and enter a spectrometer. If the quark was moving faster than average, then the electron would enter the spectrometer with a higher energy than average; if the quark were slower, then the scattered electron's energy would be relatively low. In this way, the energies of the electrons coming into the spectrometers would provide a direct measure of the energies of any quarks lurking within the proton.

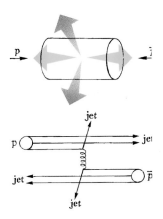

Fig. 7.21 The lower diagram shows the event in Fig. 7.20 at the level of the quarks and antiquarks. A quark and an antiquark within the colliding proton (p) and antiproton (p̄) can interact via the exchange of a gluon (curly line), and then shoot off sideways. They materialize in the detector as jets of high-momentum particles (blue arrows in upper diagram). The remaining quarks and antiquarks that formed the original proton and antiproton give rise to jets of low-momentum particles (orange arrows) that move off horizontally. The entire process takes place inside the beam pipe, in a tiny 'femtouniverse' less than 10^{-15} m across.

Fig. 7.22 The dramatic deflection of an electron when it meets a quark buried within a proton in a high-energy collision is clearly visible in this computer display of an event in the ZEUS detector at HERA, the electron–proton collider at DESY. The red outlines represent parts of the detector, in particular the cylindrical tracking chamber where tracks of charged particles are shown by yellow lines, and the 'forward' and 'rear' calorimeters to left and right respectively. The electron beam has entered from the left, the proton beam from the right, and an electron and a proton have collided at the centre. The electron has been deflected backwards, towards the top left corner where it is revealed by the large amount of energy deposited in the calorimeter (the large red spike). The quark struck by the electron has set off towards the bottom left, immediately creating new quarks and antiquarks to form a 'jet' of particles which dump their energy in the calorimeter. (The diameter of the central tracking chamber is about 1.7 m.)

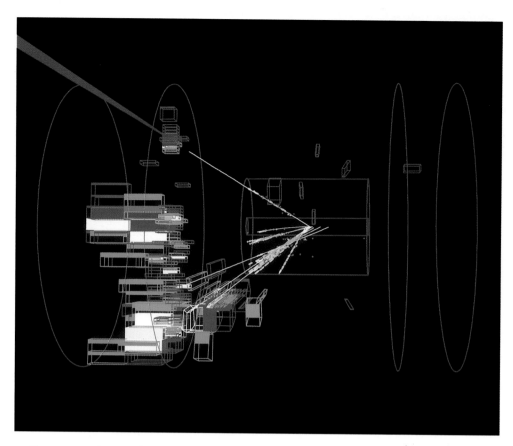

Fig. 7.23 (OPPOSITE) In these computer reconstructions of a high-energy electron–proton collision in the H1 detector at the HERA collider, a single jet of particles appears, as in Fig. 7.22, but there is no sign of the scattered electron. The upper image shows the detector as if sliced along its 10 m length, the pipe that contains the colliding beams (unseen) threading through the centre. The electron has entered the detector from the left, the proton from the right. The electron has struck a quark, which shoots towards top left, materializing immediately as a jet of particles. The lower view shows the same event seen 'end-on'. In this case, the electron and proton have come in from above and below the page, and the jet of particles shoots out to the top right. Because there was originally no momentum in this direction, something must have travelled undetected in the opposite direction, towards bottom left. It cannot be the electron, as this would have left a track, but it is in a manner of speaking the 'ghost' of the electron – an electron-neutrino.

The answer from End Station A was clear: the energies of the returning electrons varied. Then the spectrometers were repositioned so as to capture electrons bouncing at large angles instead of small angles. The variation in the energy distributions and the rate of arrival of the electrons as the angle changed showed that the proton consisted of much smaller objects. At about the same time, similar experiments with neutrinos, instead of electrons, took place at CERN. The detailed comparison of results from CERN and SLAC proved conclusively that the proton consists of three fractionally charged quarks.

In 1992, a powerful new machine for seeing into the proton came into operation at the DESY laboratory in Hamburg. In HERA, the Hadron Electron Ring Accelerator, beams of electrons and protons (and also positrons and protons) are made to collide head on. The energies involved in the collisions are much higher than at SLAC, and the results are dramatic. As the beams collide, an electron can have so much energy relative to a proton that it approaches close enough to one of the quarks to be turned almost completely back in its track, as the event in Fig. 7.22 from the ZEUS detector shows. The spray of particles from the struck quark is also clear to see.

At these ultra-high energies, the detectors at HERA also capture events where the electron seems to disappear completely, as in the event shown in Fig. 7.23, from the H1 detector. In fact, the interaction is related to the process of beta decay, where an electron and an electron-antineutrino appear when a neutron changes into a proton – or, more fundamentally, when a 'down' quark turns into an 'up' quark. At HERA, in the act of hitting a quark, the electron can turn into an electron-neutrino, which leaves no trail, to create a strikingly 'lopsided' event.

The idea of quarks began with patterns among the particles that had been found in the bubble chambers and other detectors of the 1950s and 1960s. By the end of the twentieth century, the real dynamical existence of quarks had become plain to see at particle colliders, such as HERA at DESY. At the same time, particles built from three types of quark beyond those required in Gell-Mann's theory have also emerged, in particular at colliding-beam machines. News headlines speaking of 'hidden charm', 'naked bottom', and 'top at last' have tracked the discoveries of this trio of heavier quarks, as we do in Chapter 9. But first we shall look at the development of the new kinds of accelerator and detector that made these discoveries possible.

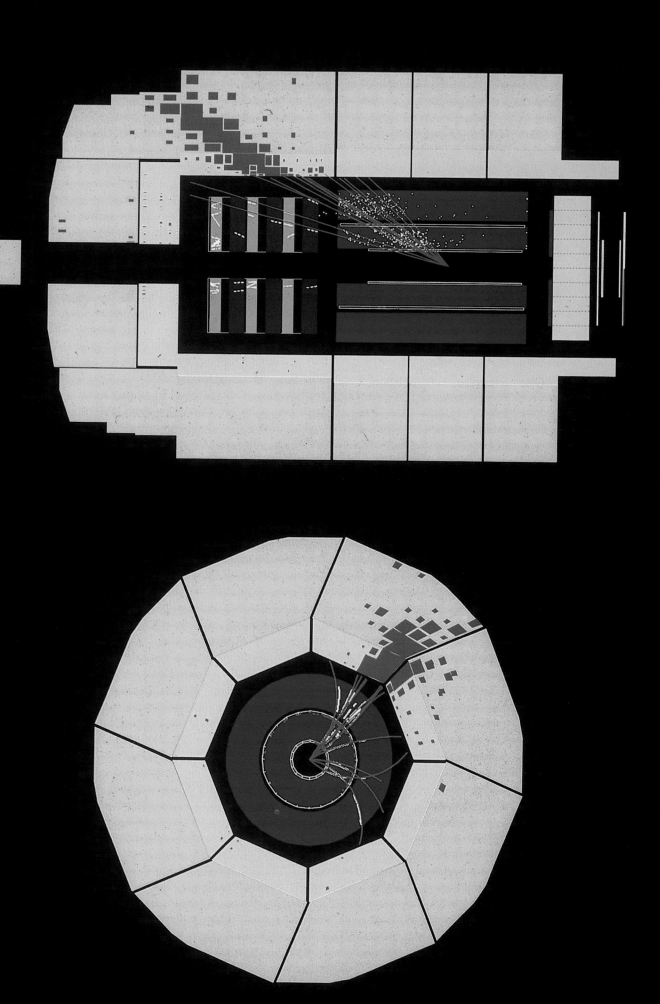

8. Colliders and Image Chambers

On 14 July 1989, as most of France prepared fireworks for Bastille Day celebrations, physicists and engineers in the Pays de Gex had excitement of a different kind on their minds. At 4.30 in the afternoon, a beam of antielectrons – positrons – travelled in a 27 km circle 100 m beneath the border between France and Switzerland. Flying along at close to the speed of light, the particles took less than a ten-thousandth of a second to make the first complete circuit of the world's newest and biggest particle accelerator – the Large Electron Positron collider, or LEP, at CERN, Geneva. Eleven days later, the first electrons circulated in the opposite direction, and the final steps in commissioning LEP began.

By mid-August the engineers were confident enough to try for the first high-energy collisions. At four points round the ring, they made small adjustments to the magnetic fields guiding the two particle beams, bringing them to meet head on at the heart of huge particle detectors. As big as a two-storey house and weighing as much as ten Jumbo jets, each detector had been built and assembled over the previous six years by hundreds of physicists and engineers from around the world. Now, at last, was the moment they had all been waiting for – the first electron–positron collisions in the marvellous new machine.

The first sight of the physicists' quarry came just before midnight on Sunday, 13 August, on the French side of the border in the detector called OPAL (which stands for OmniPurpose Apparatus for LEP). Particle tracks spraying out through the detector's multiple layers revealed the signature of a decaying Z particle, one of the carriers of the weak force. The Z, weighing in at nearly 100 proton masses, was the heaviest particle then known, and had been seen only a few hundred times throughout the world since the first observations at CERN in 1983. But now the physicists had a machine they could tune to make Z particles by the million – a veritable Z factory.

The other three big detectors – ALEPH (for Apparatus for LEP PHysics), DELPHI (for DEtector with Lepton, Photon and Hadron Identification) and L3 (because it was proposed in the third 'Letter of intent' for an experiment at LEP) – soon found their first Z particles. Together the four experiments would complement each other, as each had been designed to have different strengths, in precisely measuring electron energies, say, or in identifying particles. However, it was also vital that the experiments would be able to corroborate each other's evidence, to confirm the validity of the results from LEP in the eyes of the world.

OPAL was the most conservative of the four experiments, in general using well-understood techniques to record the decays of the Z particles. In common with the other experiments, it was constructed like a huge nest of Russian dolls wrapped around the pipe carrying the colliding beams, each layer designed to yield its own brand of information about the particles produced in the electron–positron collisions. To uncover the complexity of the assembly, you would have had to take the nest apart, one layer at a time.

As a visitor to OPAL you would take the lift from the surface 100 m down to the underground cavern, as big as a cathedral. A first glance at the detector would show its outer layer of chambers, formed into a barrel about 10 m high and 10 m long, with flat panels at the ends. These outer chambers were designed to register muons, the only charged particles able to penetrate the complete structure and deposit energy in the outer shell. If you could have peeled away this outer skin of muon chambers, you would have exposed a layer of 2500 tonnes of iron, interleaved with gas-filled tubes to pick up protons, pions, and other hadrons – in other words, particles built from quarks. This was the 'hadron

Fig. 8.1 A technician works on one of the two 'end caps' of the OPAL detector at the Large Electron Positron (LEP) collider at CERN. The complete detector was a huge layered cylindrical barrel – 10 m long and 10 m in diameter – with the beam pipe threading through its centre. The end caps in effect 'sealed' the ends of the barrel so that as few as possible of the particles created in electron–positron annihilations at the centre of the apparatus would escape detection.

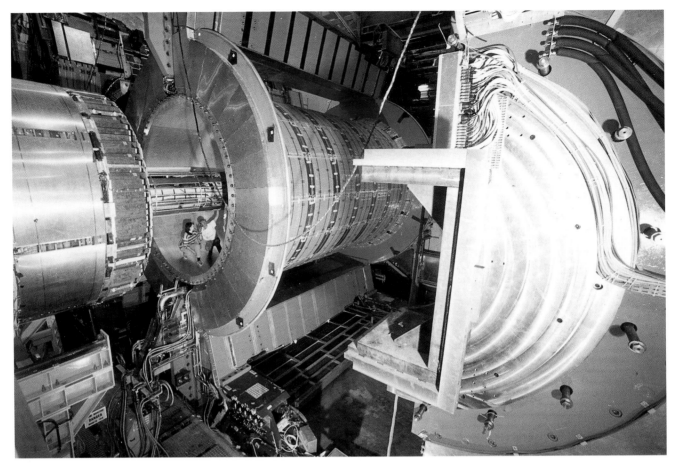

Fig. 8.2 OPAL's layered cylindrical structure, like a set of 'Russian dolls', is clearly visible here. The 8 m long bobbin-shaped structure at the centre, with the large flanges at the ends, contained the cylindrical aluminium coil of the electromagnet which provided the magnetic field to curve the tracks of charged particles for momentum measurement. To the left is the 3.7 m diameter 'jet chamber', which detected the tracks of charged particles, and fitted within the 'bobbin' of the electromagnet. The 'vertex detector' – the long cylindrical structure just above the heads of the people – formed the innermost layer, providing precise measurements of the first points on the tracks of charged particles as they emerged from the beam pipe running through its centre. To the right is one of two 'C-shaped' structures, which closed around the sides of the 'bobbin'. These contained an inner layer of lead-glass blocks that formed the electromagnetic calorimeter (to detect electrons, positrons, and photons) surrounded by the iron of the hadron calorimeter (to detect pions, protons, and other particles made up of quarks).

calorimeter', so called because it measured the energy of hadrons, just as calorimeters in other branches of science measure heat energy. The iron in the calorimeter had a dual purpose: as well as slowing down and trapping the hadrons, it formed part of the electromagnet used to bend the paths of charged particles.

Further in from the iron you would have discovered a layer of 11 488 blocks of high-quality lead glass, like the crystal of cut-glass tableware. This was the 'electromagnetic calorimeter', designed to trap electrons, positrons, and photons. Lead glass is often used as a detector because the lead in the glass – a surprising 60% or more – makes electrons and positrons radiate photons and also causes photons to convert into electron–positron pairs. The net effect is a miniature avalanche of electrons, positrons, and photons, which proceeds until all the energy of the original particle has been dissipated. The electrons and positrons travel faster in the glass than light does, and emit Čerenkov light (see p. 91) which is picked up by light-sensitive phototubes. The amount of light collected bears testimony to the energy of the original particle that entered the block.

Delving deeper into the heart of OPAL, the next layer you would have encountered, beyond the lead glass, would have been OPAL's solenoid magnet. This was an electromagnet built from a coil of aluminium 4 m in diameter and 6.5 m long. When electric current flowed through the coil it created a magnetic field to bend the paths of charged particles and provide information about their momentum, just as in a bubble chamber. But it was the next major layer that revealed the particle tracks.

The coil surrounded the 'jet chamber' – a closed cylinder 3.85 m in diameter, filled with gas and spanned by nearly 4000 wires. As charged particles passed through the gas they left ionized trails in their wake, which would produce tiny electrical signals on the nearest electrified wires. The pattern of signals recorded would reveal the tracks of the particles as they curved through the magnetic field.

Finally, having penetrated close to the centre of OPAL, you would have found the 'vertex detector'. This comprised small gas-filled cells with wires that measured particle tracks as precisely as possible, close to the point where the particles created in the electron–positron collisions emerged from the beam pipe.

Fig. 8.3 Half of the curved 'barrel' of lead-glass blocks (wrapped in reflective foil and black plastic) that formed the electromagnetic calorimeter in the OPAL detector. An identical curved section completed the barrel, surrounding the 4 m diameter cylindrical coil of the electromagnet. In all, the barrel contained 9440 blocks of lead glass to detect and measure the energy of electrons, positrons, and photons produced in the collisions at the heart of the detector. Each brick-sized block was one of 16 tapered shapes, carefully designed so as to point directly towards the collision point. The barrel was completed by an additional 2300 lead-glass blocks in the end caps.

Altogether, 14 different detector components made up the complete OPAL apparatus. They formed a hermetic system designed to trap as many particles as possible as they emerged from the electron–positron collisions at the centre. In principle, only the elusive neutrinos could escape completely, leaving no trace at all in any of the detector components. Yet even the neutrinos left a 'calling card', for they escaped with energy and momentum, both of which must be conserved in any interaction. The physicists working on OPAL knew the total energy and momentum of the colliding particles to begin with, and could add together the energy and momentum of all the particles they detected after the collision; anything that went missing had probably been spirited away by unseen neutrinos.

When LEP began running, every 22 microseconds (22 millionths of a second) needle-like bunches of electrons and positrons would pass through each other at the heart of OPAL and the other three detectors. However, the experimenters expected an interesting collision, or 'event', only about once every 40 times the bunches crossed. This was because, even though there were some million million particles in each bunch, the particles were thinly dispersed, so interactions between them were rare. But it would take milliseconds (thousandths of a second) for the computers to read all the pieces of information from the detector for one event (160 000 pieces of data in all). So, while an event was being recorded the detector would be 'dead' for many bunch crossings. The challenge was to identify and collect the interesting events, and not to miss them while recording something more mundane.

In OPAL – as in the other detectors – an electronic 'trigger' responded to the first signals from a collision to 'decide' within 10 microseconds whether something interesting had occurred. If it had not, 12 microseconds would remain to reset the system before the next bunch crossing; if it had, the process of reading out and combining the information from all the pieces of the detector would begin. After some milliseconds a display on a computer screen would recreate the pattern of particle tracks and show where energy had been deposited in the detector.

OPAL, which cost about 70 million Swiss francs (about $40 million) in all, was a cosmopolitan affair. As with the other three detectors at LEP, its components came from far and wide. There were muon chambers from the UK, parts for the hadron calorimeter from

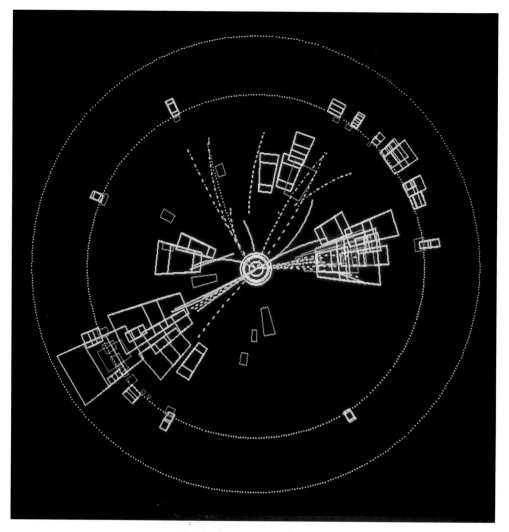

Fig. 8.4 One of the first batch of events recorded by the OPAL detector in August 1989, when the LEP collider at CERN produced its first collisions. The cylindrical detector surrounds the beam pipe where the collisions occur. In this 'end view', in effect a slice through the cylinder, the electron and positron have approached from above and below the page. On meeting, they have annihilated, their energy rematerializing as sprays of particles. The charged particles leave tracks (pale blue) in the central tracking chamber and, together with neutral particles, deposit energy in parts of the calorimeters, as indicated by the yellow blocks.

Israel, Italy, and the USA, and lead glass for the barrel from Japan. A team from Germany and CERN built the central tracking chamber, while the vertex detector inside it came from Canada. After first conception in 1981, through two years of design and then six painstaking years of prototyping and building in the different institutions, the pieces of OPAL finally came together during April 1989, just four months before the first electron–positron collisions in LEP.

This was the telling time. Each piece had been lowered into the 70 m long underground cavern, either via two lifts, or through the 10 m diameter access shaft. Japanese lead-glass and photodetectors, worth about $6 million, had to fit precisely around detectors surrounding the magnet coil from CERN; 130 tonnes of lead-glass and iron 'end-cap', with components from the UK and Israel, had to locate with the ends of the 'barrel' from CERN and Japan. The whole 3000 tonne colossus had to come together with almost millimetre accuracy. And, finally, it had all to be tilted to match the slant of the LEP collider ring itself, which had been constructed at an angle of one degree to the horizontal to avoid tunnelling through hard rock under the Jura mountains north-west of Geneva.

OPAL was typical of the particle detectors of the 1980s and 90s – an 'electronic bubble chamber', designed to cover as much of the space around the collision region as possible. With a bubble chamber, a computerized analysis of photographs from different angles allowed the full three-dimensional reconstruction of the particle tracks in space. In an electronic detector, by contrast, the tracks are recorded directly as computer data, and the three-dimensional reality of the events becomes visible when complex analysis programs paint the tracks of particles across computer screens. One device, probably more than any other, has made all this possible. This is the 'drift chamber', invented at CERN in the late 1960s, which nowadays records the tracks of charged particles in almost as much detail as bubble chambers did.

Electronic Bubble Chambers

During the 1970s, experiments in particle physics were gradually revolutionized by the inventiveness of a Frenchman at CERN – Georges Charpak. His work has led to particle detectors that combine speed with precision. In the 1960s, wire spark chambers proved valuable because they could operate much faster than a bubble chamber, although they could not provide the same amount of detailed information. Charpak's chambers challenged both these earlier devices; they operate far faster than spark chambers, while at the same time they approach the precision of the bubble chamber.

When a charged particle travels through a gas, it leaves behind a trail of ionized atoms. A whole range of particle detectors, from the cloud chamber to the wire spark chamber, depends on sensing this trail of ionization in some way. In 1968, Charpak's group of researchers discovered new ways to put the ionization to work in revealing the tracks of particles. The team developed two basic types of detector, known as the multiwire proportional chamber and the drift chamber. Both of these could work much faster and more precisely than wire spark chambers, and so deal more effectively with the copious numbers of interactions created at modern particle accelerators. The drift chamber and its variations figure in tracking charged particles in almost every experiment today.

A typical multiwire proportional chamber is superficially rather similar to a spark chamber; it is a sandwich of three planes of parallel wires (rather than the two planes of a spark chamber) fitted into a gas-filled structure. The difference between the devices lies in the way they operate. With a spark chamber, you apply a high voltage (10–20 kV) for a brief period across the closely separated planes of wires soon after a charged particle has passed through. The high voltage induces a spark to leap across the gap, but only where the gas has been ionized, along the particle's track.

A multiwire chamber, on the other hand, behaves more like the single-wire counter that Rutherford and Geiger used (see p. 26). In this case you apply the voltage (of 3–5 kV) continuously, so that the central plane of wires is at a positive electrical potential relative to the two outer planes. Immediately a charged particle passes through the gas it triggers an avalanche of ionization electrons. This avalanche grows rapidly in the intense electric field around the wire in the central plane that is nearest to the original particle's path. It is vital that these central wires are fine – 20 micrometres (millionths of a metre) or so in diameter – so that the field near to them is very strong. This means that most of the avalanche develops close to a single wire.

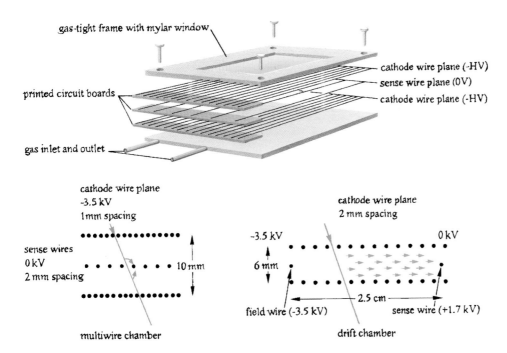

Charpak and his colleagues discovered that they could readily pinpoint the wire nearest to the ionized trail through the distinctive signal it produces, so with a series of chambers they could follow a particle's path. They also found that a chamber with wires only 1–2 mm apart within the central plane produces a signal within a few hundredths of a microsecond after a particle has passed by. Thus a multiwire chamber can handle a million particles per second passing each wire – a thousandfold improvement on the spark chamber.

Multiwire chambers in one form or another are nowadays part of many particle physics experiments, and their use has spread into astronomy and medicine, where they are particularly valuable in forming images. They come in many shapes and sizes – from flat chambers a few square centimetres in area for measuring the size of particle beams, to cylindrical sandwiches of concentric layers of wires which fit around the beam pipe at colliding beam machines.

Fig. 8.7 The wires that make up the drift chamber of the Mark II detector are seen during its construction at SLAC, in California. It is typical of modern electronic tracking chambers, which contain thousands of wires to pick up the electrons released by ionizing particles.

A multiwire chamber easily out-performs a bubble chamber in terms of the rate at which it can accept particles, but to track particles across a large volume, of a cubic metre say, requires a vast number of wires each with electronics to amplify the signals. Furthermore, it has limited precision, recording tracks at best to within a little less than a millimetre. However, the drift chamber, also developed by Charpak's group, not only provides precision, it does so with fewer wires.

The clever idea with the drift chamber is to measure time – which can be done very precisely with modern electronics – to reveal distance. The chamber again consists of parallel wires strung across a volume of gas, but some of the wires provide electric fields that in effect divide a large volume into smaller units or 'cells'. Each cell acts like an individual detector, in which the electric field directs the ionization electrons from a charged particle's track towards a central 'sense' wire. If the electrons travel at a constant velocity, then the time it takes for them to reach this wire gives a good measure of the distance of the track from the sense wire. Indeed, with this technique it has proved possible to locate particle tracks to an accuracy of some 50 micrometres.

Normally, electrons released in a gas will slow down as they lose energy in collisions with the gas molecules. Alternatively, a region of high electric field will accelerate them so that they gain energy. The drift chamber works by balancing the two effects in a carefully designed electric field, so that the energy ionization electrons gain from the field matches the energy they lose in collisions. The overall result is that the electrons travel at a constant, known velocity. The high-energy charged particles being detected travel through the chamber much faster than the electrons drift to the sense wires, so scintillation counters, which produce signals very rapidly, can be used to start an electronic 'stopwatch' as soon as they register that a charged particle has passed through the chamber. When the drifting electrons from the particle's trail reach a sense wire, the signal from the wire stops the watch, and the time can be read out by computer. Together, this time and the velocity of the drifting electrons give the distance from the sense wire, and hence a point on the ionizing particle's track.

A great advantage of the drift chamber is that the sense wires can be spaced at intervals of several centimetres or more, because it is the drift time that provides the information on position. This reduces the number of wires and ancillary electronics. However, another development, which sounds like something out of science fiction, reduces the number of wires even further. The 'time projection chamber' or TPC, invented by David Nygren from the Lawrence Berkeley Laboratory, in effect makes a complete detector out of a single large drift cell.

Nygren's idea was to have a cylinder of gas with a single electrode at high negative voltage across the middle. Electrons released along the tracks of ionizing particles drift towards the ends of the cylinder, which are positive relative to the centre. The time of arrival of the electrons at the end planes gives a measure of how far along the cylinder the electrons originated; electrons from nearer the centre take longer to reach the end. Moreover, as the electrons arrive at an end plate they create an image of a two-dimensional slice through the tracks. The measurements of the positions together with the arrival times provide enough information for a computer to reconstruct a three-dimensional image of the tracks.

Fig. 8.8 David Nygren, left, inventor of the time projection chamber, stands inside the hexagonal steel of the calorimeter that surrounded the 2 m diameter chamber built to study electron–positron collisions at the PEP collider at SLAC in the 1980s. The chamber was surrounded by the coil of a superconducting magnet, to bend the charged particle tracks.

The TPC's prowess in tracking charged particles is not the end of the story, however. The number of electrons arriving at an end plate depends on how much ionization has occurred along a track, and this in turn depends on the nature of the ionizing particle that produced the track. For example, lightweight positrons ionize less than protons of the same momentum. So the sizes of the signals at the end plates contain information which can help to distinguish between different types of particle.

The really ambitious part of Nygren's original concept was to do all this with a cylinder 2 m in diameter and 2 m long, allowing the ionization electrons to drift across distances up to 1 m. It took the best part of 10 years for the time projection chamber to come to life. Nygren's device, built for an experiment on the PEP electron–positron collider at SLAC, in California, started operating in 1983. It was a key part of a huge detector that surrounded PEP's colliding beams.

Time projection chambers have since figured in a number of experiments at different

Fig. 8.9 The world's largest time projection chamber is installed in the STAR experiment at Brookhaven National Laboratory's RHIC collider. Visible is one end of the 4.5 m long and 4 m diameter chamber as it is about to slide within the cylindrical electromagnet. The huge volume (53 cubic metres altogether) is divided into 12 segments at each end, where the ionization from charged particles traversing the chamber is detected.

laboratories. At CERN, for example, both the ALEPH and DELPHI experiments at LEP incorporated large segmented TPCs to track charged particles. With a diameter of 3.6 m and length of 4.4 m, the TPC for ALEPH was at the time the world's largest. Also at CERN, four large TPCs in the experiment code-named NA48 made visible the tracks of the multitude of charged particles produced in high-energy collisions between lead nuclei. Since 1999, the largest TPC has been a cylinder, 4.5 m long and 4 m in diameter, in an experiment at the Relativistic Heavy Ion Collider (RHIC) at the Brookhaven National Laboratory. The TPC forms a major part of STAR (for Solenoidal Tracker At RHIC), which is studying the thousands of tracks produced in high-energy collisions between nuclei of gold (see Fig. 10.13, p. 196).

Synchroclash

Electronic detectors have produced their most spectacular results in an environment that is inaccessible to bubble chambers – at colliding-beam machines where particles meet head on within the beam pipe. These machines produce more violent collisions than accelerators that fire particles at a stationary target. In a collider, the target is neither a piece of metal nor a volume of liquid, but a second particle beam travelling in the opposite direction.

Why collide beams? When high-energy particles plough into the nuclei within a stationary target, the debris of the collision is propelled forward, just as cars in a traffic queue shunt forwards when another car crashes into the back of them. From the physicists' point of view this effect is undesirable because the hard-won energy of the beam particles is being transferred largely into energy of motion – into moving particles in the target. This problem of wasted energy is overcome if we can bring particles to collide head on, so that their energy can be spent on the interaction between them. The situation then is more akin to two similar cars crashing head on. The debris flies off in all directions, and the energy is redistributed with it – none is 'wasted' in setting stationary lumps in motion.

The particles in accelerators are travelling close to the speed of light and in these circumstances the benefits of head-on collisions are even greater. As Einstein's special theory of relativity tells us, particles become heavier as they approach the speed of light and – because momentum is the product of mass and velocity – they have much more momentum to pass on to a stationary target. So the higher the energy, the more energy is wasted in moving a target, and the greater the benefits of colliding beams. These arguments were clear to accelerator builders as long ago as the 1940s, but it took 20 years for particle colliders to take shape, and another 15 years for them to become the dominant form of particle accelerator, as they still are today.

In 1943 Rolf Wideröe – whose doctoral thesis of 1928 had inspired Lawrence to invent the cyclotron (see pp. 84–85) – applied for a German patent on a scheme to store and collide particles travelling in opposite directions around the same orbit. The key word here is 'store'. If you were to fire two ordinary particle beams at each other, collisions would be few and far between – imagine firing a pair of shotguns towards each other in the hope that two pellets would collide. Wideröe's idea was to improve the odds by using a magnetic 'storage ring' to accumulate successive bursts of particles from an accelerator and so create a much denser beam. Moreover, because relatively few particles actually interact when two beams meet, the beams could circulate and intersect many times and so provide still more collisions for a given number of orbiting particles.

While nothing came directly from Wideröe's patent, the idea of colliding beams arose again in the late 1950s, and a fruitful partnership began between Gerard O'Neill from Princeton University (later to become known for his work on space colonies), and Wolfgang ('Pief') Panofsky of Stanford University. O'Neill's idea was to build two accelerator rings to store and accelerate electrons (which would be easier to work with than protons). The rings would intersect at some point where the stored beams, travelling in opposite directions, could meet head on.

Together with Panofsky, O'Neill gathered a small team of physicists to build a pair of electron storage rings at Stanford, where there already existed a 1 GeV linear electron accelerator. Construction of the two rings, which were to be side by side and joined at one common point, began in 1959. By 1965, the team had overcome all the problems and they were able to record the first physics results from colliding particle beams. Each ring stored a beam of 0.5 GeV electrons, giving a total collision energy of 1 GeV. This may not sound

Fig. 8.10 Gerard O'Neill (1927–1992).

Fig. 8.11 The first colliding-beam machine to carry out a successful program of experiments – the electron–electron storage rings built by a team from Princeton and Stanford – was working by 1965. In this picture we can see one of the rings, composed of four magnet arcs which encircle the rectangular structure. Electrons in this ring collided head on with electrons in the second ring at a point near the top centre of the picture, where the two rings shared a common straight section. Spark chambers around the intersection point recorded the tracks of particles produced in the collisions. The pipes and the curved magnet in the foreground are part of the system that fed electrons to the two rings.

much, but to free this amount of energy in a collision with a stationary target, an electron would have to be accelerated to 1000 GeV!

The Princeton–Stanford collaboration was not the only team working on storage rings. In 1959 a group of Italian physicists, under the leadership of Bruno Tsouchek, began work at the Frascati Laboratory near Rome on a small machine to collide electrons with their antimatter equivalents, positrons. Positrons have the same mass as electrons but opposite electric charge. This means that if a magnetic field bends electrons to the right, say, then it will bend positrons to the left. But suppose the electrons are moving in the opposite direction to the positrons: the magnetic field will then bend the two kinds of particle the same way. In other words, electrons and positrons travelling in opposite directions through a magnetic field will follow the same path, providing they have equal energies. The magnets that guide electrons one way round a storage ring – clockwise, say – will guide positrons the other way – anticlockwise.

The machine at Frascati was called ADA, for Annelo d'Accumulazione ('accumulation rings'), and was designed to store beams of 0.25 GeV energy. By 1962 it had stored electrons, and it was then transported to Orsay, near Paris, where a more intense electron beam was available to feed it. Towards the end of 1963, ADA's first electron–positron collisions were recorded, but the machine was never used to collect high-energy physics data. Instead, ADA was a testing ground for a breed of machine that was to change the course of particle physics in the following decades. Several similar electron–positron colliders followed – at places from Massachusetts to Novosibirsk – but it was one in California that made the greatest impact on our knowledge of fundamental particles.

In 1964, Burton Richter at the Stanford Linear Accelerator Center (SLAC) and David Ritson at Stanford University put forward a proposal to build an electron–positron collider called SPEAR, for 'Stanford Positron Electron Asymmetric Rings'. At the time, the famous 3 km long linear accelerator was still being built at SLAC, and it was not until 1970 that the Atomic Energy Commission (which at the time provided money for particle physics) gave the laboratory permission to build a simpler SPEAR, with one ring, together with a large multipurpose detector. But the money had to come from the laboratory's existing budget.

Undaunted, Richter and his team pushed ahead, and built SPEAR on a parking lot at SLAC, close to the end of the linear accelerator. Magnets to guide and focus the beams were mounted on 18 'girders' of reinforced concrete to form an oval ring, 63–80 m across, which could store particles with energies between 1.3 and 2.4 GeV. It was soon complete, and in 1972, only 20 months after approval, the first electron and positron beams were colliding. The machine had cost only $5.3 million.

How did SPEAR, the forerunner of many successful electron–positron colliders, work? First, the linear accelerator – the linac – fed the ring with successive bunches of electrons. In SPEAR these merged to form a single needle-like bunch, a few centimetres long and less than a millimetre across, containing as many as 10^{11} electrons. The positrons were created by accelerating electrons along about one third of the linac, and then firing them at a copper target. Positrons were filtered out of the resulting debris and fed back into the remainder of the linac, which now had its electric fields flipped so as to accelerate the

positively charged particles. Finally, the positrons left the linac and entered SPEAR, to circulate in the opposite direction to the electrons.

The counter-rotating bunches of electrons and positrons in SPEAR passed through each other twice per orbit when they met at points on the opposite sides of the oval ring. On one side – the 'West pit' – Richter, Martin Perl, and other physicists from SLAC, together with Willy Chinowsky, Gerson Goldhaber, George Trilling, and colleagues from the Lawrence Berkeley Laboratory, had installed the large detector they had been building while SPEAR was being constructed. This novel device – the Mark I – was destined to make great discoveries.

The Mark I covered 65% of the space around the collision zone. It was the nearest approach to an electronic bubble chamber that had yet been used. Sixteen concentric layers

Fig. 8.14 The Mark I detector at SPEAR was built by a team from SLAC and the Lawrence Berkeley Laboratory. In the mid-1970s it became famous for many discoveries, notably the J/psi particle and its relatives, and the tau lepton. The tracks of particles were recorded by wire spark chambers wrapped in concentric cylinders around the beam pipe, out to the ring where physicist Carl Friedberg has his right foot. Beyond this are two rings of protruding tubes, which are housings for photomultipliers that view various scintillation counters. The coil of the solenoidal electromagnet lies between the two layers of tubes; the magnet's iron forms the octagonal structure. To the left are rectangular magnets to guide the beams, which meet at the heart of the detector.

of cylindrical wire spark chambers – 100 000 wires in all – were wrapped around the pipe carrying the colliding beams, to track charged particles as they flew away from the collisions. Around the spark chambers, a huge coil of wire, some 3 m long and 3 m in diameter, formed an electromagnet to bend the tracks of the charged particles. Other types of detector, inside and outside the coil, helped to reveal the identity of particles, so that the physicists could later tell electrons from muons, pions from kaons, and so on.

Over the weekend of 9–10 November 1974, the Mark I's place in history became assured. As data from the detector burst onto screens in the control room, the physicists sensed they had touched on something remarkable, and soon they knew what was happening: SPEAR's collisions were providing clear evidence for a brand new particle, more than three times as heavy as the proton. Two years later, in 1976, Richter shared the Nobel prize for physics with Sam Ting from MIT, whose group had discovered the same particle at Brookhaven. The new particle became known as the J/psi (see pp. 158–159), and we now know that it is built from the charm quark bound with its antiquark. But this was not all, for by 1975, Martin Perl and colleagues working on some unusual events from SPEAR realized that they were seeing a new, third kind of lepton – a particle akin to the electron and muon, but much heavier. They named it the 'tau' (see pp. 162–163), this being the first letter of the Greek word for 'third', and in 1995 Perl also received the Nobel prize.

Fig. 8.15 (LEFT) Sam Ting (b. 1936) with members of his team in their control room at Brookhaven. A plot of the data with the peak revealing the J/psi lies on the table.

Fig. 8.16 (RIGHT) Members of the team that found the J/psi at SLAC peruse the log book. On the left is Martin Perl (b. 1927), with Burton Richter (b. 1931) in the centre, and Gerson Goldhaber on the right. The display on the screen in the background shows tracks from the J/psi in the Mark I detector.

New Particles, New Machines

During the months following the discovery of the J/psi, the Mark I detector at SPEAR collected a wealth of data on the new particle and on its excited states, which consist of a charm quark and antiquark orbiting around each other but with enhanced energies. At the same time, physicists at the Deutches Elektronen Synchrotron laboratory (DESY) in Hamburg were able to join in the pursuit of the J/psi's family using a machine called DORIS, for DOuble RIng Storage facility, which had started up during 1974. It collided electrons and positrons at a total energy of up to 7 GeV, a little below SPEAR's top energy in 1974, which modifications had increased to 8 GeV. Built with two rings, one on top of the other, DORIS was designed so that it could collide two beams of electrons or electrons and positrons. However, the discoveries at SPEAR ensured DORIS's future as an electron–positron machine. After the tau appeared at SPEAR, in 1975, experiments at DORIS provided valuable corroborative evidence for the new particle's properties.

The appearance of a new quark – charm – had at first seemed to bring with it a new natural symmetry. Physicists now knew of four kinds of quark – up, down, strange, and

Fig. 8.17 DORIS, the first electron–positron collider to be built at DESY, the accelerator laboratory in Hamburg. The machine was originally built with two rings of magnets, one on top of the other, so that it could store and collide two beams of electrons, if desired, in preference to electrons and positrons. In 1977, however, DORIS was modified and the two rings amalgamated into one, with the same radius, but now with unusually tall magnets (with blue tops). One of DORIS's important functions nowadays is to supply synchrotron radiation to experiments. High-energy electrons emit radiation as they move on curved paths. This radiation, a waste in terms of accelerating particles, provides a useful source of X-rays and UV radiation for scientists studying the structure of atoms, molecules, and materials. Here we see the pipes (heading towards top left) through which the synchrotron radiation passes out of the accelerator ring to experiments beyond the wall.

charm – and four kinds of lepton – the electron, the muon, the electron-neutrino, and the muon-neutrino. But the discovery of the tau, which presumably also had an associated neutrino, broke this neat symmetry. The possibility arose that nature might harbour still more quarks, to bring the total number of fundamental particles to a round dozen – six quarks and six leptons.

The new quarks were expected to be heavier than the charmed quark, and so should form particles still more massive than those of the J/psi's family. Storage rings such as SPEAR and DORIS had proved the ideal hunting ground for new particles, but the maximum energies of these machines precluded the discovery of anything heavier than around 8 GeV (a little more than eight proton masses). New, larger electron–positron machines were on the horizon – PEP, being built at SLAC, and PETRA, under construction at DESY, were both designed to reach a total energy of 30 GeV. But during all the excitement at SPEAR, a huge new proton synchrotron had started up at the Fermi National Accelerator Laboratory in Illinois, and it was here that another new particle made its appearance.

In 1977, Leon Lederman (later director of Fermilab) and a team of physicists from Columbia University, the State University of New York at Stony Brook, and Fermilab itself,

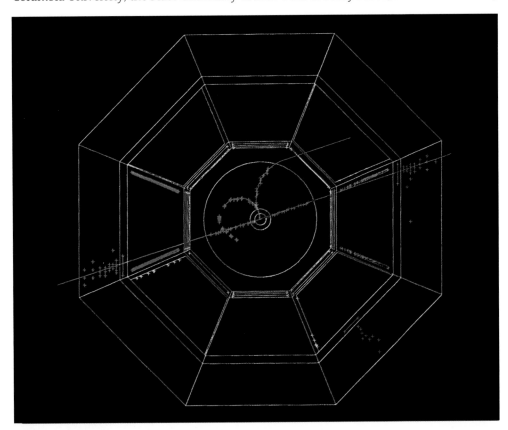

Fig. 8.18 A computer display of an electron–positron annihilation in the CLEO detector at Cornell shows the decay of an excited upsilon – a particle in which a bottom quark is bound together with its antiquark. The electron and positron have annihilated within the beam pipe – the innermost circle at the centre. The excited upsilon decays into another excited state, of lower energy, emitting an invisible photon as it does so; this new excited state then decays to the ground-state upsilon (with lowest energy) by emitting a second photon. The first photon converts to an electron (green) and a positron (red) before it enters the drift chamber, indicated by the large circle. The second photon is detected at the bottom right only when it converts (blue crosses) in the lead of the shower counter – the outermost octagonal layer. The ground-state upsilon also decays to an electron and a positron. These are of higher energy than the first pair and shoot straight off as far as the outer layer, where they produce showers. The total width of the apparatus on this display is a little more than 6 m.

discovered a particle about three times heavier than the J/psi and nine times heavier than the proton. The particle became known as the 'upsilon' and turned out to be rather like the J/psi, but this time comprising a fifth kind of quark – the bottom quark – bound with its antiquark (see p. 164).

With a mass of 9.4 GeV, the upsilon was too heavy to be made either at SPEAR or at DORIS in its original form with two rings, one on top of the other. However, by modifying the two rings of DORIS so that they formed a single ring capable of carrying much more intense beams of particles, the machine physicists at DESY were able to reach the energy region of the upsilon. In May 1978, physicists there saw the upsilon and soon began to study the various ways the bottom quark and antiquark orbit each other in the system known as 'bottomonium'. A year later, a new electron–positron collider began work at Cornell University in New York – the Cornell Electron Storage Ring, or CESR (pronounced 'caesar') – with a maximum energy of 8 GeV per beam, or 16 GeV in total. Two detectors at CESR, known as CLEO and CUSB, went on to study 'bottomonium' and other particles containing bottom quarks in great detail.

The discovery of the bottom quark made the symmetry problem of the quarks and leptons still more acute; now there really had to be a sixth, still heavier quark, dubbed 'top'. The search for this quark became one of the main priorities of particle physics in the late 1970s, and by the beginning of the 1980s a new generation of electron–positron colliders had begun the search.

In 1974, proposals had been put forward to build larger colliders both at DESY and at SLAC. The West German government seized the opportunity to help out an ailing construction industry, and plans for PETRA – Positron Electron Tandem Ring Accelerator – were approved the following year. Before the end of 1978, the new collider, which just fits into the DESY site, was complete and ready to produce its first collisions.

PETRA got off to a slow start and did not come close to its design energy of 19 GeV per beam until 1980. But when it

Fig. 8.19 The Cornell Electron Storage Ring (CESR), at Cornell University in Ithaca, New York, occupies the same tunnel as the synchrotron that feeds it with particles. The 12 GeV synchrotron (the magnet ring on the left in this picture) accelerated its first electrons in 1967. However, in the mid-1970s, the proposal to build CESR was made, and by the end of 1977 the synchrotron was successfully accelerating positrons for injection into a prototype section of the storage ring. The first electron–positron collisions in CESR, the ring on the right, occurred in June 1979, and since then Cornell has implemented a thorough study of the heavy particles that contain the bottom quark.

started up, first at 6.5 GeV and then at 8.5 GeV per beam, the total collision energy was the highest ever achieved with electrons and positrons. Moreover, there was no competition from the Positron Electron Project (PEP) at SLAC. A series of difficulties, financial and technical, delayed completion of PEP until 1980, by which time PETRA had come up trumps – not with the expected top quark, but with some remarkable results concerning the strong force between quarks.

The products of collisions at PETRA's higher energies revealed the first evidence for the radiation of gluons, the carriers of the strong force which flit between quarks and bind them together in protons, pions, and other particles that we observe. PETRA revealed that just as electrons radiate photons, so quarks can radiate gluons (see pp. 168–171).

By the spring of 1984, after various improvements, PETRA reached a new world record for positron–electron collisions of a little over 23 GeV per beam. But still there was no sign of the elusive top quark; all that the physicists at PETRA could say was that if it did exist then its mass must be greater than around 23 GeV. In turn, this implied that a machine capable of reaching higher energies would be needed if the top quark were to be found. As it turned out, the top quark would prove to be far too heavy for the electron–positron machines of the late twentieth century to produce, though hints of its existence were apparent at the Large Electron Positron collider before its direct appearance in experiments at Fermilab in 1995 (see p. 182).

Fig. 8. 20 An aerial view of DESY shows how the underground ring for PETRA just fits into the laboratory's site in a Hamburg suburb. Roads and tracks mark most of the ring's path, which passes from behind the chimney near the top centre of the picture, round by the houses at the right, close to the sports field at the bottom, and back up across the fields at the left.

The Antiproton Alternative

There is a problem with accelerating electrons and positrons in circular machines like PETRA and PEP – the lightweight particles radiate energy as they swing round bends. The effect soon becomes troublesome – double the energy of the electrons and the amount of 'synchrotron radiation', as it is known, rises sixteen times! To compensate for the energy lost in this way, electron and positron machines must use a powerful supply of the radio waves that accelerate the particles; PETRA, for example, used as much as 4.4 MW. More massive particles, such as protons, radiate far less easily, so to avoid the wasting effects of synchrotron radiation in circular electron machines, why not build proton–proton colliders, or even proton–antiproton colliders?

In the late 1950s, Gerard O'Neill had opted to build rings to collide electrons because no one at the time was sure how protons could be stored. However, a decade or so later, in 1971, engineers at CERN had succeeded in producing the first head-on collisions between protons in a machine called the ISR (for 'Intersecting Storage Rings').

The ISR consisted of two interlaced rings of magnets, with two beam pipes that crossed at eight places. It was fed by 26 GeV protons from CERN's Proton Synchrotron (PS), and brought the two beams to collide after they had each been accelerated to 31.5 GeV. The total head-on collision energy of 63 GeV was equivalent to the effect when protons strike a stationary target with an energy of 1800 GeV! In one step, the ISR had taken particle physics into a completely new energy region, and one that conventional accelerators had at the time no chance of reaching; the highest energy then planned for a synchrotron was a 'mere' 400 GeV.

Fig. 8.21 A crossing between the two beam pipes of CERN's Intersecting Storage Rings, the world's first proton–proton collider. The machine consisted of two interlaced proton storage rings, which crossed at eight places. It operated from 1971 to 1984, accelerating the proton beams to a total collision energy of 63 GeV.

Fig. 8.23 Antiprotons at CERN are created in collisions of protons with a metal target and then passed to this machine, the Antiproton Accumulator. Here successive bursts of antiprotons are added together, and 'cooled'. This is a process whereby the beam is made sufficiently well-behaved for injection into another machine – the Super Proton Synchrotron in the case of the high-energy proton–antiproton collisions studied during the 1980s. The magnets (blue) are unusually fat because they have to accommodate a beam pipe wide enough to carry the unruly antiprotons as they emerge at a variety of angles from the production target.

The physicists from around the world who first worked on the ISR were amazed by the experience, and it took several years to learn the best designs for apparatus and the most appropriate ways to analyse data. In the meantime, however, the people who built and ran proton machines were also learning. In particular, a Dutch engineer at CERN, Simon Van der Meer, thought of a way of concentrating particles in the beams in the ISR, to increase the chance of collisions when the beams crossed.

Van der Meer's idea was subtle, and relied on some ingenious manipulations of the particle beam. Put simply, you measure the positions of random samples of protons in the beam and then nudge these particles accordingly. By repeating this procedure many times, the whole beam is slowly concentrated closer to the optimum orbit. In practice, you sense the average position of protons in a 'slice' through a beam at one point in its orbit. You then use this information to send a signal across the ring to a 'kicker' that generates just the right amount of electric field to push these protons, on average, towards the ideal path. The method is called 'stochastic cooling': 'stochastic' because it operates on random samples of the beam, 'cooling' because squeezing the beam reduces its sideways motion, and smaller motions are usually associated with lower temperatures.

The machine physicists at CERN showed that stochastic cooling would work on the proton beams of the ISR in 1975. But it was in another development at CERN that Van der Meer's idea was to have an enormous impact and win him a share of the Nobel prize for physics in 1984. The technique proved vital in allowing CERN to use its 400 GeV Super Proton Synchrotron (SPS) as the world's first proton–antiproton collider.

The ISR had shown that colliding proton beams was a viable way of reaching high energies, but to reach energies much higher than the ISR would have required the construction of two machines, each equivalent to something like the SPS. In 1976, however, three physicists proposed a more cunning plan. The Italian Carlo Rubbia and Americans David Cline and Peter McIntyre suggested putting antiprotons into one of the big synchrotrons at CERN or Fermilab. The machine could then be made to operate like an electron–positron collider, with the antiprotons being bent by the same magnets and accelerated by the same electric fields while travelling in the opposite direction to the protons. It was a beautifully simple idea. The only problem was to get enough antiprotons into the machine for collisions to occur – and that is where Van der Meer's work on stochastic cooling came in.

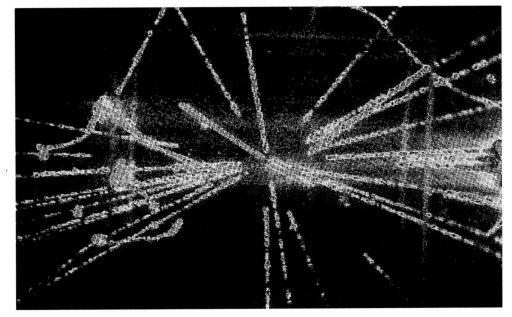

Fig. 8.24 The aftermath of a proton–antiproton annihilation at a total energy of 900 GeV. In CERN's SPS, the protons and antiprotons usually collided at a total energy of 630 GeV, but in special circumstances the beams could collide at 900 GeV. Here the tracks of charged particles produced in one of the first collisions at this high energy have been captured in the UA5 streamer chamber – a gas-filled device in which luminous streamers form along ionized trails under the influence of an electric field. This image was recorded by a TV camera and then enhanced by computer. The light intensity has been colour-coded so that the faintest areas are at the red end of the spectrum, and the brightest purple.

Antiprotons are produced in large numbers when a beam of high-energy protons strikes a metal target. The antiprotons emerge with a wide range of velocities and over a broad sweep of angles, so they cannot pass directly into a synchrotron, which operates with well-defined bunches of particles travelling at the same velocity. To tame the antiprotons before injecting them into the SPS, CERN decided to build a small machine, the Antiproton Accumulator. This would take the antiprotons from a target and use stochastic cooling to concentrate them into a well-behaved beam.

In 1978, CERN gave the official go-ahead for the proton–antiproton project and the building of the Antiproton Accumulator. Three years later, in August 1981, the accumulator delivered the first antiprotons to the SPS, and ecstatic physicists at CERN detected the first head-on collisions of protons with antiprotons at an energy of 270 GeV per beam, or 540 GeV in total. This is equivalent to a single beam of 150 000 GeV striking a stationary target.

Several experiments were soon ready to explore the new energy region. The first images of collisions at 540 GeV came from a detector known as UA5 (and four years later, in 1985, the same detector recorded the first images at a new record energy of 900 GeV).

Fig. 8.25 Carlo Rubbia (b. 1934), who played a leading role in persuading CERN to convert its Super Proton Synchrotron into a proton–antiproton collider, and who led the team that built the UA1 detector to study the proton–antiproton collisions. His profile is superimposed with part of a computer reconstruction from UA1 of the decay of a Z particle – the neutral carrier of the weak force (see pp. 176–179).

However, most of the time the proton–antiproton collisions came under the scrutiny of two larger detectors, in underground caverns at separate locations around the ring. These detectors, known as UA1 and UA2, were the predecessors of the huge complex detectors at the LEP collider, and were designed and built by large international teams of physicists and engineers. UA1, in particular, epitomized the principle of the electronic bubble chamber, with its 'image chamber', based on the drift chamber principle, to record the tracks of particles.

UA1 was masterminded by Carlo Rubbia, whose intellectual energy and physical stamina were such that he was able to be both a leading researcher at CERN and a professor across the Atlantic at Harvard. Rubbia had also been instrumental in persuading CERN to pioneer the route to proton–antiproton collisions in the SPS, and with UA1 he and his team had built a detector equal to the task of exploring the new energy region to the full. UA2, built by another consortium, was complementary to UA1, with different strengths in measuring particles. Together, the two experiments would provide vital corroborative evidence that would make their discoveries incontrovertible.

One of the first phenomena the two experiments observed was 'jets' of particles emerging from the proton–antiproton collisions. In particle physics, a jet is an individual shower of closely spaced particles, which appear to originate from a single quark (or antiquark) or gluon.

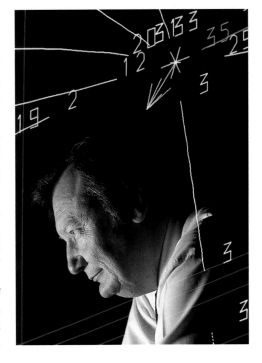

As Chapter 6 describes, quarks and gluons appear to be confined permanently within particles such as protons and pions. But when a proton and an antiproton collide at high energies, their contents – quarks, antiquarks, and gluons – can clash violently and bounce off at large angles relative to the main thrust of the debris. Even in this case, however, the quarks and gluons do not escape as free particles. Instead, they use their energy to create more quarks and antiquarks to form the clusters we detect as protons, pions, and the like. If the original quark or gluon is moving rapidly enough, it produces a 'jet' of several particles. This is the nearest we come to seeing an individual quark or gluon – the jet bears the memory of the direction in which its parent particle was originally moving.

The observation of jets at CERN provided valuable additional evidence for the existence of quarks and gluons. It reinforced the discoveries that had been made first at electron–positron colliders, such as SPEAR and PETRA, where quarks and antiquarks are produced afresh from the annihilations of electrons and positrons.

However, it was in 1983 that UA1 and UA2 finally made the discoveries for which they are most famous, and the anticipation of which in many ways had driven the whole proton–antiproton project at CERN. Early in the year the two teams announced the observation of the W particle (see pp. 172–175), the charged carrier of the weak force, with exactly the mass predicted by the electroweak theory that unifies the weak and electromagnetic forces. A few months later, they announced the discovery of the related neutral carrier, the Z particle (see pp. 176–179).

This was a triumph for the ideas of electroweak theory, a triumph for CERN with its daring proton–antiproton collider, and a triumph for Rubbia and all the physicists and engineers working on UA1 and UA2. The following year, 1984, Rubbia and Simon Van der Meer, whose ingenious antiproton cooling scheme had made the whole project possible, were rewarded with the Nobel prize.

A year later, Fermilab also began to produce proton–antiproton collisions in its Tevatron collider – initially at 800 GeV per beam, rising later to 900 GeV per beam or 1800 GeV in total. The reward for Fermilab came later, in 1995, in the form of the elusive and phenomenally heavy top quark. The 'missing' particle that had for so long been anticipated at every new high-energy machine had at last revealed itself.

Fig. 8.26 (OPPOSITE) A view down the access shaft shows the UA1 detector at the bottom of its pit, 60 m below ground. The aluminium-clad boxes visible along the top and sides of the detector contain muon chambers; they form the outer layer of UA1. In this position UA1 is not in fact in the tunnel of the SPS, where the proton–antiproton collisions occurred, but in its 'garage' where it could be worked on while the SPS was running for other experiments. When the detector was operating it was rolled into the tunnel, to the bottom left of this picture. Yellow supports hold hanks of cables, connected to the various parts of UA1, which must be long enough to follow the apparatus into the tunnel.

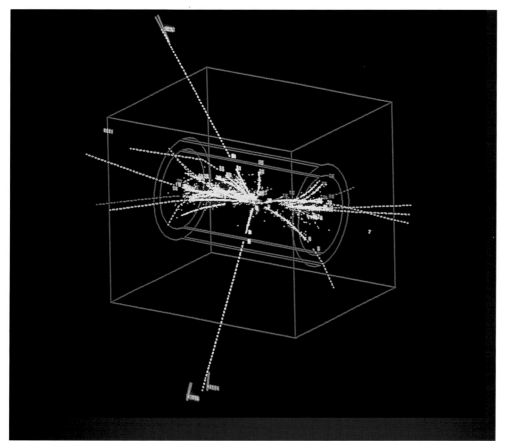

Fig. 8.27 A Z particle decays into a muon and an antimuon in this computer display from UA1. A proton and an antiproton have come in along the axis of the cylindrical Central Detector (outlined in red) and collided head on. The Central Detector reveals the tracks of the charged particles produced, which may also register in other parts of the detector. The computer has matched two of these tracks to hits in the muon chambers – indicated by the blue slashes – which lie outside the iron of the magnet (the red rectangular outline). Muons are the only charged particles that penetrate as far as this outer layer. Measurements of the momentum of the muons, from their slight curvature in the magnetic field, shows that together they add up to the expected mass of the Z particle.

The Biggest Machine in the World

Fig. 8.28 A view inside the 27 km circular tunnel of CERN's Large Electron Positron collider, which ran from 1989 to 2000. The electrons and positrons travelled in opposite directions in the beam pipe through hundreds of brown and white bending magnets (dipoles) and blue focusing magnets (quadrupoles). Originally LEP accelerated the beams to a total collision energy of around 90 GeV, but by the time of its final shut down in October 2000 it reached more than 200 GeV.

Protons may be easier to accelerate in circular machines than electrons, but they are complex objects and that means that they produce messy collisions. Moreover, in a proton–antiproton collider, the energy of each proton and antiproton is shared among their constituent quarks and gluons. The average energy carried by a single quark (or antiquark) is only about a fifth of the quoted beam energy, so the Tevatron's 1000 GeV per beam is in effect about 200 GeV per quark. Electrons and positrons, on the other hand, are fundamental particles as far as we can tell, with no discernible constituents. When an electron and a positron annihilate they become for an instant pure energy, which can then materialize as a new but equally fundamental particle–antiparticle pair. This makes it generally far easier to interpret the products of electron–positron annihilations than to sort out the debris from proton–antiproton collisions. So physicists have continued to reach for high energies with electron–positron machines, despite the penalties that come from energy losses through synchrotron radiation.

During the late 1970s, while PETRA and PEP were being built, physicists on the European Committee for Future Accelerators (ECFA) turned their thoughts to a far larger electron–positron machine – one that would reach to twice the ultimate energy of PETRA, and beyond. Their aim was to build a machine with sufficient energy to produce the then undiscovered Z particle, which electroweak theory predicted should have a mass around 90 GeV. To alleviate the problems with synchrotron radiation (recall that doubling the energy of electrons increases the synchrotron radiation losses sixteen-fold!) the plan was to make the curvature of the machine as gentle as possible. This is because doubling the radius of curvature of an electron's path halves the amount of energy lost through synchrotron radiation. There was a compromise to make, however, as a big ring would mean high capital costs in building it and equipping it with magnets.

On 16 December 1981, the CERN council made its historic decision to approve the construction of LEP – the Large Electron Positron collider, with an energy of 50 GeV per beam and a circumference of 27 km. It was possible to squeeze the ring – more than 10 times the size of PETRA – on land north and east of the laboratory's main site, between Geneva airport and the Jura mountains. Most of the ring would be in France, where the law states that people own their land down to the centre of the Earth. So although LEP would be underground, there would still be tricky negotiations ahead.

By June 1982, CERN had full authorization from the Swiss to go ahead with LEP. Agreement from the French followed in May 1983, and the massive civil engineering project could begin. Three full-face boring machines – 'moles' – were used to carve out the 3.8 m diameter tunnel as it pushed its way through 24 km of the local molasse, a sandstone-like rock. The remaining 3 km of the ring lay in harder limestone beneath the Jura mountains, and this had to be blasted out, as did the underground areas where the huge detectors were to be located.

The enormous ring was to consist of eight curved sections, each 2.8 km long, with 500 m long straight sections between them. It would be equipped with 3400 dipole magnets to bend the beams round the curves, 760 quadrupole magnets for focusing, and 512 sextupole magnets. With so many magnets, it was vital to keep down their cost, and the relatively low magnetic field required for LEP's gentle curvature allowed an innovative design for the bending magnets. They were constructed from steel laminations separated by gaps, three times their thickness, which were filled with cement and mortar. These 'concrete magnets' both cost and weighed half as much as if they had been of a more conventional design.

To reduce running costs, the radio-frequency accelerating system was also of a new design. Every 20 microseconds, particle bunches would pass through the hollow metal accelerating 'cavities' in straight sections in LEP's ring. But during the time between bunches, these structures would lose energy due to electrical currents unavoidably set up in the metal. To reduce this waste of energy, LEP's accelerating cavities were coupled to other cavities of a different design, which did not accelerate particles but did have low energy losses. Between particle bunches, the radio-frequency power was applied to these 'low-loss' cavities. Only when particles arrived in the system would the power be transferred to the accelerating cavities. With this design, LEP's maximum power load was less than that for the SPS, despite the energy lost through synchrotron radiation.

The low magnetic fields in LEP meant that it would be difficult to inject particles at very

Fig. 8.29 LEP's 27 km long tunnel was excavated for the most part by three boring machines known as 'moles'. These could bore rapidly, at about 25 m a day, through the sandstone of the plain around CERN, but blasting was necessary in some 3 km of harder rock at the foot of the Jura mountains. The ring was built with a tilt of 1.4% to keep as much as possible of the portion close to the Jura in the upper layers of softer rock.

Fig. 8.30 When LEP started up in 1989 it was powered by sixteen 1 megawatt klystrons which fed electromagnetic waves – microwaves – into 128 sets of copper accelerating cavities each time the bunches of electrons and positrons passed through them. During the intervals between the bursts of particles, the power was transferred from the accelerating cavities, where it wasted energy by heating the copper, into special storage cavities designed to lose less power. In this photograph of one of the acceleration sections at LEP, a spherical storage cavity is clearly visible above the accelerating cavities. This system of copper cavities, which was sufficient to accelerate the beams to 50 GeV each, was gradually replaced from 1996 onwards with superconducting cavities that could provide more accelerating power while using less electricity. These enabled LEP to reach more than 100 GeV per beam before it closed down in 2000.

Fig. 8.32 In a silicon strip detector, seen here in cross-section, a silicon wafer is divided into hundreds of parallel 'diode' strips formed by 'p+ silicon' channels on one surface. The p+ silicon contains additional positive 'holes' as it has been 'doped' with atoms of an element (boron) with fewer outer electrons. (By contrast the n+ silicon on the opposite surface contains atoms with more outer electrons.) Aluminium, finely divided above the strips, provides the electrodes, and oxidized silicon protects the surface between the strips. A charged particle ionizes atoms in the silicon, releasing electrons (green dots) and creating positive 'holes' (red dots), which move to the positive and negative electrodes respectively. The charge collected creates a signal in the readout on the strip nearest to the track, providing a measure of its position to within around a hundredth of a millimetre.

low energies, so the aim was to inject electrons and positrons already accelerated to 20 GeV. Initially, the idea was to build a new synchrotron for this purpose, but the accelerator experts at CERN realized they could save money by using two existing accelerators – the PS and SPS – to feed LEP. However, two new machines had to be built to provide the initial electron and positron beams: the LEP Injector Linac (LIL) and the Electron Positron Accumulator. By July 1987, the new injector and accumulator were finished, positrons had been sent through the system as far as the SPS, and installation of the huge detectors had begun in the completed experimental halls. But it was not until the following February that the full circle of the tunnel was complete when the last blast under the Jura allowed both ends to meet – to within a centimetre! Five months later, the first positrons entered the LEP ring and were accelerated through the first completed octant – one eighth of LEP, or 3.5 km of tunnel.

The positrons started life when 200 MeV electrons in the first linac in LIL struck a small tungsten target. Positrons produced in the collisions were separated off and accelerated in the second linac to 600 MeV. Successive pulses of positrons were then stored in the new accumulator ring until there was a reasonable number, and they were ready for feeding into the 30 year old PS. The PS took the positrons to 3.5 GeV and then passed them on to the SPS for acceleration to 20 GeV and the final leg of their journey before injection into LEP. The first successful injection at 23.53 on 12 July 1988 demonstrated that all the components of this amazing subatomic pinball machine worked correctly in unison. The challenge was now on to complete the remainder of the ring and fulfil the promise of the first collisions by the summer of 1989.

In May 1989, the final magnet was installed, and in June the beam pipe was completed – the longest ultra-high vacuum system ever built – and the final kilometres were pumped down to a pressure lower than that on the Moon. On 14 July a positron beam made its first full lap around the machine, and on 13 August the first electron–positron collisions occurred. The world's largest scientific instrument – and 'Z factory' – was in business.

Silicon Microscopes

The new particles of the 1970s live much longer than the resonances first observed in the previous decade, but they do not survive as long as the strange particles, which can be 'seen' through measurable tracks (or gaps in the case of neutral particles) in cloud and bubble chambers. Particles containing charmed or bottom quarks live typically for only 10^{-13} s, decaying a thousand times more rapidly than their strange counterparts. The lifetime of the tau lepton is also about 10^{-13} s.

When particles with short lifetimes are made in collisions between a particle beam and a fixed target, they do not have time to move very far from the general 'forward' direction of the beam that created them. A particle with a lifetime of 10^{-13} s, for instance, strays no more than 300 micrometres from this forward direction. This makes distinguishing the 'vees' – the points at which the particles decay – very tricky. In colliding-beam experiments, the problem is even more severe: such short-lived particles do not even have time to escape from the beam pipe.

The solution in both cases is to ensure that the part of the detector closest to the collision point has as high a resolution as possible. Often, this means using a high-precision wire chamber. Some fixed-target experiments, however, have seen the return of a material that initiated the particle explosion of the 1940s and 1950s – photographic emulsion. But just as silicon chips began to replace film in video and still photography in the 1990s, so silicon has taken over from emulsion in particle detection. Nowadays, almost every experiment has a silicon 'vertex' detector, which can reveal the 'vertices' where tracks

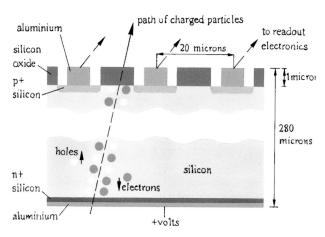

Fig. 8.33 (TOP) The silicon strip detector that lies at the heart of the BaBar experiment at SLAC's PEP-II collider contains five concentric layers made up from 52 doubled-sided silicon strip detectors, with a total of 150 000 strips.

Fig. 8.34 (BOTTOM LEFT) A silicon 'strip' detector for the D0 experiment at the Tevatron at Fermilab. The detector, fabricated from a silicon wafer rather as silicon microchips are, is divided lengthwise into 400 strips, each with microcircuitry attached at one end.

Fig. 8.35 (BOTTOM RIGHT) Technicians install the new 'silicon microvertex detector' at the centre of the ZEUS experiment at DESY. The detector measures charged particle tracks to 0.01 mm and allows the experiment to pinpoint the decays of particles containing heavy quarks, which have been created in the high-energy electron–proton collisions at the HERA collider. The detector was installed in 2001 in readiness for an upgraded HERA to restart with a much higher collision rate.

diverge as short-lived particles decay to those with longer lifetimes.

The most common technique with silicon is to divide its surface during fabrication into fine parallel strips which act as diodes. These are simple electronic components that conduct electric current in one direction only. When a charged particle passes through the silicon it ionizes the atoms, liberating electrons and positive ions. Only the strips through which the particle has passed will collect this liberated charge and produce a signal. In a sense the detector is like a wire chamber, with strip-like diodes instead of wires, and silicon instead of gas. However, the power of the silicon detector is such that, with modern fabrication techniques, the strips are typically spaced 25 microns (millionths of a metre) apart, yielding a precision on measuring particle tracks of only 10 microns.

Silicon strip detectors have come into their own at colliders, providing high-resolution 'microscopes' to see back into the beam pipe, where the decay vertices of particles can occur close to the collision point. They have proved particularly important in identifying B particles, which contain the heavy bottom quark. Bottom quarks prefer to decay to charm quarks, which in turn like to decay to strange quarks. Particles containing either of these quarks decay in about 10^{-13} s, and travel only a few millimetres, even when created at the highest-energy machines. Yet the silicon 'microscopes' constructed at the heart of detectors can often pinpoint the sequence of decays, from bottom to charm to strange particles. At the Tevatron at Fermilab, the ability to 'see' bottom particles in this way was critical in the discovery of the long-sought top quark, which likes to decay to a bottom quark (see Figs. 9.34–9.35, pp. 182–183).

The charge-coupled device, or CCD – well known now in cameras – provides another opportunity for particle physicists to exploit the silicon chip in their attempts to develop detectors with high resolution. A CCD consists of a two-dimensional array of 'picture elements', or 'pixels', each about 0.02 mm square. Electrons released by light, or by an ionizing particle, collect in the pixels, and can then be fed into electronic circuitry in a way that 'remembers' the locations of the relevant pixels. The numbers of electrons liberated by

Fig. 8.36 A computer reconstructs the tracks of charged particles through a barrel of silicon 'ladders' in the SLD experiment at SLAC. Each ladder supported two charge-coupled devices (CCDs), one on each side. The CCDs, 8 cm long and about 1 cm wide, were silicon chips divided into 3.2 million picture elements or 'pixels'. A charged particle passing through a pixel would ionize the silicon and produce a signal that could reveal the position of the particle to within about 0.004 mm. The inner diameter of the barrel, viewed here from one end, was about 5 cm, only slightly larger than the beam pipe in which electrons and positrons collided head on. The CCDs lay at the heart of the 4000 tonne SLD experiment, and provided precision measurements of the tracks of charged particles the moment they emerged from the beam pipe. This precision enabled the computer to extrapolate the tracks back inside the beam pipe to locate their point of origin, as Fig. 8.37 shows.

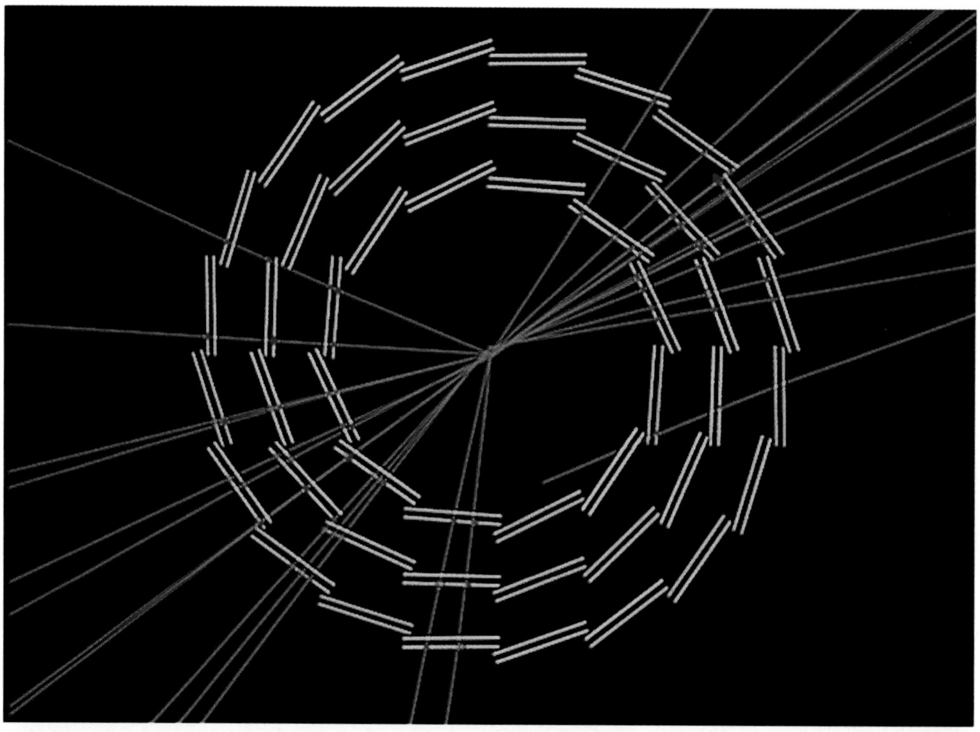

Fig. 8.37 This enlarged view of the centre of Fig. 8.36 shows how some of the tracks detected by the SLD's barrel of CCDs clearly originate from points – 'vertices' – some distance apart. An electron and positron have come from above and below the page and annihilated at the central point from which tracks shoot off in different directions. Several more tracks originate from two vertices to the left and right of the central interaction point. These points are only a few millimetres apart and indicate where short-lived particles – B mesons containing bottom quarks – decay. The bottom quarks – in fact a bottom quark and a bottom antiquark – were themselves created in the decay of a Z particle produced in the initial electron–positron annihilation.

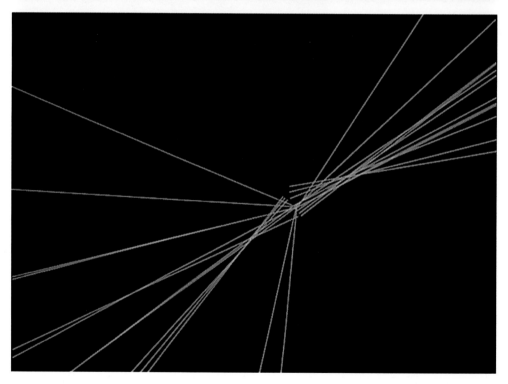

charged particles are small, so if CCDs are to be used as particle detectors they must be kept cool – at temperatures of −100 C or so – to minimize the effects of heat, which jiggles atoms and shakes electrons free. However the devices have the advantage over strip detectors of providing inherently two-dimensional information, and this makes CCDs more accurate.

The most complex CCD 'microscope' built so far formed part of the SLAC Large Detector (SLD), which focused in on the decays of Z particles produced during the 1990s by the linear electron–positron collider at SLAC (see pp. 153–154). Installed at the heart of the 4000 tonne SLD in January 1996, the CCD detector was a mere 15 cm in diameter, but it contained 96 CCDs, mounted in three concentric shells. The total of 307 million pixels tracked particles to a precision of about 0.004 mm, and helped the SLD team to make several 'world's best' measurements of the production and decay of Z particles.

All Kinds of Collider

It took just over six years and about 1300 million Swiss francs ($800 million) to build LEP, once the civil engineering had begun in 1983. The project was completed to time and to cost, and it provided the thousands of physicists who come to CERN with a world-class machine for the 1990s. At the same time, other leading laboratories were building new particle colliders, sometimes by converting existing accelerators.

CERN had pioneered the idea of the proton–antiproton collider, with the conversion of the Super Proton Synchrotron in 1981, and the physicists there were rewarded two years later with the discovery of the W and Z particles. Fermilab followed suit in 1985, when the first proton–antiproton collisions occurred in the superconducting, high-energy Tevatron (see p. 102). The Tevatron accelerates its proton and antiproton beams to 1000 GeV and then brings them to meet head on in the world's highest-energy collisions, at a total energy of 2000 GeV or 2 TeV. At these energies, the collisions can make W and Z particles in greater abundance than at CERN's proton–antiproton collider, and the physicists at Fermilab were able to improve on the original measurements from CERN. The Tevatron made history when it discovered another long-sought particle, the sixth and heaviest kind of quark – 'top' – in 1995 (see p. 182).

Meanwhile, the Stanford Linear Accelerator Center, SLAC, had converted its 3 km long linear accelerator into an ingenious collider, which could produce Z particles in electron–positron collisions, as at LEP. The trick was to make the 'linac' feed electrons and positrons into two intersecting arcs, the electrons travelling one way round the arc and the positrons the other so that they would collide where the arcs met. By also upgrading the

Fig. 8.38 The CDF detector at Fermilab, which together with the D0 detector found the first evidence for the top quark. In this picture, the pieces of the detector are pulled apart. Normally the complete detector is closed around the beam pipe of the Tevatron, where protons and antiprotons collide head on. The red sections are part of the steel support structure. The blue and black components, to the right and left, are parts of the electromagnetic and hadron calorimeters. These stop electrons and photons, and hadrons (protons, pions, etc.), respectively, and measure their energies.

linac so that its maximum energy was increased from 30 GeV to 50 GeV, the new SLAC Linear Collider (SLC) could reach collision energies of 100 GeV, enough to make the Z particle. In April 1989, the Mark II detector – successor to the famous Mark I (see Fig. 8.14, p. 139) – detected its first Z particles, three months before LEP.

During this time another collider began to take shape in Europe, quite different from the others. SLAC's linear accelerator had gained fame in the early 1970s, when its electrons delved deep into protons to reveal the quarks hiding there. Higher-energy electrons would in principle delve deeper, perhaps deep enough to reveal structure within quarks themselves. One way to produce higher-energy collisions between electrons and protons would have been by building an even longer linac, but a better alternative was to follow the collider route to high energies. Imagine taking the linac at SLAC and turning it round in a circle. Then, instead of having the target protons in a stationary lump of metal or a tank of hydrogen, as at SLAC, introduce a moving target of protons circulating in the opposite direction. This is the principle of the Hadron Electron Ring Accelerator, HERA, at the DESY laboratory in Hamburg.

Electrons and protons are very different, so the beams must be accelerated in separate machines before they are collided – quite simply, a proton synchrotron and an electron synchrotron are both required! Because lightweight electrons radiate away energy as they curve round an accelerator ring, electron synchrotrons are built with as gentle a curvature as possible – hence LEP's huge circumference of 27 km. For protons, nearly 2000 times heavier than electrons, the problem is much less severe, so much higher energies can be reached in relatively small rings.

To fit a proton machine and an electron machine in the same tunnel, the engineers at DESY had to compromise. First they designed as big a ring as they could, bearing in mind both costs and the fact that they would be tunnelling under suburban Hamburg. The size of this ring – with a circumference of 6.3 km – in effect set the highest electron energy they could reasonably reach at 30 GeV. They then designed the highest-energy proton machine that could be fitted into the same tunnel. By opting for the high magnetic fields generated

Fig. 8.39 HERA, the Hadron Electron Ring Accelerator, consists of two particle storage rings – one for electrons (or positrons) and one for protons. In this view inside the 6.3 km tunnel, the yellow electromagnets of the electron ring are visible below the white pipe of the cryostat that contains the superconducting magnets of the proton ring. The superconducting magnets, which are cooled by liquid helium to their operating temperature of −269 C (4.4 K), can run at higher electric currents than conventional copper electromagnets without losses due to heating, and so create stronger magnetic fields. This means that the protons can be accelerated to energies as high as 820 GeV, while being kept by the magnets in the 6.3 km ring. The electron beam on the other hand easily loses energy (as synchrotron radiation) as it bends, and the higher the energy, the more the radiation losses are. So the energy of the electron beam is kept relatively low, at 30 GeV, and normally conducting electromagnets easily keep it on its path round the ring.

Fig. 8.40 Various layers of the H1 detector can be seen here before they are closed together in position at the HERA collider. The large grey cylinder is the 5.8 m diameter coil of the superconducting magnet, which is kept at a temperature of about 4 degrees above absolute zero (4 K). This will be moved to the left to line up with the beam line which passes through the centre of the smaller grey superconducting magnet visible at far left between concrete shielding blocks (dull red). The iron sections of the main magnet (blue), which will move in to surround the large superconducting coil, are layered with detectors to register particles such as muons that will pass straight through them.

by superconducting magnets – as pioneered at Fermilab (see p. 102) – they were able to build a machine where the protons had a maximum energy of 820 GeV. The head-on collisions between 30 GeV electrons and 820 GeV protons are equivalent to electrons of 52 000 GeV striking protons in a stationary target, and that would have required a linear accelerator 3000 km long, four times the length of Germany!

As at other accelerators, the beams for HERA pass through several stages before they reach the maximum energies in the final ring. And as at CERN, the engineers at DESY made good use of existing machines in the early stages – in effect 'recycling' the older accelerators. Both the electron and proton beams start off in separate linear accelerators and are then fed into separate small synchrotrons. One of these is DESY's original electron synchrotron, converted to accelerate protons. The other ring is a completely new electron machine with a smaller radius, which fits inside the ring of the old synchrotron. The third stage for both kinds of beam is acceleration in the modified PETRA (see p. 142), and by the time they leave this machine the electrons have reached 12 GeV and the protons 40 GeV. They are now ready for injection into the separate rings of HERA for the fourth and final stage of their acceleration prior to head-on collision.

Fig. 8.41 Bjorn Wiik (1937–1999), in the HERA tunnel at DESY. Having worked on TASSO, one of the experiments that first saw gluons (see p. 168) at DESY, he became responsible for the construction of HERA's superconducting proton ring. He became Director of DESY in 1993, a position he still held at the time of his untimely death in 1999.

HERA's beams collide at four points around the ring, at two of which sit huge detectors called H1 and ZEUS. These are designed both to measure precisely the electron scattered in the collision and to collect as much as possible of the debris from the proton. Two specialized experiments, which started up after H1 and ZEUS, occupy the other two collision points. HERMES (for HERA Measurement of Spin) has been designed to investigate the precise origins of the spin, or intrinsic angular momentum, of the proton, while HERA-B's speciality is the study of B particles, which contain the heavy bottom quark.

DESY proved its ability to make proton machines as good as its electron machines when the first electron–proton collisions were seen on 19 October 1991. Seven months later, on 31 May 1992, H1 and ZEUS were ready to record their first collisions – and the journey into the proton was set to go deeper than ever before.

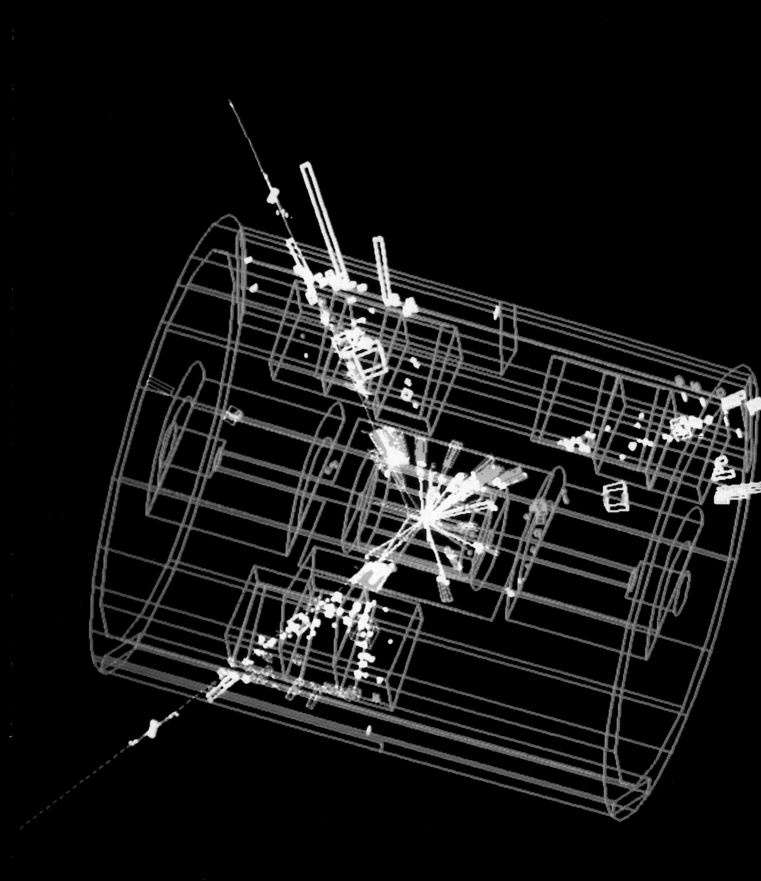

9. From Charm to Top

Matter is built from quarks and leptons, held together by fundamental forces, which in turn are mediated by particles known collectively as gauge bosons. This statement summarizes what particle physicists call the 'Standard Model' and it represents the state of our understanding of the nature of matter at the start of the twenty-first century. Perhaps surprisingly, much of the basis of this Standard Model became established in a relatively short time, in a remarkably fruitful decade from 1974 to 1984.

A series of important discoveries began in November 1974, with the first observations of the J/psi particle. This proved to contain a new, fourth type of quark ('charm'). Then, within another three years, evidence emerged for a fifth type of quark ('bottom'), and for a third kind of electrically charged lepton, the tau, to add to the electron and the muon.

All this might have confused the picture had it not been for important advances in understanding the forces between particles, which occurred over the same period. The idea of unifying the weak and electromagnetic forces within one framework began to seem more and more to fit with the reality of nature. More than merely accommodating a fourth quark, the 'electroweak' theory demanded it. This put increased pressure on experimenters to find the predicted carriers of the electroweak force – the W^+, W^-, and Z^0 particles.

At the same time, a new theory for the strong force, modelled on the same concept of force-carrying particles, made great headway. This was based on a novel property known as 'colour', which is analogous to electric charge. In other words the strong force is a 'colour force'. According to the theory, known as quantum chromodynamics, quarks carry colour and are bound together by carrier particles called gluons. By 1984, experimenters had not only found convincing evidence for gluons but also for the elusive Ws and Z.

While the underlying similarity between the fundamental forces was emerging, there were also signs of a pattern among the quarks and leptons. They seemed to occur in pairs, and the discovery of the fifth quark and the tau – the third charged lepton – strongly implied that three pairs of each occurred in nature, suggesting six quarks and six leptons in all. But it was to be the mid-1990s before experimenters caught the first glimpses of the sixth kind of quark, 'top', and the beginning of the twenty-first century before the sixth lepton – the tau-neutrino – was finally detected. In the 1970s and 1980s, experimenters continued to use bubble chambers and counter experiments just as in the previous two decades. But most of the major discoveries – and the images in this chapter – came from electronic detectors set up in a new configuration: in a 'barrel' surrounding the site of a head-on clash between two particle beams travelling in opposite directions. The earliest successes with a colliding-beam machine, or 'collider', came at SLAC, where a small magnet ring called SPEAR was set up to collide electrons with positrons. The physicists there were rewarded not only with some of the first evidence for the J/psi and its charmed relatives, but also with the tau. Later, a larger electron–positron collider at DESY, in Hamburg, claimed the discovery of the gluon. Then, in the early 1980s, CERN stole the scene with the discovery of the W and Z particles at its proton–antiproton collider.

The 1990s saw the epitome of both kinds of collider, with the Large Electron Positron (LEP) collider at CERN, and the Tevatron, a proton–antiproton collider at Fermilab. LEP's collisions allowed a detailed study of the unity between the forces and in the pattern of quarks and leptons, while the Tevatron yielded the sixth quark and the sixth lepton, the neutrino that partners the tau.

Fig. 9.1 Modern computer reconstructions of particle tracks in electronic detectors can be rotated, coloured, and depicted in whatever way makes the underlying physics clearer to see. This view of the aftermath of an electron–positron annihilation in the L3 detector at the Large Electron Positron (LEP) collider at CERN has been rotated to an unusual angle, clearly separating the three jets of particles. Elements of the basically cylindrical detector are indicated by the red lines, the particle tracks and energy deposits are shown in various other colours. The original electron and positron are unseen but would have come in from left and right, along the axis of the apparatus, and annihilated at the centre of the detector to form a single Z particle. The Z decayed almost instantly into a quark and an antiquark, one of which has radiated a gluon. The quark, antiquark, and the gluon are the source of the three jets of particles.

Charmed Particles

On the morning of Monday, 11 November 1974, members of the Program Advisory Committee at SLAC were assembling for one of their regular meetings. When one of the committee members, Sam Ting from the Brookhaven National Laboratory, met Burt Richter, a leading experimenter at SLAC, he said, 'Burt, I have some interesting physics to tell you about.' Richter responded immediately, 'Sam, I have some interesting physics to tell *you* about.' Neither realized that they had each discovered the same particle in entirely different experiments, nearly 5000 km apart. Richter's team had already called the particle after the Greek letter Ψ (psi); Ting had opted for 'J', the Chinese character for Ting. To this day it is known as the J/psi, an unwieldy name for a short-lived particle that was to open a new era in particle physics.

The J/psi revealed itself as a resonance. At SLAC's electron–positron storage ring, SPEAR, it produced a sharp spike in the number of charged particles emerging from the electron–positron annihilations (Fig. 9.2). In Ting's experiment at Brookhaven, it was responsible for a similar spike in the number of pairs of electrons and positrons produced in the collisions of high-energy protons with a beryllium target (Fig. 9.3). In both experiments the spike occurred at a total electron–positron energy of 3.1 GeV – the mass of the new particle, more than three times as heavy as the proton.

By 1974 resonances were nothing new, but the J/psi was remarkable because its spike was very narrow. In the quantum world of subatomic particles, the narrower a resonance, the longer its lifetime, and the width of the J/psi corresponded to a life of 10^{-20} s. This does not sound very long, but it was a thousand times longer than expected for a particle as heavy as the J/psi, which should have decayed even more rapidly to lighter particles.

Within 10 days of their first discovery, Richter's team at SPEAR found a second spike at a slightly higher energy, just below 3.7 GeV. This had to be due to another new particle – the Ψ' (psi-prime) – again with a narrow width and therefore a relatively long lifetime. The physicists were stunned. It was as if anthropologists had stumbled on a tribe of people who lived to an age of 70 000 years. What could be prolonging the life of the new particles?

The most likely possibility was that the J/psi and the Ψ' possessed some new property, which they could not easily discard and which prohibited a rapid decay. In the months following these discoveries, an enormous number of papers were published offering explanations of the new particles, until one theory began to emerge head and shoulders above the others.

In 1970, theorists Sheldon Glashow, John Iliopoulos, and Luciano Maiani had been considering how to incorporate the behaviour of quarks into a single 'unified' theory of electromagnetic and weak forces. They discovered that the way was clear to realizing such a theory if a fourth type of quark existed, which they called 'charm'. With the discovery of the J/psi, the idea of a charmed quark came right to the forefront of theoretical wisdom. The properties of the J/psi and its heavier relative could easily be explained if they were each built from a charmed quark bound with its antiquark – a system known as 'charmonium'.

A charmed quark and a charmed antiquark make a particle that contains charm within it but which has no net charm overall; the charm carried by the quark and the 'anticharm' of the antiquark cancel out. The quark and antiquark move around each other – rather as the electron and proton do in a hydrogen atom – in a variety of orbitals with differing energies. If the quark and antiquark orbit with high energy they form a relatively heavy particle, because mass is equivalent to energy. This heavier particle can emit energy (in the form of photons, pions, muons, or electrons) as the quark and antiquark move to a state of lower energy, and so form a lighter particle.

However, once the quark and antiquark are in the lowest energy state possible they remain, surviving until they come close enough to annihilate one another, their total mass-energy rematerializing as lighter particles. The J/psi corresponds to this lowest energy state in the charmonium system, while the Ψ' is the second lowest.

In Fig. 9.4 we see the results of the decay of a Ψ' to a J/psi accompanied by the emission of two pions (one positive, one negative) which carry away the excess energy. This is followed almost immediately by the J/psi's decay, as the charmed quark and antiquark annihilate, their energy appearing as a positron and an electron. Appropriately, the particle tracks write out the Greek letter Ψ.

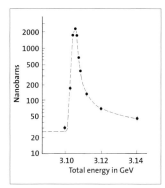

Fig. 9.2 This dramatic hundredfold increase in the number of hadrons produced in electron–positron annihilations at a total energy of 3.1 GeV signalled the production and decay of the J/psi particle in the Mark I detector at SPEAR.

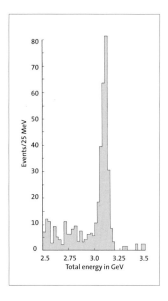

Fig. 9.3 The spike in the number of electron–positron pairs produced in collisions of a proton beam with a beryllium target at Brookhaven also revealed the existence of the J/psi.

With a mass of 3.1 GeV, the J/psi can decay in many ways to lighter particles. Nearly 90% of the time it decays to particles containing the lighter quarks – pions and kaons in a whole variety of combinations. More rarely it decays to an electron and a positron, as in Fig. 9.4, or to a muon and an antimuon. As muons are penetrating particles, this mode of decay can be particularly distinctive, as Fig. 9.5 shows. Here an electron and a proton have collided at high energy in the HERA machine at DESY, in Hamburg, at the heart of the detector called H1. The electron and proton travel through the beam pipe, unseen, but a J/psi produced in the collision betrays its presence when it decays to a muon and an antimuon, which leave symmetrically divergent tracks in the detector.

The charm quark can also combine with any of the lighter quarks (up, down, and strange) to form either baryons built from clusters of three quarks, or mesons formed from quark–antiquark pairs. Just as there is a world of strange matter, as physicists discovered in the early 1950s, so there is a whole range of charmed matter, containing charmed quarks and antiquarks, where the overall charm does not cancel out as it does in charmonium.

We saw in Chapter 5 that in particle collisions the strong force always makes strange particles in pairs, because the creation of a strange quark is always balanced by the creation of a strange antiquark. Precisely the same rule applies to charmed particles: when the strong or electromagnetic forces produce a charmed quark, they must also produce a charmed antiquark. Sometimes the quark and antiquark form charmonium; but if they are moving fast enough they can escape from one another and associate instead with other quarks or antiquarks emerging from the same collision. In this latter case the charmed quark and antiquark appear 'wrapped up' in different charmed particles. As they fly apart they decay individually in a process analogous to the beta decay of a neutron, in which one type of quark transmutes into another variety. Thus the charm created by the strong force in the initial collisions leaks away slowly through the agency of the weak force responsible for beta decay.

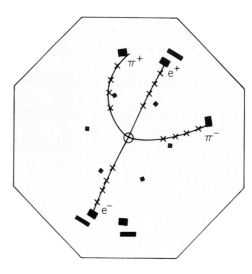

Fig. 9.4 A psi-prime (Ψ') particle writes its Greek name in the Mark I detector as it decays to two pions (the curved tracks) and a J/psi, which immediately decays to a positron (e^+) and an electron (e^-). The octagon outlines the detector, approximately 2 m from the centre. The crosses mark hits in four layers of concentric cylindrical spark chambers, and the dark bars indicate scintillation counters that have fired. The Ψ' was formed in the annihilation of an electron and a positron, which collided at the centre of the detector.

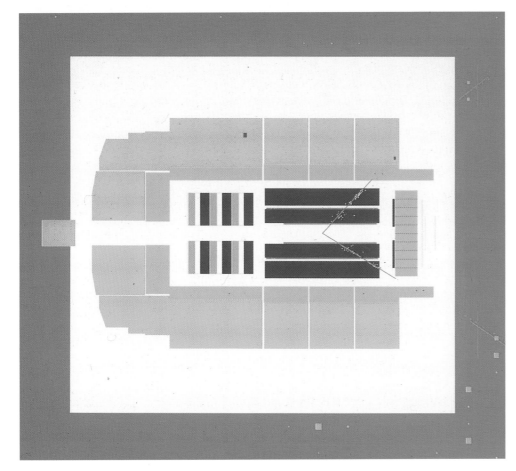

Fig. 9.5 Two lone symmetric tracks reveal the decay of a J/psi particle in the H1 experiment at the HERA collider, DESY. The J/psi has been produced in a high-energy glancing collision of an electron and a proton (both unseen in the beam pipe along the central axis of the apparatus). The various colours mark out H1's different layers, from the central tracking chamber (dark blue), through the lead and steel calorimeter (green and orange) to the iron of the magnet (outer blue rectangle), which is layered with detectors to reveal any particles that penetrate this far, in particular muons. The only indication that an electron, from the left, has interacted with a proton, from the right, is the presence of the two tracks, which are almost undeflected by the magnetic field and which link with signals (orange) in the outer iron layer. These must be due to energetic, penetrating muons. Together their energies add up to the equivalent of the mass of the J/psi particle. (Note the asymmetric appearance of this 'side view' of H1, with more detector components to the left. This is because the protons, coming from the right, have much higher energies (820 GeV) than the electrons (30 GeV), so after a collision most of the particles produced normally tend to travel towards the left.)

Fig. 9.6 One of the first examples of an event consistent with the production and decay of a charmed baryon (three-quark particle), photographed in the '7 foot' (2.1 m) bubble chamber at Brookhaven in 1974. A neutrino (unseen) enters the picture from below and collides with a proton in the chamber's liquid. The collision produces five charged particles – a negative muon (red), three positive pions (blue), and a negative pion (green track at left) – and a neutral lambda, which leaves no track. (Note how the muon and one of the pions knock electrons (yellow) out of the liquid, which spiral round in the chamber's magnetic field.) The lambda produces a characteristic 'V' when it decays to a proton (purple) and a pi-minus (green). The momenta and angles of the tracks together imply that the lambda and the four pions produced with it have come from the decay of a charmed sigma particle, with a mass of around 2.4 GeV. But the decay happened too quickly – within 10^{-12} s – for the original charmed particle to leave an observable track in this chamber.

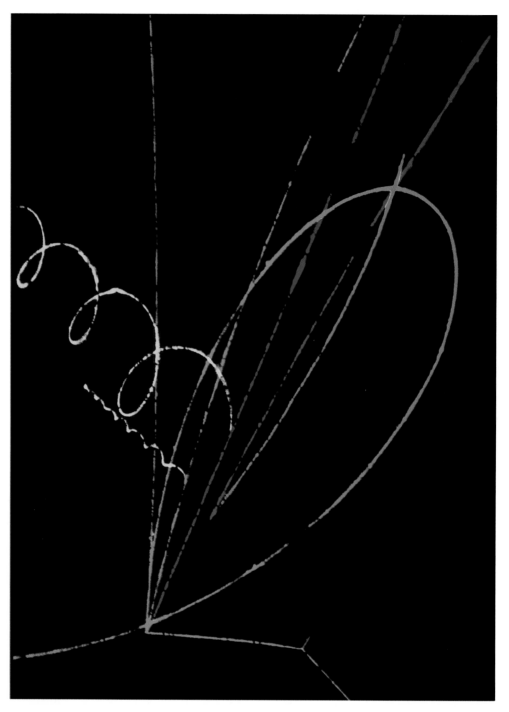

Figure 9.6 shows one of the first images consistent with the decay of a charmed baryon, taken in the '7 foot' bubble chamber at Brookhaven in 1975. We cannot see the track of the charmed particle, but we can infer its existence from the information contained in the tracks of its decay products. An invisible neutrino enters the bubble chamber and interacts with a proton, producing a muon and a 'charmed sigma' – a particle resembling the sigma particle but with a charmed quark in place of the strange quark. The muon moves swiftly from the scene, leaving a characteristic long, almost straight track; the charmed sigma decays before it can leave a discernible track.

Theory indicates that when a charmed quark decays it most often becomes a strange quark, which means that strange particles such as kaons should bear witness to the brief existence and subsequent decay of a charmed particle. The first clear evidence for charmed particles appeared in this way in the spring of 1976 in data from electron–positron collisions at SPEAR. A detailed analysis showed that combinations of pions and kaons were produced more frequently when their energies added up to a certain amount. The implication was that at this energy the pions and kaons were emerging from the decays of a specific particle, with a mass equal to the total energy. This was identified as the charmed particle known as the D^0 – a combination of a charmed quark and an up antiquark with a mass of 1.863 GeV. A few months later, further analysis had revealed more states, in particular the D^+, comprising a charmed quark bound with a down antiquark, with a mass of 1.868 GeV.

These discoveries, made through spikes in energy distributions, are reminiscent of the way we observe the J/psi. However the lifetimes of the charmed particles are some 10 million times longer – 10^{-13} s as compared with 10^{-20} s for the J/psi. In the mid-1970s, such lifetimes were close to the borderline of what could be detected directly in bubble chambers, but the advent of high-resolution techniques in the 1980s made it possible to see charmed particles directly by their trails. These observations reinforce the notion that particles we identify through spikes in energy plots are just as real as those we observe through their trails in detectors.

In Fig. 9.7, a photon provides the energy to create a pair of charmed particles as it hits a proton in a bubble chamber at SLAC. Two charmed mesons are produced, one charged and one neutral. The charged meson leaves a visible track and then decays into three charged particles whose trails are also seen. The neutral charmed meson decays into two charged particles, one negative and one positive, whose tracks form a 'vee'. The beauty of this picture is that we can measure the distances that the charmed particles travel before they decay, which provides useful information on lifetimes.

The discovery of charmed particles showed that nature exhibits a symmetry between quarks and leptons. The electron and its neutrino are matched by the up and down quarks; the muon and its neutrino are matched by the quarks bearing charm and strangeness. Our everyday world, and indeed the Universe we inhabit, comprises matter whose nuclei contain up and down quarks. However, we can imagine a 'Mark 2 Universe' built from strange and charmed quarks. Such matter probably existed fleetingly after the Big Bang, alongside up and down matter, but today we are left only with glimpses of strangeness and charm, and we have yet to understand the precise significance of these qualities, which seem unnecessary in our everyday world.

Fig. 9.7 The 'footprint' of the associated production of charmed particles in the Hybrid Facility bubble chamber at SLAC. An invisible photon has come in at the left to collide with a proton in the liquid hydrogen in the bubble chamber, producing two charmed particles – one neutral and one charged. The neutral charmed particle, probably a D^0, leaves no track but decays, forming the 'V' at bottom right. The charged charmed particle travels about 2 mm before it too decays, to three charged particles. The bubbles in this chamber were allowed to grow only to about 0.055 mm before they were photographed. This allowed the tiny tracks (or gaps) due to the short-lived charmed particles to be observed, unlike the example in Fig. 9.6.

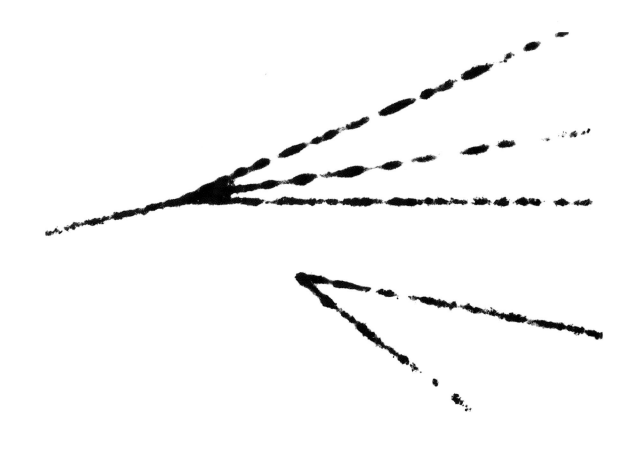

The Tau

Particle physicists were still congratulating themselves on having discovered the symmetry between four types of quark and four leptons when they were thrown into disarray. The same experiment that had caught some of the first glimpses of charm unearthed an unexpected guest at the feast: the tau. The tau is an electrically charged lepton, a much heavier version of the electron and muon. It weighs about twice as much as a proton, 20 times as much as the muon, and a staggering 4000 times as much as the electron. The law of nature that determines this bizarre numerology is one of the unsolved puzzles in particle physics today.

Like the electron and muon, the tau has negative charge and exists in an antimatter version with positive charge. It is not affected by the strong force, but it does take part in electromagnetic and weak interactions. And just as the electron and muon are partnered by their own neutrinos, the tau is partnered by a third variety, the tau-neutrino, bringing the total number of leptons to six.

When an electron and a positron annihilate in a head-on collision, they can rematerialize as new forms of matter, provided the total energy is high enough to create

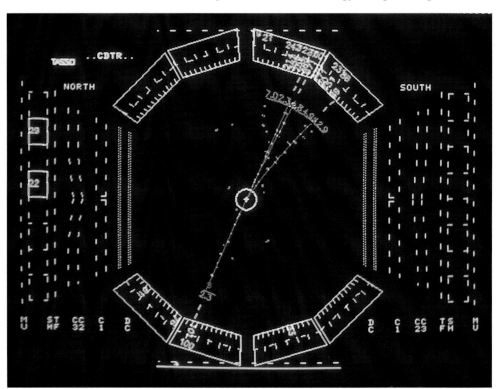

Fig. 9.8 The distinctive '3 + 1' decay of a tau-plus and a tau-minus in the TASSO detector at DESY. An electron and positron have annihilated at the point marked by the yellow cross. A tau-plus and a tau-minus materialize from the annihilation and shoot off in opposite directions, but decay while still within the beam pipe (the yellow circle). The tau-plus decays to two invisible neutrinos and a positive muon, which travels towards bottom left. The muon can be identified because it penetrates the argon-lead shower counters (purple blocks) to score a hit in one of the muon chambers (indicated by the blue line, with the cross showing the location of the hit). The tau-minus, on the other hand, has decayed to an invisible neutrino and three charged pions, which travel towards top right. The pions and the muon all leave tracks in the drift chamber, which fills most of the space between the beam pipe and the shower counters. Hits on the drift chamber wires are marked by blue bars; the particle tracks, as calculated by the computer, are in red. The drift chamber extends to about 1.3 m from the centre of the detector.

the appropriate antiparticle along with a new particle. If the total energy is above about 3.6 GeV, a tau and an antitau can emerge back to back. Four times in every hundred the negative tau decays into an electron and two neutrinos while the positive antitau decays into a positive muon and two neutrinos. Alternatively, the tau and antitau can produce a negative muon and a positron together with unseen neutrinos. These are very distinctive reactions, because an annihilation between an electron and a positron appears to yield an electron and an antimuon (or a positron and a muon). It was events such as these that gave the first hints of the tau to Martin Perl and his colleagues working on the Mark I detector at the SPEAR electron–positron collider in 1974, though it was 1975 before the researchers were sure of what they were seeing.

The tau, together with the muon and the electron, is a member of the lepton family of particles. It might seem rather surprising therefore that it can also decay to particles such as pions and kaons, which contain quarks. This is because the tau is so heavy – heavy enough to produce quarks and antiquarks *in matching pairs* to form pions and kaons. In these decays, the tau in effect turns into a tau-neutrino, which preserves its 'lepton-ness', while the remaining energy emerges initially as a quark and an antiquark, rather than an electron and neutrino, or muon and neutrino, as in the decays Perl's team first observed. Moreover, because the tau is heavy and has many ways to decay, its lifetime is brief – about 3×10^{-13} s – and any track it leaves is correspondingly minute. So when taus are made in low-energy electron–positron annihilations, it is difficult to distinguish decays to pions and kaons from other annihilation reactions that have created quark–antiquark pairs. At higher energies, however, quarks and antiquarks created directly in annihilations usually generate tight 'jets' of many particles, which contrast with the relatively few particles produced when a tau decays.

A particularly clean example of the tau's decay to quarks is the so-called '3 + 1' decay, as shown in Fig. 9.8. Here a positive and a negative tau have been produced together at the centre of the TASSO detector, which studied electron–positron collisions at the PETRA collider at DESY. One of the taus decays into a neutrino and three charged particles (pions), which leave tracks in one direction; these are balanced in the opposite direction by the track of a lone charged particle (a muon) that emerges from the other tau (together with neutrinos). This contrasts with the '3 + 3' decay shown in Fig. 9.9, recorded in the ALEPH detector at the higher collision energies of the LEP collider at CERN. In this case, both taus have decayed to three charged pions, together with unseen neutrinos.

Fig. 9.9 A '3 + 3' decay of a tau-plus and a tau-minus produced in the decay of a Z particle, formed in an electron–positron annihilation at the centre of the ALEPH detector at LEP. In each case the tau has decayed to three charged pions, which leave tracks first in the silicon vertex detector (first black ring outside the tiny circle of the beam pipe), then in the inner tracking chamber (the next black ring) and then in the time projection chambers (large black ring). These particles also deposit energy and stop in the surrounding electromagnetic (yellow) and hadron (red) calorimeters. Unseen are the tau-neutrino and tau-antineutrino that must have been produced in the decay of the tau-minus and tau-plus respectively.

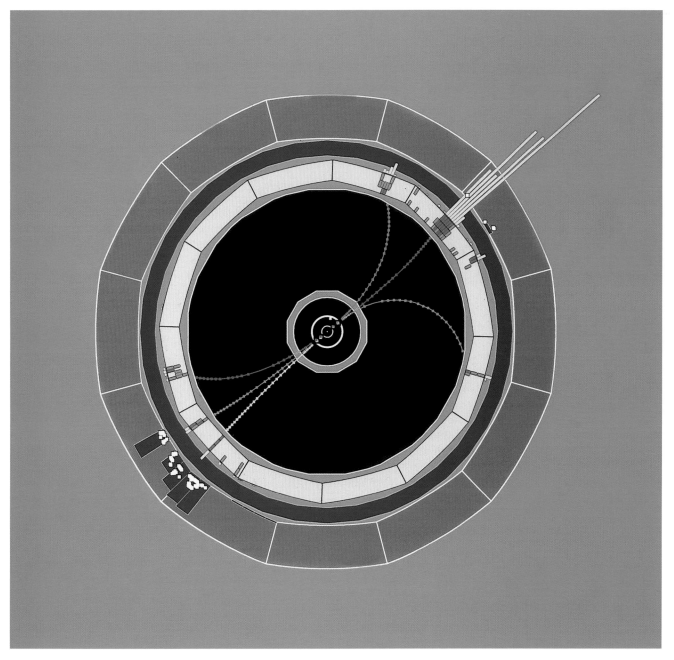

Bottom Particles

The discovery of charm in 1974 had begun to suggest a symmetry between quarks and leptons. In the everyday world up and down quarks are matched by the electron and its neutrino. At higher energies, a second 'generation' of quarks and leptons appears, with the strange and charmed quarks matched by the muon and its neutrino. But then came the tau lepton. The similarity of the tau to the muon and electron naturally suggested that there is a third 'generation' of quarks. So confident were many physicists in the symmetry between quarks and leptons that they took the discovery of the tau as a prophecy of further quarks, which would restore the symmetry.

The prophecy was fulfilled – at least in part – on 30 June 1977, when Leon Lederman announced an historic discovery at Fermilab. His team had found a new particle, which they called the upsilon particle, and it provided the first indication of a fifth variety of quark. Although sometimes known as 'beauty' – after charm came beauty! – it is more often called, rather prosaically, 'bottom', in analogy with 'down', for like the down quark the bottom quark has charge −1/3. Lederman's upsilon particle consists of a bottom quark bound with its antiquark to form the lowest energy state of 'bottomonium', with a mass of 9.46 GeV.

Lederman had performed a similar experiment to the one that had produced the J/psi for Ting. (Indeed, some years earlier in an experiment at Brookhaven, Lederman's group narrowly missed discovering the J/psi.) Ting's team detected electron–positron pairs produced when a high-energy proton beam smashed into a beryllium target. Lederman and his colleagues chose instead to detect pairs of negative and positive muons created in a similar way by Fermilab's high-energy protons. They were rewarded with the discovery of peaks in the production of muon pairs at particular energies, just as if the muons were the decay products of new particles. The team found first two, and later three, members of the bottomonium family.

In Fig. 9.11 an upsilon particle decays to two muons, just as Lederman's team first observed. In this case, the particle has been produced in rather different circumstances, in the high-energy annihilation of an electron and a positron at the LEP collider at CERN. Most of the energy of the annihilation has gone into a quark and an antiquark that create two jets of many particles. However, the tracks of two muons emerge together in a different direction, and the total energy of these particles is compatible with the mass of the upsilon.

Fig. 9.10 Leon Lederman (b. 1922).

Fig. 9.11 (OPPOSITE) The tracks of two muons (to the right, in red) reveal the decay of an upsilon particle in an image that echoes the decay of the J/psi in Fig. 9.5. The similarity is wholly appropriate as the upsilon consists of a bottom quark and antiquark bound together, while the lighter J/psi is a 'bound state' of the lighter charm quark and its antiquark. In this case the upsilon has been made in the annihilation of an electron and a positron at the centre of the DELPHI detector at LEP. The annihilation probably gave rise initially to a high-energy quark and antiquark, which initiated the two jets of particles that travel to the left. The upsilon would have been created as either the quark or the antiquark radiated a gamma ray, which then formed the bound state of the bottom quark and antiquark.

The analogies between the upsilon and its bottomonium family on the one hand, and the J/psi and its charmonium family on the other, are strong. In bottomonium, the heavy bottom quark and antiquark orbit each other with increasing energies, which correspond to particles of increasing mass. Today, details of several members of the family are well known, having been studied extensively, particularly in experiments at the Cornell Electron Storage Ring or CESR (see Fig. 8.19, p. 142). This electron–positron collider started up in 1979 and continues to be a major centre for the study of upsilon particles.

As in the case of charm, where there exist charmed particles as well as charmonium, so there exist bottom particles as well as bottomonium. Bottom particles consist of a single bottom quark or antiquark accompanied by quarks or antiquarks of the other varieties. The bottom quark has a charge of −1/3, which is the same as the strange quark (as well as the down quark). Thus the varieties of bottom particles should mirror the strange particles, though the bottom particles are five to ten times heavier than their strange analogues. For example, the negative kaon, built from a strange quark and an up antiquark, has a mass of about 0.5 GeV; its bottom analogue, known as the B⁻, consists of a bottom quark and an up antiquark, and weighs in at about 5 GeV.

The relatively heavy bottom quarks soon decay to lighter charmed quarks, which in their turn also decay rapidly, so the bottom particles quickly disintegrate into more familiar particles. The lifetime of bottom particles is typically some 10^{-12} s. This is ten times longer than the charmed particles, yet studying bottom particles has still challenged

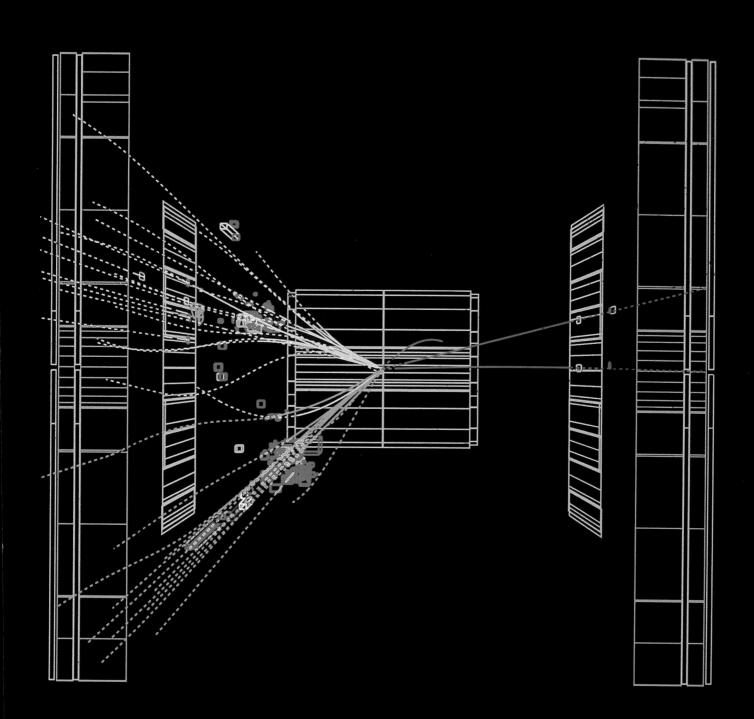

the ingenuity of experimenters. The basic problem is the large mass of the bottom particles. Because they are much heavier than even charmed particles, they can decay in many ways and to many particles. So experimenters must look for clues such as the fast lepton released when a bottom quark decays to a charmed quark, or the production of charmed particles.

In Fig. 9.12, a B^0 reveals its presence when it decays to a J/psi and a K^{0*} (a more massive, excited form of the neutral kaon). The J/psi leaves a classic signature in the form of a pair of muons that shoot off to the outer limits of the CDF detector at Fermilab. Closer

Fig. 9.12 Two muons (green), a positive kaon (red), and a negative pion (yellow) shoot out towards the top of this computer display of particles produced in a proton–antiproton collision in the CDF detector at Fermilab. They emanate from the decay of a short-lived B^0 meson, made from a bottom antiquark bound with a down quark. The B^0 in fact decays first to two short-lived particles – a J/psi and a K^{0*} – which decay swiftly in their turn. The pion and kaon are the products of the K^{0*}, while the muon pair are the remnants of the J/psi. Tracks of other particles created in the initial proton–antiproton collision are shown in dark blue. Notice that the tracks of the muons no longer bend towards the edges of the detector, which is 10 m across in total, as there is no magnetic field in this outer region.

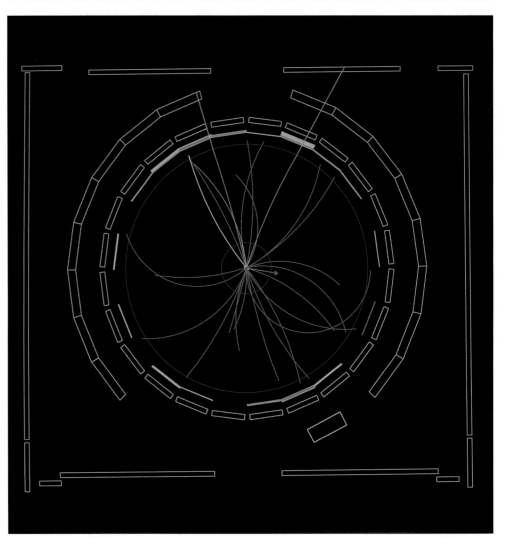

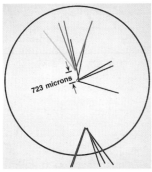

Fig. 9.13 This diagram is a stylized view of some of the particle tracks in Fig. 9.12 close to the interaction point. Precision measurements of the tracks as they emerge from the beam pipe into the surrounding silicon tracking detector allow the extrapolation of the tracks back into the beam pipe. Coloured here as in Fig. 9.12 are the yellow, red, and green tracks associated with the particles from the decay of a B^0 meson, originating 0.723 mm (723 microns) from the collision point. The blue tracks originating near the bottom of the diagram are particles produced in the decay of another B meson; this contained the bottom quark produced in the original annihilation together with the bottom antiquark of the B^0. The blue tracks originating at the collision point itself are due to other particles produced in the collision. (The diameter of the outer circle shown here represents a distance of 1 cm.)

inspection of the tracks shows that the two muons originate at the same point as two other charged particles, which are consistent with being a negative pion and a positive kaon, produced in the decay of a K^{0*}. The four tracks appear together 0.7 mm from the point where a proton and an antiproton have collided at the centre of CDF, as indicated in the close-up of the interaction point (Fig. 9.13); their measured energies and momenta add together to give the mass of a B^0 meson.

The bottom quark can combine with any of the four lighter varieties of quark to form a large number of different bottom particles. Figure 9.14 shows several steps in the decay of an anti-B_s^0, the particle formed when a bottom quark binds together with a strange antiquark. In this event, from the ALEPH experiment at CERN, precise measurements of charged particle tracks have been made in pieces of silicon wafer subdivided into fine strips. These silicon-strip detectors were packed as closely as possible to the beam pipe where it threads through ALEPH, and where the initial electron–positron collision occurred. The resulting measurements were so precise that the computer analysing the event could extrapolate the tracks back to reveal the point where the short-lived anti-B_s^0 decayed *inside* the beam pipe (Fig. 9.15).

'Microscopes' such as ALEPH's silicon detector are helping physicists to pinpoint precisely the production and decay of short-lived particles containing bottom and charmed quarks. The properties of these particles, in particular their lifetimes, provide stringent tests of modern theories of the decays of quarks and also yield insights into the nature of the weak force, which is responsible for the decays. In this way, the measurements of bottom particles test the Standard Model of particles and forces, and help to pave the way to an understanding of the three different generations of quarks and leptons.

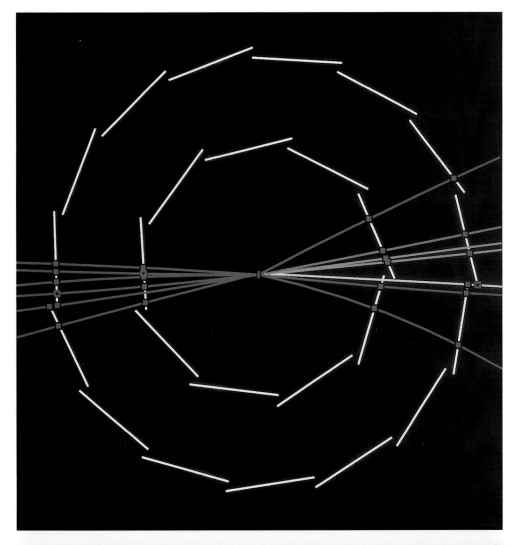

Fig. 9.14 (LEFT) Charged particle tracks from the decay of a B meson are pinpointed by the innermost detector in the ALEPH experiment at CERN. In this display the 'ladders' of silicon strip detector are seen end-on, with strips that have detected particles highlighted by red dots. An electron and positron (unseen) have annihilated at the centre. The 22 cm diameter detector measures the location of the tracks to an accuracy of around 0.1 mm.

Fig. 9.15 (BELOW) This enlarged view shows the detail revealed by the silicon detector. Tracks to the right show the sequential decay of heavier to lighter quarks. The interaction produces a $\bar{B}^0_s$ meson, consisting of a bottom quark bound with a strange antiquark. The $\bar{B}^0_s$ decays within the beam pipe, but its presence is inferred because it decays away from the main interaction point. The heavy bottom quark in the $\bar{B}^0_s$ changes into a lighter charm quark, emitting an electron (green) and an unseen neutrino. The charm quark combines with the strange antiquark to form a D^+_s meson, which itself decays where the blue and yellow tracks originate. Here the charm quark turns into a strange quark, together with an up quark and a down antiquark which form a positive pion (yellow). The strange quark and the original strange antiquark form a negative kaon (lower blue track) and a positive kaon (upper blue track) by combining with an up antiquark and an up quark created from the strong force field around the quarks.

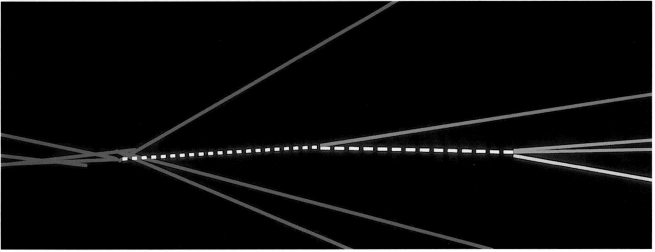

Gluons

The most powerful force we know of – the strong force – binds together the quarks from which the protons and neutrons and all the other hadrons are made. The inter-quark force is so strong that it is apparently impossible to prize a single 'naked' quark out of a hadron. It is as if quarks are stuck together by a kind of superglue. Elucidating the nature of this glue was one of the major achievements of particle physics in the 1970s.

Quantum theory implies that all the fundamental forces of nature are transmitted by carrier particles, which physicists call gauge bosons. In the case of the electromagnetic force, the carrier particle is the photon. In the case of the strong force, the carriers are gluons, and according to theory, there are eight varieties of them. The gluons are massless bundles of strong radiation just as the photon is a massless bundle of electromagnetic radiation. But whereas the photons are free to travel indefinitely through space, gluons appear to be free only within the confines of a 'femtouniverse' – a region some 10^{-15} m, or 1 femtometre, in radius. This is the typical size of a particle such as a proton or a pion. Gluons are confined within hadrons much as quarks are; and like the quarks, they can advertise their presence indirectly by generating jets of particles in energetic collisions.

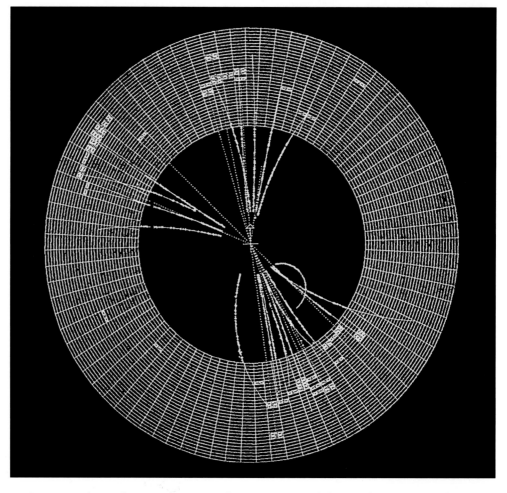

Fig. 9.16 Three jets of particles in the JADE detector reveal the signature that provided the first conclusive evidence for gluons in several experiments at the PETRA collider, DESY. This computer display shows an 'end view' of the tracks of particles in the cylindrical wire chamber that surrounded the electron–positron collider's beam pipe. The tracks have been extrapolated beyond the wire chamber to link with energy deposits in the surrounding calorimeter. An electron and positron have annihilated at the centre of the detector to create a quark and an antiquark, each of which instantly creates a jet of particles that emerges into the tracking chamber. However either the quark or the antiquark has radiated a gluon, to create the third jet of particles.

One example is when an electron and a positron annihilate and create a quark and an antiquark. If the energy is high enough the quark and the antiquark fly apart from each other and shower into two jets of hadrons, such as pions and kaons. This violent separation may shake loose one or more gluons. If the gluon comes off with enough energy, it will produce its own jet of particles distinct from those created by the quark and antiquark. In 1979, 'three-jet' events, with one jet from a gluon, were observed in experiments at the PETRA electron–positron collider in Hamburg. The events provided the first clear visual evidence for the radiation of gluons, just as predicted by the theory of the strong force.

Figure 9.16 shows a striking three-jet event from the JADE detector at PETRA, although at PETRA such clean examples were rare; often the gluon was emitted closer to the path of the quark or antiquark and the jets merged into each other. At higher energy colliders, not only do clear three-jet events become more frequent, but so do events where more than one gluon is radiated. In Fig. 9.17, four jets of particles shoot out from an electron–positron collision at the heart of the SLD, a major detector at the SLAC Linear Collider, which reached energies more than double those of PETRA.

The radiation of gluons studied at PETRA and subsequent colliders has provided ample confirmation of the theory of quark behaviour and the strong force that physicists have developed since the early 1960s. The central idea is that quarks carry a form of charge in addition to their electrical charge. This additional charge is known as 'colour'. The theory is known as quantum chromodynamics (QCD) and is closely modelled on the theory of quantum electrodynamics (QED) which was developed in the 1940s to explain how the electromagnetic force applies to electrically charged subatomic particles, such as electrons.

Colour is considered to be a new kind of charge, analogous to electric charge. Electric charge can be positive or negative and so can each type of colour charge. In this analogy, if we choose to call the colour charge that quarks carry 'positive', then antiquarks have 'negative' colour charge. Identical electric charges repel each other, and opposite charges attract and neutralize each other. The same is true of colour charge, and this is why quarks

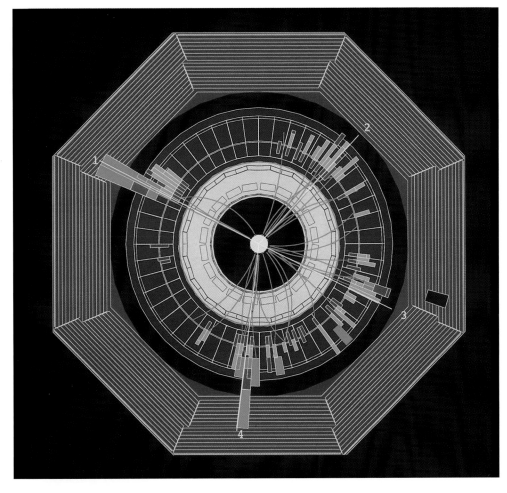

Fig. 9.17 This computer reconstruction from the SLD experiment at SLAC shows four jets of charged particles, bursting out sideways from an electron–positron annihilation at the centre. Such events are consistent with the radiation of two gluons by a quark and antiquark produced by the annihilation.

attract antiquarks to form mesons such as pions, kaons, the J/psi , and the upsilon.

The great difference between electric and colour charge is that while the former comes in only one type, which can be positive or negative, colour charge effectively comes in three types, each of which can be positive or negative. The three types of colour charge are called red, blue, and yellow (or green), by analogy with the three primary colours. Not only do positive and negative colour charges attract each other, so do the different colours themselves: unlike colours attract, like colours repel. Red attracts blue and yellow, for instance, but repels red.

The result is that it takes three quarks of different colour charge to form the particles we call baryons – for example, the proton, the neutron, and the omega-minus.

Baryons may consist of any combination of the six available quark 'flavours' – such as the two ups and one down of the proton, or the three strange quarks of the omega-minus – but they always consist of three differently coloured quarks. Thus the omega-minus must consist of one red-coloured strange quark, one blue-coloured strange quark, and one yellow-coloured strange quark. In baryons, the three colours combine to form a white particle. Mesons, on the other hand, consist of a quark bound with an antiquark, where the antiquark has the negative version of the quark's colour charge. Again, the overall colour charge is neutral. It seems that nature allows only neutral, or white, particles to exist. This is the reason why we do not observe particles that consist of two quarks, say, or two antiquarks.

If colour is hidden within baryons and mesons, how have we discovered its existence? The clue lies in particles such as the omega-minus, which at first sight appears to contain three identical quarks. According to a basic rule of quantum theory – Pauli's Exclusion Principle – a particle should not be able to contain more than one quark in a given quantum state. To overcome this paradox, American theorist Oscar Greenberg proposed in 1964 that quarks not only come in different 'flavours' – up, down, strange, and so on – but also in different 'colours'. If the three strange quarks forming the omega-minus each have a different colour, then they are no longer identical and Pauli's principle is not violated.

Back in the 1940s and 1950s, theorists thought that pions were the transmitters of the strong force. But experiments later showed that pions and other hadrons are composite particles, built from quarks, so the theory of the strong force had to be revised completely. We now believe that it is the colour *within* the proton and neutron that attracts them to each other to build nuclei. This process may have similarities to the way that electrical charge *within* atoms manages to build up complex molecules. Just as electrons are exchanged between atoms bound within a molecule, so are quarks and antiquarks – in clusters we call 'pions' – exchanged between the protons and neutrons in a nucleus.

Gluons participate in this colour world as 'carrier' particles. They transmit the colour force between one quark and another, in the way that the photon transmits the

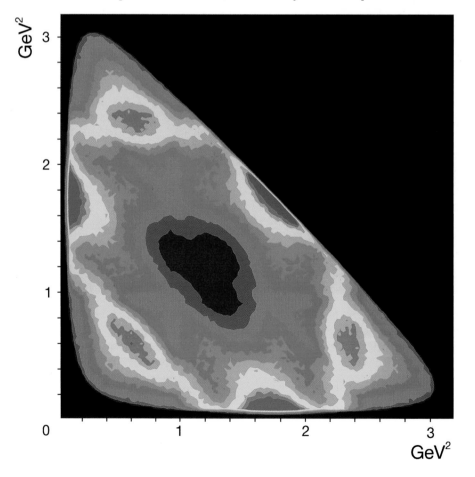

Fig. 9.18 This example of a Dalitz plot (see Fig. 7.11, p. 117) from the Crystal Barrel experiment at CERN shows a resonance that could be a glueball – a particle formed only from gluons bound together. Each point on the graph corresponds to different combinations of the energy of two pairs of neutral pions, and the colours depict the number of events with that combination of energies. Red shows locations with the most events, green and blue the least. The red 'high spots' reveal resonances (very short-lived particle states) and one of these – at the top of the plot – is the state that could be a glueball.

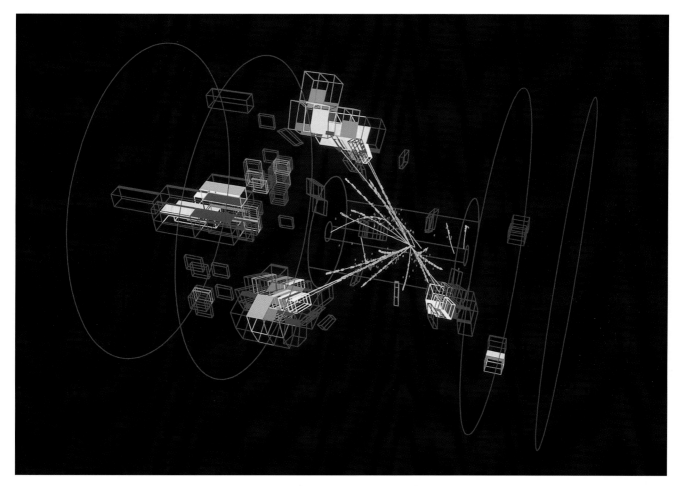

electromagnetic force between electrically charged particles. But there is one crucial difference between the photon and the gluons. The photon is not electrically charged itself, it is neutral; and because it is neutral it does not interact with other photons. Gluons, on the other hand, *are* colour charged and they therefore interact strongly with each other as well as with the quarks. As a result 'glueballs' can form, where gluons bind strongly to one another without need for quarks. The glueballs are short-lived, their gluon content quickly turning into quark–antiquark pairs that form mesons – in particular, neutral pions. By studying neutral pions produced in the annihilation of protons and antiprotons, the Crystal Barrel experiment at CERN has produced the clearest evidence yet for the ephemeral glueballs, which appear as resonances in the data collected (Fig. 9.18).

The neutrality of photons enables them to transmit the electromagnetic force through space, the strength of the force decreasing with distance. The gluons, by contrast, tend to pull on each other because they are colour charged. This appears to be why the quarks and gluons are confined to their femtouniverse.

Meanwhile, physicists continue to explore the femtouniverse of quarks and gluons contained within one of the commonest of particles, the humble proton. Experiments at the HERA collider at DESY minutely probe the proton's deep interior with energetic electrons – the ultimate microscope. The three quarks within a proton are continually exchanging gluons, and the gluons in turn can split into quark–antiquark pairs which exist only fleetingly before recombining to form gluons. The electrons at HERA probe this complex mixture, so that it spits out quarks and gluons, which instantly form their characteristic 'jets'. In Fig. 9.19, three jets of particles emerge in the ZEUS detector at HERA, while a fourth clump of energy in the detector, close to the beam pipe, is due to the debris from the proton that has been smashed up by the probing electron. The jets here are most likely due to a quark, an antiquark, and a gluon. Here the proton is proving to be the ultimate laboratory for studying the details of quantum chromodynamics – the theory of the strong force, and the theory of gluons.

Fig. 9.19 A computer display from the ZEUS experiment at the HERA collider shows three jets of particles created in the high-energy collision of an electron and a proton. The electron and proton (unseen in the beam pipe) have come in from the left and right respectively, and interacted at the centre of the detector. Four clusters of blocks of colour show where particles have deposited energy in the uranium calorimeter that surrounds the central tracking detector, indicated by the red outlines. Three of the clusters have tracks of charged particles leading to them. These jets are all at large angles to the beam pipe, and are directly due to the interaction between the electron and a quark or antiquark within the proton. A likely possibility is that one jet is due to a gluon, knocked out of the proton, while the other two jets are due to a quark and an antiquark created in the interaction. The fourth cluster of energy (the coloured blocks at far left) surrounds the beam pipe in the direction in which the proton was heading. This is due to particles created by the remainder of the proton that did not interact directly with the electron. These particles are produced at shallow angles, and remain initially within the beam pipe, so they leave no tracks in the tracking detector.

The W Particle

Fig. 9.20 An image of one of the first observations of the W particle – the charged carrier of the weak force – captured in the UA1 detector at CERN in 1982. UA1 detected the head-on collisions of protons and antiprotons, which in this view came in from the left and right to collide at the centre of the detector. The computer display shows the central part of the apparatus, which revealed the tracks of charged particles through the ionization picked up by thousands of wires. Each dot in the image corresponds to a wire that registered a pulse of ionization. As many as 65 tracks have been produced, only one of which reveals the decay of a W particle created fleetingly in the proton–antiproton collision. The track – white, with a pink arrow at the end – is due to a high-energy electron. Adding together the energies of all the other particles reveals that a relatively large amount of energy has disappeared in the direction opposite to the electron, presumably spirited away by an invisible neutrino. Together, the neutrino and electron carry energy equivalent to the mass of the short-lived W particle.

In the hot, dense conditions at the heart of the Sun, hydrogen nuclei – single protons – collide continuously, making brief contact as in some kind of eternal dance. Only very rarely do two protons stay together, and for that to happen one of them must actually change into a neutron, so that together they form a deuteron, the nucleus of 'heavy hydrogen'. The probability for this is so low that a proton has a 50:50 chance of surviving for 5 billion years in the Sun's core before becoming bound in a deuteron, even though the centre of the Sun is 10 times as dense as lead.

The force that allows the proton to change into a neutron also triggers the beta decay of the neutron and other particles. It was aptly named the 'weak force' when theorists first considered it, for its ability to cause these reactions, in other words its 'strength', seemed so much weaker than the well-known electromagnetic force. We know now, however, that this weakness is an illusion apparent only at relatively low energies. (Surprisingly, energies at the centre of the Sun are low on a cosmic scale!)

The weak and the electromagnetic forces have intrinsically the same strength. The apparent differences arise because the particles that carry the weak force are heavy, while the photon, which carries the electromagnetic force, has no mass at all. From the start, physicists suspected that the weak force carriers – called W for 'weak' – had to be heavy, because the weak force is limited in its range; it is not felt beyond distances greater than 10^{-15} m. This link between mass and range is a consequence of the 'fine print' in nature's book of rules. In our everyday macroworld energy is conserved, but in the quantum microworld of subatomic particles weird things can happen. The balance of energy need not be exactly maintained, provided any imbalance occurs over a very short time and hence over a very short distance – so short that we are oblivious to it with our macroscopic senses.

When a particle such as a neutron decays by the weak force, in effect it draws on its energy account for just long enough to make a W particle. We know that the energy account is overdrawn for only a very short time, because the force is very short-ranged, so we can infer that the amount borrowed is large – in other words, the W particles are heavy. But how heavy? The answer came in the 1970s with electroweak theory, which forged the link between the weak and electromagnetic forces. When information about the weak force gleaned from experiments with neutrinos was fed into the theory, it indicated that the mass of the W particles must be about 80 GeV, or nearly 90 times heavier than a neutron.

We can never capture an 'overdrawn' W particle of the kind that triggers the decay of a neutron, or changes a proton to a neutron at the heart of the Sun. This inability is not due to a lack of technology; it is forbidden by a principle of nature. The W in these cases is like Lewis Carroll's Cheshire cat, and we can detect only its grin – namely the electron and the antineutrino that the W leaves behind when a neutron decays, or the positron and neutrino that it produces in the Sun. To succeed in making a real W requires an input of the full mass-energy from elsewhere, so that the energy account has enough credit to pay for the W's mass; even in the centre of the Sun energies fall far short of this.

In the 1970s, encouraged by the successes of electroweak theory, physicists began to think of ways of providing enough energy to create W particles. In particular, the theory had created a second challenge in the form of another weak force carrier, called the Z. While the W exists with either positive or negative charge, the Z is neutral and gives rise to a different kind of weak process, known as a 'neutral current' (see next section). Both particles live for less than 10^{-24} s before the weak force destroys them, which set a challenge to the detectives on their trail. However, the daunting task was made easier by the fact that electroweak theory indicated precisely into what particles the Ws and Zs would decay and how frequently each kind of decay would occur.

So the stage was set for the proposal to convert CERN's large proton synchrotron, the SPS, into a proton–antiproton collider that would reach energies high enough to yield a few W and Z particles. Two teams of physicists began to build large detectors, known as UA1 and UA2, to catch the debris from the collisions. In January 1983, the teams triumphantly announced success: they had seen a few W particles decay, and the masses of those particles agreed with the expectations of electroweak theory, with an average value of 81 GeV and an uncertainty of 5 GeV either way. Figure 9.20 shows one of the first W decays captured in the UA1 detector at CERN. Tens of charged particles have been thrown out by a violent proton–antiproton collision at the centre of the detector, but only one signals the W's decay – a high-energy electron, which together with an invisible neutrino, carries the energy that was once contained in the mass of a W particle.

Over the next seven years, the UA1 and UA2 experiments collected more examples of W particles, and were able to measure the mass of the W with increasing precision. The UA2 experiment, in particular, increased its tally of Ws to 3559, reducing the error on the mass measurement to less than 0.5%. However, during 1988, another machine began to produce W particles at a higher rate. The Tevatron at Fermilab generated proton–antiproton collisions at a total energy of 1.8 TeV – six times greater than the energy of CERN's collider. This higher energy helped the Tevatron to produce many more W particles. By 1997, when the Tevatron shut down for the construction of the new Main Injector (see p. 82), the two experiments there had recorded the decays of more than 160 000 W particles, giving a mass of 80.454 GeV, to an accuracy of less than 0.1%. This is like knowing your own body weight to about 50 g (or 2 ounces).

In Fig. 9.21, a W particle has left its characteristic mark in the D0 detector at the Tevatron. The W was formed in a proton–antiproton interaction at the centre, and recoiled against a jet of particles that emerge into the lower part of the detector. But the W almost immediately decayed into an electron, which shoots off to the right, and a neutrino. The neutrino leaves no track, but its presence is registered by the 'missing' energy and momentum that it carries away, towards the top of the detector.

W particles can be produced in any particle collision, provided the energy is high enough. Figure 9.22 shows the decay of a W particle produced in an electron–proton collision at the HERA machine in Hamburg. The track of a single electron, with no charged particle to balance its momentum, provides a remarkably clear example in the ZEUS detector. The accompanying neutrino has left no trace at all as it has flown through the various layers of the detector, and the debris from the initial proton has disappeared unseen along the beam pipe that passes through the detector's heart.

W particles have also been produced in high-energy collisions at the Large Electron Positron collider, LEP, at CERN. When particles and antiparticles collide, their net charge is zero, so if a W particle is produced, its electric charge must be neutralized by an accompanying particle with opposite charge. In proton–antiproton collisions, two complex

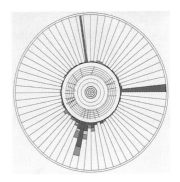

Fig. 9.21 A W particle decays in the same way as in Fig. 9.20, but in this case in the D0 detector, which monitors collisions between protons and antiprotons at the Tevatron at Fermilab. This computer display shows an end-on view of the detector, with the beam pipe at the centre. A jet of particles points downwards, towards blocks of energy deposited in the calorimeter surrounding the central tracking layers. The W particle has recoiled and set off in the opposite direction only to decay immediately into an electron and a neutrino. The electron has gone to the right, to deposit energy (the red block) in the electromagnetic calorimeter. The path of the neutrino is shown going almost directly upwards. Although the neutrino is not directly detected, its direction is inferred from an imbalance in the energy and momentum of the detected particles.

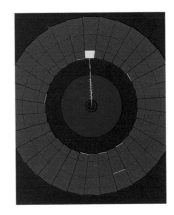

Fig. 9.22 A single electron marks the demise of a W particle in the ZEUS detector at DESY. The W has been produced in the glancing collision of an electron and a proton, which have come in from above and below the page, meeting within the beam pipe at the centre of the image. The electron from the W's decay is revealed through its track and the deposit of energy in the first layer of the calorimeter that surrounds the tracking chamber. To balance momentum, the neutrino also produced in the decay must have moved away in the opposite direction to the electron, escaping undetected.

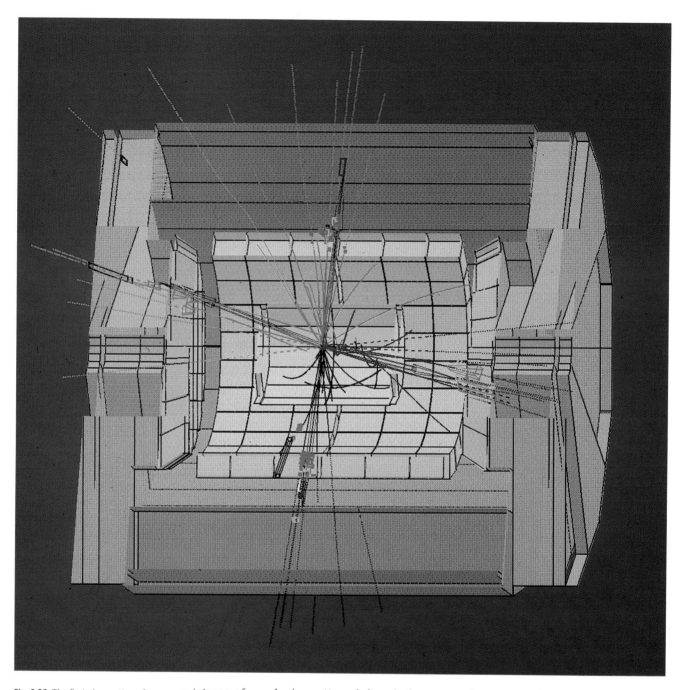

Fig. 9.23 The first observation of a pair of W particles, W⁺ and W⁻ – produced simultaneously in the annihilation of an electron with a positron. The computer display shows a cut-away view of the DELPHI detector at the LEP collider, CERN. The electron and positron have come in from left and right, unseen within the LEP beam pipe, and annihilated at the centre to produce four distinct jets of charged particles. These are due to the decays of the two W particles, each to a quark and an antiquark which in turn initiate a jet of particles, to give four jets in total.

mixtures of quarks (or antiquarks) and gluons are thrown together. From the ensuing melange, a single W particle can emerge accompanied by something quite different to balance its electric charge – for example, the recoiling jet of particles we saw in Fig. 9.21. In collisions between electrons and positrons, however, the initial particle and antiparticle annihilate each other completely. If a W⁺ is formed from the annihilation then it must be accompanied by an exactly opposite W⁻. So electron–positron collisions *can* produce W particles, provided that the collision has enough energy to produce a pair of them, a W⁺ together with a W⁻.

In July 1996, LEP produced its first W pairs when it began to use new superconducting radio-frequency cavities to accelerate its beams to higher energies than before. The first W⁺W⁻ pair emerged in the DELPHI detector – a symmetric burst of four jets of particles, as seen in Fig. 9.23, created as each W particle converted into a quark–antiquark pair.

The W particles in a pair do not have to decay the same way – each can decay in any way open to a W particle. In an example from OPAL, another detector at LEP, Fig. 9.24 shows the decay of one W particle to a quark–antiquark pair, while the other has produced a muon and an unseen muon-neutrino. With a host of W decays of all kinds, the experimenters at

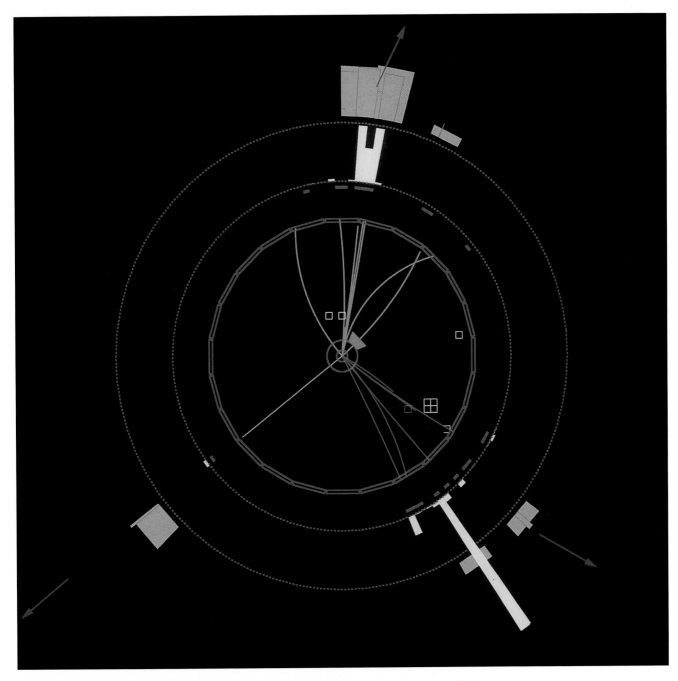

LEP were able to add their measurements to the world average for the mass of the W, bringing it to 80.436 GeV with an error of less than 0.05%.

The precise 'weighing' of the W particle is important because the particle's mass is not directly predicted by the electroweak theory. The key word here is 'directly'. Electroweak theory does provide a mass for the W particle – predicting that it should be about 80 GeV – but this is based on other numbers that must be measured. In particular, it depends on the ratio between the strengths of 'charged weak currents', which involve the charged W particles, and 'neutral weak currents', where the neutral Z particle participates. In a similar way, the theory says nothing about the mass of the heaviest of the quarks, the top quark. But the theory does relate the masses of both the W particle and the top quark to a third mass, namely the mass of the particle dubbed the 'Higgs boson'. This particle is believed to be responsible for giving mass to the W and Z – and indeed to the other fundamental particles (see pp. 192–193) – but it has yet to be observed. The more precisely the masses of the W and the top quark are measured, the tighter the constraints on the mass of the Higgs particle – and the more exciting the search for it as physicists home in on the predicted mass.

Fig. 9.24 A pair of W particles created in an electron–positron collision at the centre of the OPAL detector at LEP reveal two of the ways that a W particle can decay. Tracks leading to energy deposits in the layers of calorimeters are clustered mainly into two jets of particles. These are due to a quark and an antiquark produced by the decay of one of the W particles. A lone straight track, which points towards the bottom left, is due to a muon, the only charged particle that will penetrate to the outer layers of OPAL, as indicated by the arrow. Balancing the muon is a muon-neutrino (unseen), produced with the muon in the decay of the other W particle. (Notice that both jets of particles also contain a muon, again indicated by red arrows.)

The Z Particle

When James Clerk Maxwell united electric and magnetic effects in the 1860s in his theory of electromagnetism, he discovered more than he had bargained for: he found that his theory also described light, demonstrating that it is an electromagnetic wave. A century later, Sheldon Glashow and Steven Weinberg at Harvard University and Abdus Salam at Imperial College, London, independently put forward ideas that united the electromagnetic and the weak forces. They too found that 'unification' brought something extra – in this case, an unexpected interaction involving the weak force, but with no change of charge.

Glashow, Weinberg, and Salam found that their unified 'electroweak' theory required three 'carrier' particles for the weak force: two charged particles (positive and negative) and a neutral particle. The existence of the charged carriers, W^+ and W^-, had been considered since the first theories of the weak force in the 1930s, because a flow of electric charge occurs in the processes of beta decay. However, the neutral carrier, the Z^0, was completely new, and evidence of its existence would be crucial in confirming the correctness of the electroweak hypothesis.

The theory predicted that the Z^0 would transmit a form of the weak force never seen before: it would cause weakly interacting neutrinos to bounce off electrons or quarks, knocking them into motion as in a game of billiards. This became known as the 'neutral current' interaction, because it is mediated by the neutral Z rather than the charged W.

In 1972–73, inspired by the theory, physicists working on the huge Gargamelle bubble

Fig. 9.25 A photograph from the Gargamelle bubble chamber at CERN shows the first observation of a 'neutral current' event, due to a Z particle – the neutral carrier of the weak force. A beam of neutrinos has entered from the bottom of the image but, being neutral, the neutrinos leave no tracks in the liquid of the chamber. However, near the bottom a track begins where a charged particle, set in motion by a neutrino, makes its way up the picture. But as the particle moves along, its track fragments and branches, curling left and right – the characteristic signature of photons as they create electron–positron pairs (see Fig. 5.2, p. 66). The initial charged particle is an electron, which radiates the photons as it interacts with the electric fields of the atomic nuclei in the liquid. This 'billiard ball' collision between a neutrino and an electron, in which neither particle changes type, can only be due to the weak force and to the exchange of a neutral force carrier – the Z particle.

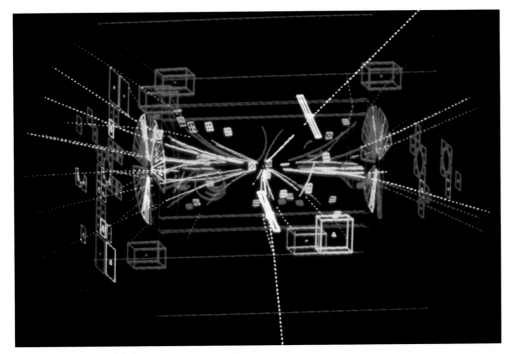

Fig. 9.26 This photograph of a computer display from the UA1 experiment at CERN reveals the production and immediate decay of a Z particle, created in the collision of a proton and an antiproton at the heart of the detector. The proton and antiproton have entered from the left and right, and most of the tracks of charged particles produced continue towards the right and left respectively. Two tracks, however, go in very different directions at a wide angle to the others. These are the two pale blue tracks – the blue indicating a high energy. The tracks lead to pale blue blocks indicating energy deposited in the layer of UA1 that reveals electrons and positrons. The two tracks are due to an electron and a positron, which share energy equivalent to the mass of the Z^0 that created them when it decayed.

chamber at CERN pored over 290 000 photographs of interactions produced when beams of neutrinos or antineutrinos had entered the chamber. They found just 166 examples of the new type of interaction. Figure 9.25 shows the first image ever where a neutrino in the beam knocks an atomic electron from the bubble chamber liquid into flight. The neutrino remains unseen, while the electron leaves an unmistakable track, with curling branches that form when the electron radiates a photon, which in turn converts into an electron–positron pair.

The discovery of neutral current interactions – circumstantial evidence for the neutral Z particle – strengthened confidence in electroweak theory, and in 1979 Glashow, Salam, and Weinberg were awarded the Nobel prize. All that remained was for actual Z particles, and the related Ws, to be shaken loose in high-energy collisions. As the previous section described, the observations of the first real W particles at CERN's proton–antiproton collider were announced in January 1983. News of the rarer Z particles came from the same experiments a few months later. The Z appeared with the mass expected from the theory, and at the rate the theory predicted.

Finding the W and Z particles in the proton–antiproton collisions was by no means straightforward. When we collide protons with antiprotons we are in fact colliding a cluster of three quarks (and attendant gluons) with a similar cluster involving three antiquarks. Only one of the quarks and one of the antiquarks may make a Z (or W) particle, but the attendant maelstrom from the other quarks and antiquarks can mask the vital signature of the particle being sought.

Figure 9.26 shows the first Z particle found in the UA1 experiment at CERN. We do not see the colliding proton and antiproton, which have entered from the left and right, only the debris that they create. Most of the debris consists of relatively low-energy pions and kaons, whose trails bend in the magnetic field surrounding the collision zone. But two straighter tracks stand out from the throng. They belong to higher-energy particles, which are identified as an electron and a positron. Together their energies add up to 93 GeV – close to the mass of the Z particle calculated from electroweak theory.

For six years, CERN's proton–antiproton collider was the only machine capable of making Z particles, but by 1989 Zs had begun to appear elsewhere. In the USA, the Tevatron could produce both Z and W particles in its proton–antiproton collisions, just as at CERN, but in much greater numbers. Physicists at SLAC also began to make their own Z particles at the new Stanford Linear Collider, or SLC (see pp. 153–154). In April 1989, the Mark II detector – successor to the famous Mark I – revealed for the first time the rise and fall in the reaction rate as the total energy of the colliding beams was increased through the mass of the Z particle. The Z, with its brief lifetime of 10^{-25} s, had made its presence felt as a resonance.

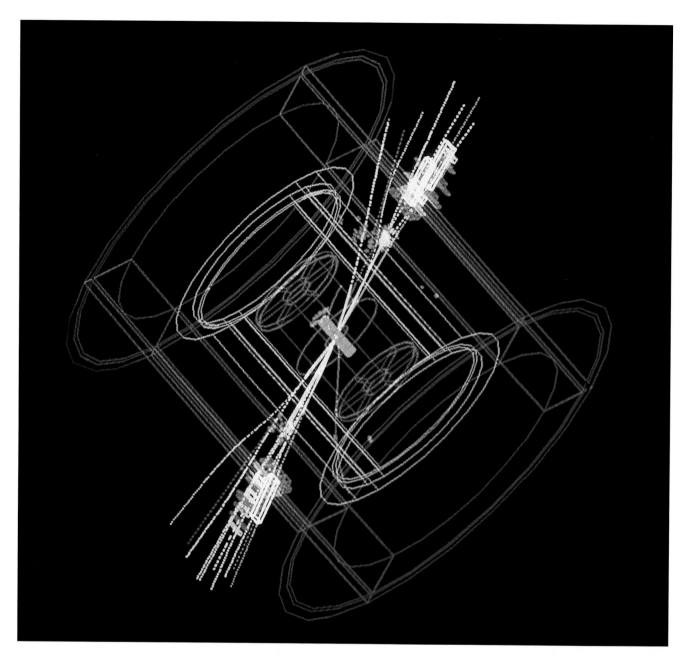

Fig. 9.27 In the DELPHI detector at CERN, the process of Fig. 9.26 is reversed, as an electron and a positron annihilate to produce a Z particle, which then decays into a quark and an antiquark. This computer display shows the two distinct jets of particles produced by the quark and the antiquark as they shoot out back-to-back from the decay of the Z particle, made at rest at the centre of the detector. The particle tracks measured in DELPHI's time projection chamber (or TPC, see p. 135) are shown as solid lines, while the dotted portions indicate the calculated extensions to these tracks. The coloured blocks show where the particles have deposited energy in the hadron calorimeter, beyond the TPC, thereby revealing themselves as particles made from quarks.

Then in August 1989, the machine that was to dominate the production of Z particles came into operation. CERN's Large Electron Positron (LEP) collider was designed to make Zs by the million, and it lived up to all expectations. Figure 9.27 shows an event from DELPHI, a detector at LEP, in which a Z decays into a quark and an antiquark. The strong force prevents these particles from emerging on their own into the apparatus. Instead, the quark and antiquark swiftly gather newly minted quarks and antiquarks from the surrounding strong force field, to form two sprays – or 'jets' – of particles which shoot out from the annihilation point in opposite directions.

In the first few weeks of operation alone, the four experiments at LEP collected more than ten thousand Zs – and they began to answer questions that extended beyond the simple vital statistics of the Z particle itself. In particular, the results indicated that there were no more neutrinos like the electron-, muon-, and tau-neutrinos to be found! For the first time in history a limit to the number of at least one kind of elementary particle had been found.

At LEP, as at the SLC, the Z particle is made in reactions that mirror exactly the way it was first seen in the UA1 experiment. The total energy of the colliding beams is tuned to the mass of the Z particle, so that the electrons and positrons annihilate to make Zs. Once made, each Z retains no knowledge of its creation, and can decay to any possible particle–antiparticle pair. The only proviso is that the total mass of the particle–antiparticle pair is less than the mass of the Z particle, the remaining mass turning into the kinetic energy of the pair and often into the production of other particles.

The lifetime of the Z is determined in part by the number of ways in which it can decay; the more ways that it can decay, the shorter its brief life. As quantum theory tells us, the shorter a particle's lifetime, the broader its resonance peak, so a measurement of the breadth of the Z's peak allows us to establish its lifetime – and hence the number of different particles into which the Z can decay. In particular, the breadth of the peak holds the key to just how many kinds of neutrinos there are, for the lightweight neutrinos must figure among the possible decays of the Z. The only problem is that these are the very decays that the experiments cannot see, for when a Z decays to a neutrino and an antineutrino the two particles escape the apparatus completely undetected!

By comparing the measured width of the Z's resonance peak for the visible decays with calculations based on electroweak theory, the physicists at LEP and SLC could test their measurements against theories with different numbers of neutrino types. The results from the SLC strongly suggested that there are three types of neutrino. With its higher intensity, and consequent larger number of Z particles, LEP soon produced the same result, but with much greater precision – Fig. 9.28 shows the measurement by the ALEPH experiment. The chance that there are four types of neutrino was now less than one in a thousand. Ten years of LEP running and more than 10 million Z particles later, the match between the data and the curve for three neutrinos is indisputable.

The vast number of Z particles at LEP meant that the particle itself became a 'laboratory' in which the physicists could study electroweak theory, probing many different aspects with great precision. Figure 9.29 shows the decay of two Z particles produced together in 1997, when LEP was running at double its earlier energy. In a quarter of a century, from the first evidence for neutral currents in the Gargamelle bubble chamber to 'ZZ' events like this at LEP, the Z became one of the best measured of fundamental particles. By the time that LEP ceased running in 2000, the mass of the Z was known to a precision better than one part in ten thousand.

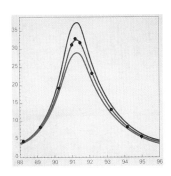

Fig. 9.28 The Z particle shows up as a resonance (see pp. 115–117) as the total energy of the electron and positron beams is increased from 88 GeV to 95 GeV at the Large Electron Positron collider at CERN. The black dots – data from the ALEPH experiment – show how the production of particles made from quarks (hadrons) rises and then falls again, as the energy passes through the mass-energy of the Z particle, with a peak just above 91 GeV. The width of the peak arises from the intrinsic uncertainty in the Z particle's mass, which is inversely related to its lifetime – the broader the peak, the shorter the lifetime. Although the data are from collisions where the Z decays to a quark and an antiquark, the lifetime also depends on decays to leptons. While the decays to charged leptons (electrons, muons, and taus) can also be measured, decays to neutral leptons – the neutrinos – are not detected in the experiment. However, the lifetime of the Z particle, and hence its width, can be calculated for different numbers of neutrino types. The three curves shown here correspond to two, three, and four neutrino types (going from top to bottom). Only the middle curve fits the width of the data peak correctly, showing that there can be only three types of neutrino.

Fig. 9.29 A computer display from the L3 detector reveals the decay of two Z particles produced simultaneously in an electron–positron annihilation at the LEP collider. One of the Z particles has decayed to a quark and an antiquark, to create two sprays of hadrons (particles made from quarks), which deposit their energy to the left and right in the hadron calorimeter, as indicated by the outer purple blocks. The other Z particle has decayed to an electron and a positron, each of which deposits energy in the electromagnetic calorimeter, as shown by the green 'towers' pointing towards the top and bottom. In each case, the energies of the particles involved add together to give a total that corresponds to the mass of the Z particle.

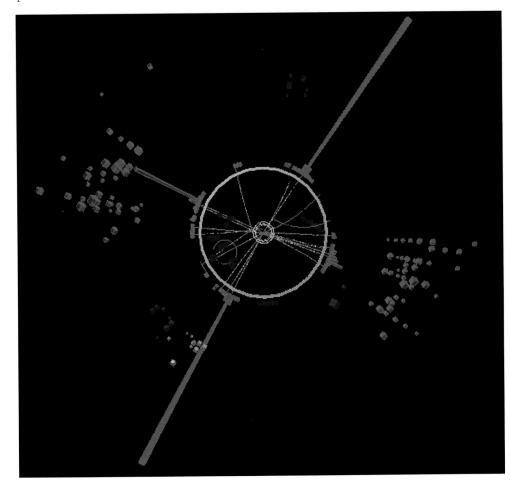

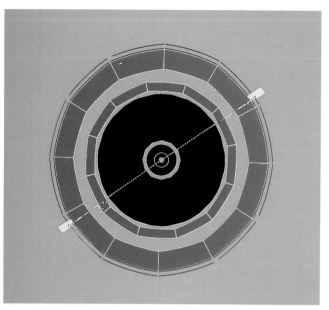

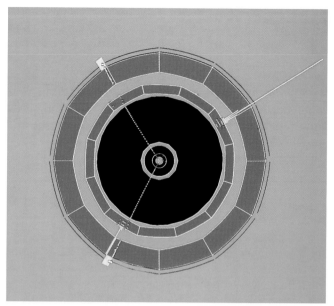

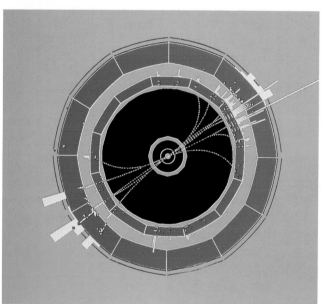

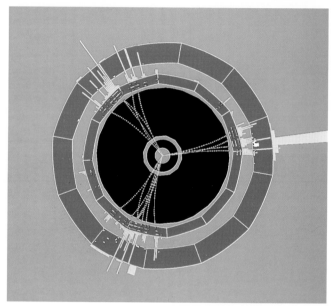

Unity

By the end of the 1970s, there were clear signs of deep relationships among the basic particles of matter and between the fundamental forces. The discovery of the tau lepton hinted that there might be 'matching sets' of six quarks and six leptons, while the existence of gluons revealed an underlying similarity between the fundamental forces. This underlying unity became much clearer through the 1980s, as different experiments consolidated the role of gluons as carriers of the strong force, and revealed the W and Z particles that carry the weak force just as predicted by the theory that links the weak and electromagnetic forces. But it is in the stark beauty of the electron–positron collisions at LEP in the 1990s that this unity can most easily be seen.

In this section we have chosen events from one experiment at LEP – ALEPH – and displayed them all in the same way to allow direct comparisons between different forces and different particles. In these events the total energy of the electron and positron beams in LEP is set to the mass-energy of the Z particle. The annihilating electron and positron produce a Z, which almost immediately decays to a new particle and matching antiparticle, often of a completely new variety. In each case the electron and the positron have approached from above and below the page to annihilate within the beam pipe at the centre of the detector. The electron and the positron collide head on with equal but oppositely directed momentum, so their total momentum is zero, and the Z particle is made at rest.

A particularly simple example is the production of a muon and an antimuon, as in Fig. 9.30. The muon and antimuon must emerge back to back, as dictated by the conservation of momentum. The original electron and positron have travelled at the same speed but in opposite directions before their head-on collision, so the particle and antiparticle that result must likewise emerge with equal speeds in opposite directions. However, the line of their flight need not be the same as that of the colliding electron and positron; indeed, the directions differ more often than not. Electroweak theory, which unites the electromagnetic and weak forces, predicts the chance that these two directions will be similar, or at 30 degrees, or at any other angle. The experiments at LEP were able to measure this 'angular distribution' by observing thousands of events like Fig. 9.30, and recording how many occur at each angle. The result is precisely as electroweak theory predicts.

Electroweak theory encompasses quantum electrodynamics, or QED, the quantum theory of electromagnetic effects, and this theory implies that events at LEP can be more complicated. QED predicts that the electromagnetic field surrounding electric charges can appear as the field carriers – photons. As a negatively charged muon and its positively charged antiparticle flee from their point of creation, one or more photons may be shaken loose from the field.

The simplest possibility is shown in Fig. 9.31, where a single photon emerges together with the muon and antimuon. Now the muon and its antiparticle are no longer back-to-back, as one has recoiled sideways to balance the momentum carried away by the photon. The muons have left tracks, as before, but the photon has no electric charge and so leaves no trail in the central tracking detector. Instead, the photon's presence is revealed as it dumps its energy in the electromagnetic calorimeter – the layer specially designed to capture photons as well as electrons.

The weak force acts equally on quarks as on leptons, so the Z particle can just as readily decay into a quark and its corresponding antiquark. Initially, these emerge back to back just like the muon and antimuon of Fig. 9.30. However, the quark and antiquark are 'free' for less than 10^{-23} seconds before their separation has disturbed the intervening space so much that it becomes filled with further quarks and antiquarks. The result (Fig. 9.32) is two jets of hadrons – particles made from quarks and antiquarks – which emerge back to back as the relics of the original quark and antiquark. Furthermore, the angular distribution of the line connecting the jets, relative to the original electron–positron direction, turns out to be identical to the angular distribution for muons. This is direct evidence that quarks and leptons, although otherwise so different, respond in exactly the same way to the weak force carried by the Z particle.

Like muons, quarks carry electric charge, so quarks can radiate photons just as muons do. However, when quarks are involved something more dramatic can happen, which provides visible evidence for the unity underlying the electromagnetic and strong forces. Quarks carry 'colour', the 'charge' of the strong force (see p. 169), as well as the more familiar electric charge, and it seems that the laws describing the behaviour of colour charges are very much like those that control electric charge. In particular, in the same way that the electromagnetic field can appear as photons, so the colour field around quarks can be manifest as gluons, the carriers of the strong force (see p. 168). In Fig. 9.33 a quark has shaken out a gluon, just as the photon was shaken out in Fig. 9.31, but unlike the photon the gluon is not able to emerge 'free' into the detector. Instead, it immediately creates a third jet of particles which carry the balance of momentum in the event – and the resemblance between Figs. 9.31 and 9.33 becomes clear.

Comparisons of the processes at work in Figs. 9.31 and 9.33 confirm the similarities between QED, the quantum theory of the electromagnetic force, and QCD or quantum chromodynamics, the analogous theory of the strong force. Together with the established link between the electromagnetic and weak forces, these observations have inspired the belief that at much higher energies than at LEP – at around 10^{15} GeV – the particles and forces must be united in a still more profound symmetry. Such energies are far beyond anything we can achieve in artificial accelerators, but they would have been typical of the very early Universe and the unity between particles and forces would then have been of prime importance.

Figs. 9.30–9.33 (OPPOSITE) Four examples of the decay of the Z particle illustrate the unity that underlies particle interactions. These computer displays all show a 'beam's-eye view' of the cylindrically symmetric detector layers within the ALEPH experiment at CERN's LEP collider.

Fig. 9.30 (TOP LEFT) Here the Z particle has decayed to a muon and an antimuon, which fly off back-to-back to conserve momentum (which must remain zero). The particle and antiparticle leave tracks in the central tracking chambers, and then pass through the calorimeters, shown as red polygons. Finally, they pass through the muon chambers, indicated by the outer red lines. Muons are the only charged particles that reach this far.

Fig. 9.31 (TOP RIGHT) A muon and an antimuon are again revealed by the tracks leading to the white dots in the muon chambers, but the tracks are not back-to-back. The total momentum is not conserved by the muon and the antimuon alone. The balance of momentum is held by a third particle, which is revealed when it strikes the inner red ring and deposits all its energy. This is the electromagnetic calorimeter, so the particle must be a photon, radiated from the muon or antimuon via the electromagnetic force.

Fig. 9.32 (BOTTOM LEFT) Here the Z particle has produced several charged particles which form two back-to-back 'jets'. The particles leave energy in the inner red ring of the electromagnetic calorimeter and in the outer red ring of the hadron calorimeter. The particles are mainly hadrons – particles built from quarks. The two jets have emerged from a quark and an antiquark formed from the Z particle. The quark and antiquark have set off in opposite directions, just like the muon and antimuon in Fig. 9.30, but have immediately 'seeded' new particles – hadrons – to form the two jets seen in the detector.

Fig. 9.33 (BOTTOM RIGHT) In this example, an additional jet of particles appears, compared with Fig. 9.32, and none of the jets is back-to-back. There is, however, a resemblance to the directions of the muon and the antimuon in Fig. 9.31 which is more than superficial. In this case, either the original quark or antiquark has radiated a gluon through the strong force. The gluon, like the quark and the antiquark, has emerged into the detector as a jet of hadrons, to give three jets in all.

The Top Quark

'Where is top?' – the question haunted particle physicists after the discovery in 1977 of the bottom quark, hidden in the upsilon particle. Everyone knew that the sixth quark – if it existed – had to be heavier than the bottom quark, but just how much heavier was at first far from clear. As machines successively reached for higher energies, experimenters diligently searched for signs of top quarks. In the end, they had to wait 18 years until the top quark finally weighed in – a great bruiser of a fundamental particle, with a mass as great as that of an entire atom of gold!

Precision measurements of the Z particle at LEP had hinted that the top quark was out there with a mass of around 150–200 GeV. At such a mass it could influence the Z in its short life but was nevertheless far too heavy to materialize directly at LEP. The production of this most heavy of elementary particles would require energies that in the late twentieth century were only available across the Atlantic, at Fermilab.

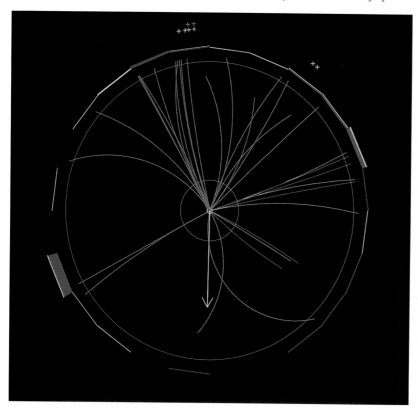

The previous two quarks to be found – charm and bottom – had both been discovered in their 'hidden' states, charmonium and bottomonium, in which they are bound to their antiquarks. Only later were particles found in which charm and bottom combine with other quark 'flavours'. However, the top quark turned out to be so heavy that it decays before the strong force has time to bind it to anything. Its lifetime is calculated to be as short as 0.4 yoctoseconds – 0.4 million million million millionths of a second, or 0.4 million millionths of a picosecond. Yet in 1995 two experimental teams at Fermilab were at last able to announce with confidence that they had established the existence of the top quark.

The first hints that top had been found had come the year before from one of the experiments, the Collider Detector at Fermilab, or CDF. This 5000 tonne apparatus had been built to study the world's highest-energy collisions in the Tevatron, Fermilab's superconducting accelerator, which started colliding protons and antiprotons head on in 1987 with 900 GeV per beam – and a total collision energy of 1800 GeV. The Tevatron would appear to have the potential to create particle–antiparticle pairs with a total mass equivalent to 1800 GeV. However, this energy has to be shared among the various quarks, antiquarks, and gluons that form the colliding particles. So it turns out that collisions are rare where a quark and an antiquark collide with sufficient energy to make something as heavy as a top quark and its antiquark. In practice, only after several years of improvements both to the Tevatron and to CDF were the conditions right for the top quark shyly to emerge, in events collected by CDF between August 1992 and April 1993.

During this time, the CDF team – by now comprising about 440 physicists from 35 institutions, ranging from Argonne to Yale, and from Bologna to Tsukuba – had accumulated some 16 million events, each with information from around 100 000 individual detector elements that comprise the complete apparatus. Within this wealth of data were 12 events bearing the hallmarks of the top quark. But how did the physicists know what to look for?

Theory implied that the extremely short-lived top quark would prefer to decay to a bottom quark and, as it does so, also emit a W particle – the charged carrier of the weak force that governs the decay process. So if the violence of the collision produces a top quark and a top antiquark, the immediate products as these two die are a bottom quark, a bottom antiquark, and two W particles (of opposite charge). The bottom quark (or antiquark) lives

Fig. 9.34 This computer display shows the tracks of particles produced when a proton and an antiproton collided at the centre of the CDF detector at Fermilab to produce a top quark and antiquark. The top quark has decayed to a bottom quark, which creates one jet, and a W⁺ particle, which decays to a positron and a neutrino. The positron deposits its energy in the electromagnetic calorimeter (the bright pink block); the neutrino's direction is indicated by the yellow arrow. The top antiquark has decayed into a bottom antiquark, which produces a second jet, and a W⁻ particle, which turns into a quark and an antiquark. These two then give rise to a third and fourth jet (see Fig. 9.35 for more detail).

for only a tenth of a picosecond, but this is a million million times longer than the top quark's brief life, and time enough for it to bind with an antiquark (or quark) newly minted from the local strong force field. The resulting B particle – which is far lighter than top, and therefore carries much of the original mass-energy as energy of motion – travels several millimetres before it too decays to produce a 'jet' of particles. The W particles, by contrast, have several options open for their decays, some of which are easier to interpret than others.

The cleanest case is when both Ws decay either to an electron and its neutrino, or to a muon and its neutrino. The electrons and muons are relatively easy to identify, while the neutrinos disappear out of the apparatus leaving only the 'signature' of the missing energy they spirit away. So one sign of the production of top and antitop is an event with two jets (from the bottom quark and antiquark) and electrons and/or muons together with missing energy (all from the two W particles). In their data from 1992–93, the CDF team found two examples of this kind.

A more common occurrence is for one of the Ws to decay to a muon or electron and a neutrino, while the other decays to a quark and an antiquark, so producing two more jets of particles to add to the two from the bottom quark and antiquark. Figure 9.34 shows a particularly clear example in the CDF detector, in which a positron emerges together with four jets – a 'golden candidate event' for the production and decay of top and antitop. Close inspection of the jets in the silicon detectors close to the beam pipe in CDF show that two jets originate a small distance from the interaction point, indicating that they are jets formed from short-lived B and anti-B particles (Fig. 9.35). CDF found 10 events of this kind in their first data, making a total of 12 candidates for top decays – all of which indicated a mass for the top quark of around 175 GeV, or nearly 200 times heavier than a proton!

Although these candidates for top were found by August 1993, it took the team a further nine months to consider all the possible ways that nature could be tricking them. Could more prosaic processes masquerade as top decays? Week after week, through the autumn and winter, the CDF physicists battled with the complexities of the data – often with as many as 100 particles per event – as they tried to convince one another that they were really seeing top events. Rumours began to spread that the top quark had been found, and by April 1994 the researchers were in a position to confirm them. On 26 April they announced their conclusions to a packed colloquium at Fermilab. There was only about a 1 in 400 chance that the 12 events they had found were due to effects other than top – but the CDF physicists remained cautious and stopped short of claiming discovery. 'We have not yet observed enough examples of top quark production to establish the particle's existence beyond question,' announced Melvyn Sochet, the team's spokesman.

At this stage, D0, the other large experiment at the Tevatron, had even fewer examples of top quark candidates. But with the results from CDF now hinting strongly at what mass the quark must have, the D0 team could focus their analysis. In doing so they found stronger hints for the existence of top, but they still did not have enough data to prove conclusive. Figure 9.36 (overleaf) shows a 'top event' from D0. This is similar to the example from CDF (Fig. 9.34), but in this case one of the W particles has decayed to a muon and a neutrino rather than an electron and neutrino.

By 1995, both experiments had collected four times as much data, and CDF had installed an improved silicon 'vertex detector' for identifying the all-important decays of B particles. The evidence for the top quark was now much stronger, and on 2 March 1995 the two teams announced the 'observation of the top quark'. In both experiments the odds were less than 500 000 to 1 that the supposed top quark decays could be due to something else. Around the world, particle physicists breathed a sigh of relief. The physicists at LEP were especially pleased, as the Fermilab teams had found the top quark with just the mass that the data from LEP had indicated.

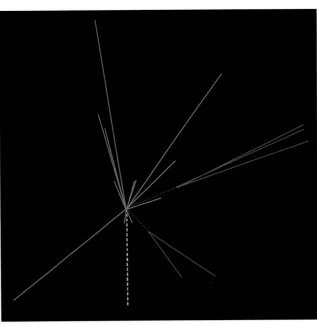

Fig. 9.35 This close-up view shows how the tracks associated with the bottom quark and antiquark produced in Fig. 9.34 could be identified. Here the charged particle tracks from the main tracking chamber, nearly 3 m in diameter, are extrapolated back towards the proton–antiproton interaction point *within* the beam pipe. The red tracks clearly intersect some distance away from the collision, indicating that they have been produced in the decays of two short-lived particles, which have travelled only 4.5 mm and 2.2 mm in their brief lifetimes. These particles are most probably two B mesons, each of which contains a bottom quark (or antiquark) bound with a lighter antiquark (or quark). They give rise to two jets towards the right in Fig. 9.34. The other two jets, towards the top of both images, come from the decay of the W⁻ produced in the top antiquark's decay. Such precise extrapolation was made possible through measurements in the layer of silicon detectors (see p. 150) surrounding the beam pipe, which located tracks to within 0.015 mm.

Fig. 9.36 An artistic rendering of the decay of a top quark and antiquark, produced in a proton–antiproton collision at the heart of the D0 experiment at Fermilab's Tevatron. The top and antitop decay to a bottom quark and antiquark respectively, each of the decays producing a W particle, which also changes into other particles. One of the W particles turns into a muon and a neutrino. The muon penetrates the different layers of D0 and leaves the blue track to the top right. The neutrino escapes undetected, but its direction and the energy it took with it can be calculated and are indicated by the long pink block at far right. The bottom quark (or antiquark) that accompanied this W particle has produced a jet of particles, revealed by the calorimeters (the coloured blocks to the right of top centre). The other W particle from the decays of the original top and antitop has decayed into a quark and antiquark, which in turn create the two jets in the left half of the image. The fourth jet, towards the bottom right, is due to the bottom quark (or antiquark) from this decay. Here the jet includes a muon (the blue track) produced among the decay products of the bottom quark.

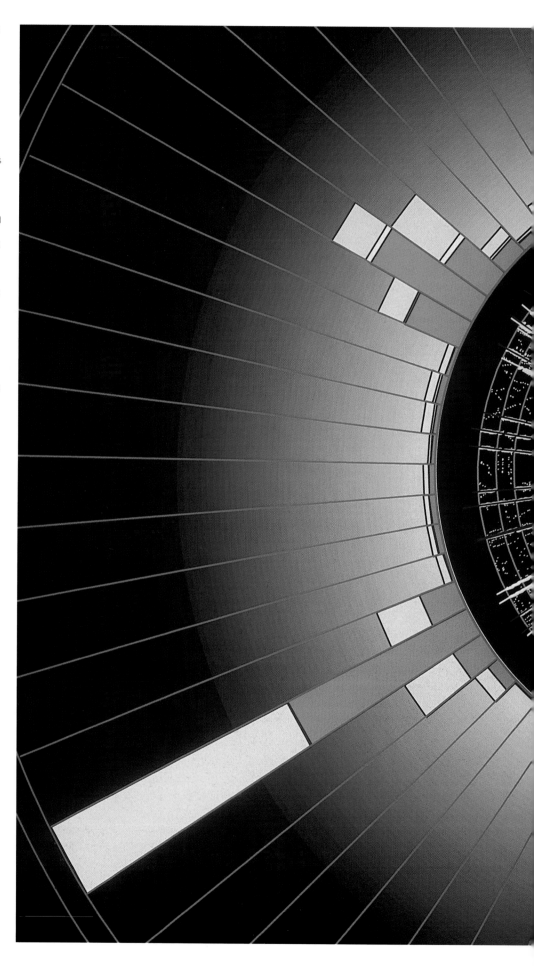

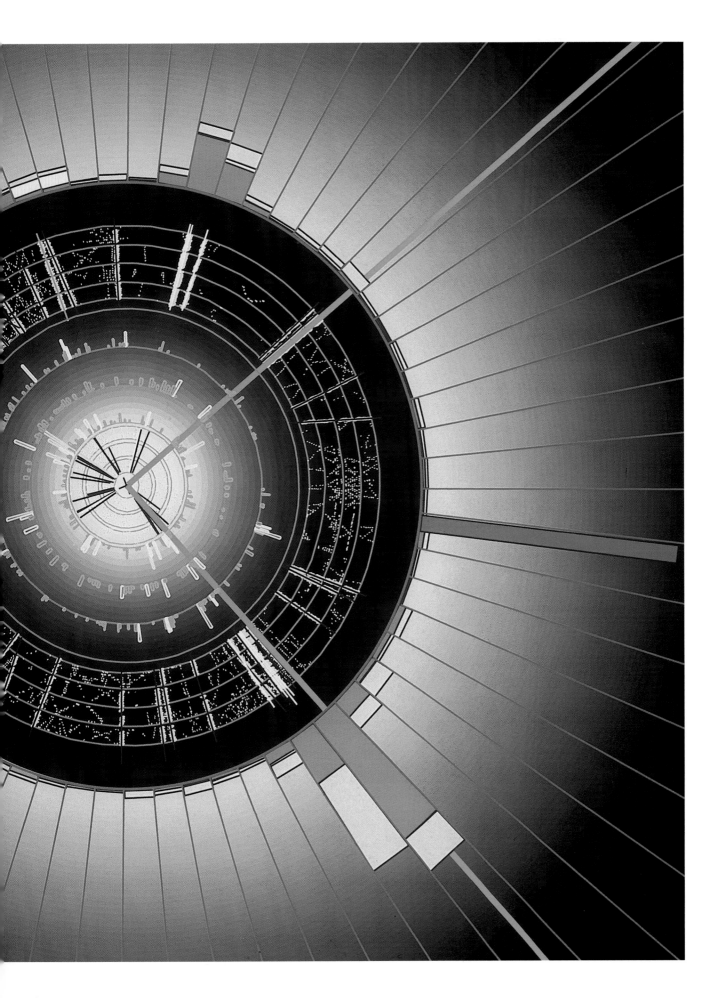

10. Future Challenges

Egocentricity – ourselves at the centre of the Universe – was the received wisdom for more than a thousand years, until Nicolaus Copernicus realized in the sixteenth century that the Earth and the planets all revolve around the Sun. During the twentieth century our place in the grand scheme of things has been pushed ever further from the centre of the action. We have found that the Solar System lies in the remote backyard of a galaxy of a billion suns, which is itself but one of innumerable galaxies. As Edwin Hubble discovered in 1929, these galaxies are rushing away from one another in a way that implies that the Universe is expanding. This observation has become embodied in one of the foundations of modern cosmology – the hot Big Bang. According to this theory, in the first instants of the Universe, some 15 billion years ago, all the material in the Universe today was created in a dense fireball smaller than a clenched fist. In recent years the suspicion has grown that the Big Bang may also have created exotic forms of matter, not present on Earth today, in amounts that may outnumber the stuff that we presently know. The nature of this 'dark matter' is one of the puzzles of modern physics, and it leaves us with the sobering thought that we may be nothing more than flotsam on a sea of dark matter.

Although Hubble's discovery of the expanding Universe – which ranks among the most far-reaching discoveries in science – further displaced us from the centre of things, it has offered the tantalizing possibility that an understanding of the origins of the Universe may be attainable. The symbiosis of the cosmology of Hubble's expanding Universe and the discoveries of particle physics is now beginning to realize this dream.

The link between cosmology and particle physics arises because in the immediate aftermath of the Big Bang, the highly compressed Universe would have been incredibly hot – a fireball of radiation and matter. Under these circumstances the particles of matter would have been colliding at high energies, similar to the conditions that we can create today in a small region of space within a particle accelerator. In this way the study of the fundamental particles is making contact with the large-scale Universe; Earth-bound high-energy experiments are mimicking the condition of the young Universe.

As the Universe expanded in space and time, it cooled. Today, 15 billion years later, the Universe is bathed in microwave radiation, whose temperature is −270 C, a mere 3 degrees above absolute zero. This is the chilly remnant of the original fireball and it provides us with a single benchmark temperature in the history of the Universe. If we imagine playing the film of this history back in time, we come to the conclusion that during its first 10^{-33} s the Universe was hotter than 10^{32} degrees. At these temperatures, equivalent to energies far higher than anything achieved at an accelerator on Earth, fundamental particles of matter and antimatter emerged and annihilated continuously. The Universe was an expanding froth of quarks, antiquarks, leptons, antileptons, photons, W particles, Z particles, gluons, and maybe other particles as yet unknown to experiment or undreamed of by theorists.

While much of the evidence fits with this picture, we immediately see one of the major questions at the frontiers of our ignorance: where is the antimatter today? Or, to put it

Fig. 10.1 (OPPOSITE) The Hubble Space Telescope reveals a myriad of distant galaxies within a tiny piece of sky that would be obscured by the width of a thread held at arm's length. The telescope is named after Edwin Hubble who realized in 1929 that galaxies are rushing apart from each other, so providing evidence that the Universe is expanding. This in turn led to the modern view that the Universe began with a hot and energetic Big Bang. Now particle · physicists have joined forces with cosmologists and astrophysicists in the endeavour to understand the origins of matter and ultimately of · the Universe itself.

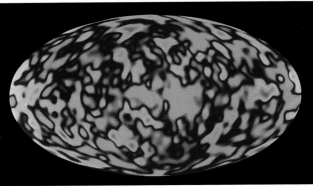

Fig. 10.2 This false-colour map of the whole sky shows variations in the cosmic microwave background radiation as measured by the Cosmic Background Explorer (COBE) satellite. This background radiation, which has an average temperature of 2.73 degrees above absolute zero (2.73 K), is the remnant of the enormous heat of the Big Bang, cooled as the Universe has expanded over the past 15 billion or so years. The map shows variations from the average temperature. Pink and red areas are warmer, while blue areas are cooler – red corresponds to a difference of +0.27 mK (millionths of a degree), pale blue to −0.27 mK.

another way, why do we exist at all? If matter and antimatter were produced equally in the early Universe, why is the Universe now asymmetric, favouring matter in bulk to the exclusion of antimatter? Why didn't the matter and antimatter annihilate? It seems likely that early in the history of the Universe, possibly within 10^{-33} s, a minutely small excess of matter – about 1 part in a billion – arose to lead some billions of years later to all the matter in the Universe.

The next important epoch came a billionth of a second (10^{-9} s) after creation. By then the Universe was cool enough for the massive W and Z particles to slow down the weak interactions they mediate relative to the electromagnetic interactions mediated by the massless photons. It was due to this difference that weak interactions – as in beta decay, for example – began to seem 'weak' in comparison to electromagnetic effects. In 1983, when CERN's proton–antiproton collider produced the first 'man-made' W and Z particles, it recreated the conditions in the Universe when it was a mere billionth of a second old.

The discovery that the weak force appears feeble because the W and Z particles are heavy, whereas the stronger force of electromagnetism is carried by the massless photon, brings us to another of the frontier questions in particle physics. Although mass as a concept is familiar to all of us, its true nature remains a mystery, albeit a mystery that particle physicists hope one day to resolve.

Returning to our movie of the evolving Universe, we can now fast forward until it is a hundredth of a second old (10^{-2} s). The quarks and gluons had remained almost free particles until this time, but as the Universe cooled further the colour forces (see p. 169) on the quarks strengthened and gathered them into clusters to form protons and neutrons. This is the epoch that was recreated in the experiments in the early 1970s at the Stanford Linear Accelerator Center, where the first signs of quarks in the proton were seen. More recent experiments are attempting to recreate the primordial quark–gluon 'plasma', which would have congealed into the protons and neutrons that form matter today.

After 100 s the Universe was cool enough for the protons and neutrons to form nuclei. The temperature everywhere was like that at the centre of the Sun today. After 300 000 years (10^{13} s) the average temperature had dropped to a mere thousand degrees – cooler than the present Sun's surface. The heat could no longer prevent the attraction of the negative electrons to the positively charged nuclei and the first neutral atoms formed.

Once atoms had formed, matter could begin to cluster together and eventually form galaxies. Today these galaxies rush apart driven by the continuing expansion of the Universe. What will become of it all? If the kinetic energy of the separating galaxies is greater than their mutual gravitational attraction, the expansion will continue forever. How the Universe ends will then depend to some extent on whether protons, the stuff of matter, are permanently stable or whether they erode slowly into radiation. The other possibility is that the Universe is overweight, so that it will ultimately stop expanding and collapse together again under the pull of its own gravity.

We cannot say with certainty which way things will turn out because the Universe appears to be very close to the critical dividing line between collapse and continual expansion. The visible Universe, which we observe through its radiations, contains significantly less than 10% of the matter necessary to reverse the expansion. However, there are reasons to believe that this visible matter is outweighed by mysterious dark matter, which does not show up in telescopes of any kind.

The nature of the dark matter is another of the big questions facing particle physics and cosmology. The mystery matter could consist of clusters of hitherto unknown particles. Another possibility is that it consists, at least in part, of neutrinos. The hot Big Bang theory predicts that there should be about 300 neutrinos in every cubic centimetre of the Universe. This is some 1000 million times the density of protons, so the visible galaxies are but islands in a sea of neutrinos. If just one of the three types of neutrino weighs even as little as 30 ev (one 30 billionth of a proton's mass) then neutrinos would dominate the Universe's gravity and the dark matter sums could be explained. The electron-neutrino's mass is the best studied of all, but experiments so far only yield an 'upper limit' which indicates that these neutrinos cannot have a mass greater than about 3 eV. But do neutrinos have any mass at all? There are exciting indications emerging that perhaps they do.

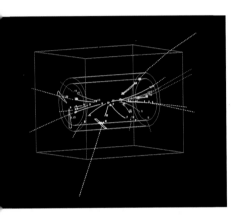

Fig. 10.3 Collisions at modern particle accelerators mimic conditions that must have prevailed in the very early Universe. This computer display from the UA1 experiment at CERN shows one of the first observed examples of the 'signature' of a W particle, created in the head-on collision of a high-energy proton and antiproton at the centre of the detector. An outline of the detector is shown in red, the cylinder marking out the central tracking chamber. Most of the particles created in the collision have gone left and right, in the general direction of the initial proton and antiproton. But one high-momentum track (blue) heads towards the bottom of the apparatus and connects to a cell in the electromagnetic calorimeter (white bars). This indicates that the track belongs to an electron. It appears to have nothing travelling in the opposite direction to balance it, suggesting that a neutrino must have passed unseen that way. Calculating the missing energy indicates that the neutrino and electron together possess the right energy to have come from the decay of a W particle (mass 83 GeV). Such collisions echo the state of the Universe a mere billionth of a second after the Big Bang.

What Happened to the Antimatter?

Theories and observation all suggest that in the first moments of the Universe particles of matter and antimatter were created in equal abundance. Yet today the Universe is not like this at all: everything we know in the large-scale cosmos consists of matter. How this asymmetry came about is one of the great puzzles of both particle physics and cosmology.

How can we be so sure that the Universe is dominated by what we define as matter – based, as our world is, on a positive proton rather than its antithesis, the negative antiproton? The critical feature is that when antimatter and matter touch they annihilate one another in a characteristic burst of gamma rays. In particular, electrons and positrons annihilate most easily when they are more or less at rest, and emit gamma rays of a very precise energy, equivalent to their total mass. The joyful return of astronauts from the Moon and the successful landing of probes on Mars proved that there is no antimatter up there. The solar wind, a continuous breeze of subatomic particles from the Sun, hits the Moon all the time. Had the solar wind been made of antimatter, its collision with the Moon would have led to sharp flashes of gamma rays; their absence confirms that the solar wind and by inference the Sun are also made of matter. Nor are any gamma rays with energy characteristic of electron–positron annihilation emitted when the solar wind hits asteroids, planets, or comets. As a result, we infer that our immediate neighbourhood is made entirely of matter.

There are clouds of gas permeating the Galaxy and the absence of the critical gamma rays extends our search for antimatter throughout the Milky Way. The largest amount of antimatter that there could be in the Galaxy, given that we do not detect these tell-tale signals, is staggeringly tiny – less than one part in a thousand million million. Through the most powerful telescopes we can see millions of galaxies distributed throughout the heavens, some of which are in close encounters and distended as the tidal forces tug on their

Fig. 10.4 Our Galaxy consists not only of billions of stars but also of clouds of gas and dust, which can be seen obscuring stars in this image of the Milky Way, taken from Mount Graham in Arizona. If any of the stars or the dust between them were antimatter, we would observe the tell-tale gamma rays produced by annihilations where this antimatter met with the matter of our own immediate neighbourhood.

Fig. 10.5 Andrei Sakharov (1921–1989) in the 1960s.

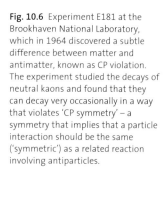

Fig. 10.6 Experiment E181 at the Brookhaven National Laboratory, which in 1964 discovered a subtle difference between matter and antimatter, known as CP violation. The experiment studied the decays of neutral kaons and found that they can decay very occasionally in a way that violates 'CP symmetry' – a symmetry that implies that a particle interaction should be the same ('symmetric') as a related reaction involving antiparticles.

individual stars. If any of these colliding galaxies were made of antistars their boundaries would be delineated by characteristic gamma rays, corresponding to the annihilation of electrons and positrons. But none are seen. The gamma rays that are observed out in the Universe – including the dramatic gamma-ray bursts – do not have the energies that would come from electron–positron annihilation.

It seems that we inhabit a volume of matter that is some 10 billion light years in diameter. It is tempting to extrapolate from this huge volume to conclude that the entire Universe is made from matter; however, this observed region represents only a tiny fraction of the 15 billion years the Universe has existed, and in terms of volume it is minuscule. So we are left with one of two possible puzzles. We have to explain why the Universe settled into huge distinct clusters of matter and antimatter (where we happen to live in one of the domains of matter), or why matter has won out overall. In either case, this is one of the great mysteries confronting modern physics.

Most physicists favour the idea that there is some subtle asymmetry between matter and antimatter, and that soon after the Big Bang this tipped the balance in favour of a Universe dominated by matter. This suspicion is based on subtle differences in how matter and antimatter behave at the level of the fundamental particles. The challenge now is to study these differences in detail in order to identify their origins and, perhaps, the source of the asymmetry between matter and antimatter in the cosmos.

In the first instants of the Big Bang, equal quantities of particles and antiparticles would have continuously formed and annihilated in the cauldron of high-energy radiation. Then, as the Universe expanded and cooled, annihilation would begin to dominate as energies became too low to create particle–antiparticle pairs. Eventually, if the symmetry between matter and antimatter were perfect, only radiation would have been left – a Universe full of photons. Our very existence shows that this cannot have been quite true, although it was almost true. We can estimate from observations of the present Universe that photons outnumber particles of matter – specifically, protons – by as much as a billion to 1. It seems that while most of the protons and antiprotons formed in the early Universe annihilated to leave us with a sea of radiation, one proton in a billion had no antiproton with which to annihilate, and the stars and galaxies we observe are the consequence.

In 1967, the Russian theorist Andrei Sakharov – who had earlier won the Lenin Prize for

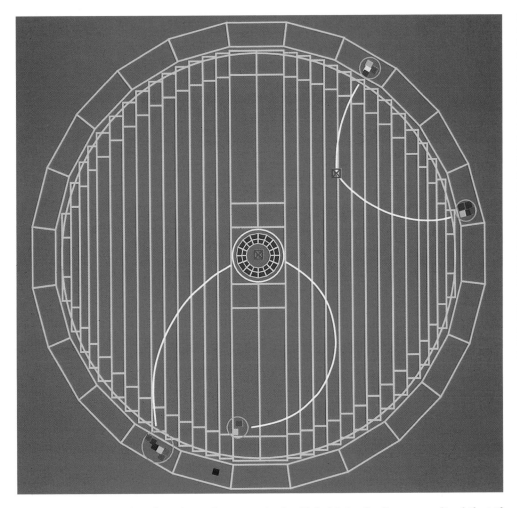

the Soviet hydrogen bomb and was later to win the Nobel Prize for Peace – realized that if protons were to outnumber antiprotons even in small numbers, then there had to be some small difference in the rate at which protons and antiprotons participated in certain reactions in the early Universe. In particular, he showed that nature must exhibit a breakdown of what is known as 'CP symmetry'.

'C' (for charge conjugation) refers to replacing quarks with antiquarks, or vice versa, in any interaction. You can think of it as seeing an interaction reflected in a 'mirror' that has the effect of changing quarks into antiquarks and antiquarks into quarks. 'P' (for parity) refers to the ability to invert an interaction completely in space, like a mirror that switches an image not only from right to left, but also top to bottom and back to front. If CP symmetry is perfect, then the combination of these two 'reflections' yields an interaction proceeding at the same rate as the original one, and exactly like it in all respects except that particles are replaced by antiparticles, and vice versa.

However, in 1964 James Christenson, James Cronin, Val Fitch, and Rene Turlay, working at the Brookhaven National Laboratory in New York, observed an interaction that would be impossible if CP symmetry is always valid. The interaction occurs in the bizarre behaviour of neutral kaons. Kaons (see pp. 74–75) are made of a quark and an antiquark and as such are an equal mixture of matter and antimatter. The neutral kaon (K^0) consists of a down quark and a strange antiquark, while its antiparticle (written $\overline{K}^0$) consists of a down antiquark and a strange quark. The K^0 and $\overline{K}^0$ are thus different particles, but they are intimately related through the weak force which, rather surprisingly, allows a K^0 to change to a $\overline{K}^0$, and vice versa, via interactions between their quarks and antiquarks. What this effect means is that once a neutral kaon or neutral antikaon is created some quantum mechanical 'mixing' begins to occur. As time passes an initially pure K^0, say, can evolve into a pure $\overline{K}^0$ via two different particle states that are in-between 'mixtures' of K^0 and $\overline{K}^0$. These in-between mixtures are known as the K_S ('S' for short) and the K_L ('L' for long). The K_L lives about 600 times longer than the K_S.

Fig. 10.8 The CPLEAR experiment at CERN not only studied the matter–antimatter asymmetry known as CP violation in kaons, but also observed the related asymmetry between certain interactions of kaons and the same time-reversed interactions. CPLEAR studied kaons produced in the collisions of antiprotons with protons. Here the antiproton beam line appears from the bottom of the picture, just left of centre, and curves to the right through focusing and bending magnets before reaching the CPLEAR detector, much of which is painted bright yellow. To the left is another antiproton beam line which later splits to serve two other experiments. To the right are some of the concrete blocks that shield detectors from scattered radiation.

The important feature is that the states K_L and K_S behave differently in the combined 'mirrors' of CP. The two states decay in different ways, the K_S to two pions, the K_L to three pions. If CP symmetry were perfect this pattern of decay would always be true. The K_L, for example, would never decay to two pions. However, as Cronin and Fitch and their colleagues first observed, in about 0.3% of cases the K_L does decay to two pions.

Initially, the reason why CP symmetry is violated in the neutral kaons remained an enigma. Then in 1973, two Japanese theorists realized that CP violation could occur in kaons, if there were more kinds of quark. Only three quarks were known at the time, although a fourth had been hypothesized by some theorists. However, M. Kobayashi and T. Maskawa required a third pair of quarks in addition to the two pairs of up/down and charm/strange. Building on work by the Italian, Nicola Cabibbo, they realized that they could summarize the responses of the different types of quark to the weak interaction by numbers expressed in a 3 by 3 matrix – now known as the CKM matrix after the three theorists. If certain of these numbers are complex (involve the square root of minus one) CP violation can arise.

This work would have remained a theoretical curiosity had it not been for the discovery of a fifth quark, bottom, in 1977, followed eventually by the sixth quark, top, in 1995, so giving the third pair or 'generation'. The realization that there really are three generations of quarks galvanized interest in the somewhat arcane idea of the CKM matrix. The question now in the minds of many physicists is whether the 'accident' of three generations is what has led to the dominance of matter in our Universe. Theory implies that CP violation should be a large effect in the case of B mesons, which are similar to kaons but with the strange quark replaced by a bottom quark. In the late 1990s the B-meson system became the subject of intensive experimental investigation, as specialized 'B factories' became available, and in July 2001 the first clear evidence for CP violation in B mesons was announced, as Chapter 11 describes (see p. 211).

The CP violation observed in kaons and B mesons is a breakdown in the symmetry of their interactions, but an overall symmetry in the interactions can be restored theoretically – at a price. The interactions must also embody an 'arrow of time', a small but clear distinction between forwards and backwards in time. In our familiar, everyday world, such a concept hardly seems surprising at all; indeed, it meshes completely with our experience. But until the observation of CP violation in the 1960s, physicists had always believed that at the fundamental level there should be a symmetry between a particle interaction and the same process run backwards in time. For example, a pion and a proton can make the resonance known as the delta; and, conversely, the delta can decay into a pion and a proton. However, the discovery of CP violation suggested that the asymmetric effects of CP violation could be related to a difference between kaon interactions moving forwards and backwards in time. In other words, if you were to look at the interactions in a CP 'mirror' that also reversed time, all the asymmetries would cancel out, with the result that you would see a symmetry between the interactions on either side of the mirror.

This implied that it should be possible to measure a small difference between some neutral kaon interaction and the equivalent time-reversed process. In 1998, the team on the CPLEAR experiment at CERN announced that they had made such an observation. For the first time, an arrow of time had been revealed at the level of fundamental particles.

What is Mass?

The action of a magnet and the burning of the Sun appear so different that it seems remarkable that the processes are both manifestations of a single fundamental force – the electroweak force. This is the force carried by the familiar photon of electromagnetism, and by the W and Z bosons, which are responsible for the weak interactions that not only initiate solar burning but also underlie certain types of radioactivity. Yet if these effects are so closely intertwined, why do they appear so different in our daily experiences, that is, at relatively low temperatures and energies? One reason is that the particle associated with electromagnetism, the photon, is massless, whereas the W and Z bosons, which are associated with the weak force, have huge masses and each weighs in at nearly 100 proton masses, or as much as an atom of silver.

Experiments at CERN's Large Electron Positron (LEP) collider and other laboratories have measured the masses of the W and Z bosons to great precision: one part in a thousand for the W and better than one part in ten thousand for the Z. Moreover, if theorists insert these experimentally measured masses into their equations, they are able to describe the detailed interactions of the bosons with remarkable accuracy. However, theory alone provides no clue as to what the masses of the W and Z particles should be.

Indeed, the enigma goes deeper still, for we do not really know what mass is, although it is familiar as a concept that permeates all of science. Mass, according to Newton and decades of experience, is the source of gravity, which controls the Universe. Einstein, on the other hand, taught us that mass is the 'm' in the most famous equation of physics, $E = mc^2$, so mass is a form of energy. But what actually is mass? Where does it come from?

The Standard Model of the fundamental particles and the forces that act among them explains mass by proposing that it is due to a new field, named the Higgs field after Peter Higgs who in 1964 was one of the first to recognize this theoretical possibility. Gravitational and electromagnetic fields are not the only things that fill the Universe: the Higgs field also permeates all of space. Were there no Higgs field, according to the theory, the fundamental particles would have no mass. What we recognize as mass is, in part, the effect of the interaction between particles and the Higgs field. Photons do not interact with the Higgs field and so are massless; the W and Z bosons do interact and thereby acquire their large masses. The building blocks of matter, the quarks and leptons, are also presumed to gain their masses by interacting with the Higgs field.

Just as electromagnetic fields produce the quantum bundles we call photons so should the Higgs field manifest itself in particles, which are given the name of Higgs bosons. In the simplest theory there is just one type of Higgs boson, but in more complex theories there are more. What do we know about these new bosons? Precision measurements made at LEP and other accelerators, when combined with the mathematics of quantum theory and the Standard Model, enable theorists to determine the energies at which the Higgs boson – or whatever it is that gives rise to mass – should be revealed. These calculations imply that the origins of mass were frozen into the fabric of the Universe just a millionth of a millionth of a second after the Big Bang, when the temperature had 'cooled' to below ten thousand million million degrees. It is just possible that proton–antiproton collisions at Fermilab's upgraded Tevatron might catch the first signs of a Higgs boson. However, to make a dedicated exploration of this energy region, where the puzzle of mass should be revealed, requires a machine that will access higher energies. The machine to do this is the Large Hadron Collider, or LHC, being built at CERN, which should start up in 2006 (see Chapter 11).

Fig. 10.9 Peter Higgs (b. 1929) in 1988. In the 1960s, Higgs and others proposed that particles acquire their masses through interactions with a field that pervades the Universe – the stronger this interaction, the more massive a particle appears to be. The blackboard shows equations and a diagram describing this 'Higgs field'. An important consequence of the theory is that, associated with the field, there must exist one or more 'Higgs particles'.

Does Quark–Gluon Plasma Exist?

In the searing heat of the Big Bang the quarks and gluons, which in today's cold Universe are trapped inside protons and neutrons, would have been too hot to stick together. Instead, these fundamental particles would have existed in a dense, energetic 'soup' of quarks and gluons. Matter in this state is called 'quark–gluon plasma' or QGP for short. It is analogous to the state of matter known as plasma, such as is found in the heart of the Sun, which consists of intermingled swarms of electrons and nuclei too energetic to bind together to form neutral atoms. Physicists believe that QGP might still exist today in the hearts of neutron stars, the residue of the spectacular stellar explosions known as supernovae. Neutron stars are so dense that a piece the size of a pinhead would weigh more than the Eiffel Tower. However, even if QGP does exist in these remote exotic places, we can look at it only across the vast distances of space. If we are to study it more closely, to learn more about the first moments of the Universe, we need to recreate QGP in the laboratory.

Fig. 10.10 The collision of a high-energy sulphur ion with a nucleus in a target of gold produces a myriad particle tracks in the NA35 experiment at CERN. This image, from 1991, involves a sulphur ion, which has come in from the left, with an energy of 6.4 TeV (6.4 tera electron volts, or 6400 GeV).

Physicists are attempting to make QGP by smashing large atomic nuclei into one another at such high energies that the protons and neutrons squeeze together. The hope is that the nuclei will 'melt' – in other words, that the quarks and gluons will flow throughout the nucleus rather than remaining 'frozen' into individual neutrons and protons.

It is easy to knock atoms out of molecules, or electrons out of atoms, or even neutrons and protons from atomic nuclei. But no one has ever managed to liberate an individual quark or gluon from within its neutron or proton prison. However, theorists believe that when nuclei collide at high enough energies the volume within which the quarks and gluons are trapped will increase – in the theorists' jargon, the quarks and gluons become 'deconfined'. This is believed to be the first step towards QGP.

In the QGP state the enlarged volume of free-flowing quarks and gluons lasts long enough for them all to reach the same average energy or, equivalently, temperature. This is analogous to what happens when you pour hot water into a cold bath. At first there will be local hot spots in among the cold but after a time the temperature evens out; the bath is said to be 'thermalized'. Similarly, with deconfinement in the atomic nucleus the hot spots must be big enough for thermalization to happen, and for QGP to be formed.

The first claims that experiments were probably making QGP in the laboratory came in January 2000 from several teams at CERN. In the 1980s, the machine physicists at CERN converted the Super Proton Synchrotron (see Fig. 6.36, p. 102) into a 'heavy ion' accelerator. Ions are atoms with missing electrons, which means they are no longer electrically neutral. Instead, they are positively charged and so can be accelerated. If all the electrons are missing, what remains is the bare atomic nucleus. (Negative ions can also be made, by attaching extra electrons to an atom.)

CERN began by accelerating relatively small nuclei, such as oxygen (8 protons and 8 neutrons) and sulphur (16 protons and 16 neutrons), but by 1994 had progressed to producing its first beam of high-energy lead nuclei (82 protons, 126 neutrons). Initially, the lead beam would consist of ions with 27 electrons missing (lead 27+), knocked out in multiple collisions with electons in a plasma. The remaining 55 electrons were gradually stripped off as the beam progressed from the special heavy-ion linear accelerator through the Booster accelerator and the Proton Synchrotron (see Fig. 6.28, p. 97) to the SPS. By the time the lead ions reached the SPS they were naked nuclei – with their 82 protons giving them a charge of +82 – and they were accelerated to 33 TeV per nucleus, or 160 GeV for each of the 82 protons and 126 neutrons they contained. Then they were ready to be extracted from the SPS and smashed into a target – and the search for QGP could begin.

The challenge for the physicists was not only to make QGP, but to know that they had made it! A variety of effects should occur as nuclear matter heats up, turns into QGP, and then cools down again so that ordinary matter condenses out rather like water condensing from steam. In particular, there are three specific signals that experimenters can look for when they smash nuclei together at very high energies.

The first clue could be an increase in the production of particles containing strange quarks. Strange quarks are heavier than the more common up and down varieties, so they are harder to produce from normal matter. However, theory shows that QGP is more stable if it contains strange quarks in similar amounts to the up and down varieties. If QGP is made in a nuclear collision, then as it cools the quarks will cluster together again to form familiar composite particles. An increased number of strange quarks in the QGP should lead directly to an increase in the number of strange particles, such as kaons, emerging from the collisions. The experiments at CERN have seen hints that the number of strange quarks does increase as collisions get hotter (more energetic), but this alone was not enough to convince the researchers that they had made QGP.

Charmed quarks (and their antiquarks), which are even heavier than strange quarks, can also be produced when nuclei collide, and a charmed quark and antiquark can combine together to make the J/psi particle. When nuclei collide at relatively low energies, the number of J/psi particles that emerge is higher for large nuclei, such as lead, than it is for smaller ones. This is sensible: the increased number of neutrons and protons in a large nucleus enables more of everything, in particular J/psi particles, to be produced. However, experimenters at CERN in 1997 noticed a dramatically different effect when they directed lead beams into lead targets at very high energies. It transpired that the production of J/psi particles in the collisions of lead nuclei was less than expected in comparison with collisions of lighter nuclei, contrary to previous experience at lower energies. Furthermore, the number of J/psi particles was reduced when the lead nuclei smashed head on into the target nuclei, rather than in more glancing collisions.

This is just what would be expected if QGP had been formed. In the melting pot of QGP the charmed quark and antiquark have difficulty in binding to one another to form a J/psi particle. Instead they pair with the other lighter varieties of quark to make charmed particles. Thus a reduction in the production of J/psi particles could be a signal for QGP. Moreover, theory indicates that QGP should be formed in head-on collisions of heavy nuclei such as lead, rather than in glancing collisions, or in those involving lighter nuclei. It is in these very circumstances that the number of J/psi particles appears to be reduced in the experiments at CERN.

Finally, there is an analogy between particles escaping from the heart of QGP and those escaping from the high-temperature conventional plasma that exists at the centre of the Sun. Neutrinos are the only particles that can easily escape from the solar plasma, as they have no electric charge and interact only weakly with the charged electrons and nuclei that form the plasma. QGP, on the other hand, consists of strongly interacting quarks and gluons, so not only neutrinos but also electrons and positrons can escape unhindered, as they do not interact through the strong force. Several kinds of particle, including pions and kaons, produce electrons and/or positrons when they die, and according to theory some of these

Fig. 10.11 A collision between a high-energy lead ion and a lead nucleus in a target creates a dramatic spray of hundreds of charged particle tracks in the NA49 experiment at CERN's Super Proton Synchrotron (SPS). The tracks have been detected by four large time projection chambers (TPCs, see p. 135), which are outlined in grey. The first two TPCs, near the bottom and centre of the image, are located in large superconducting magnets, which bend the tracks of the particles according to their momentum. The higher momentum particles continue farther up the page to two large TPCs, each nearly 4 m x 4 m in area, and about 1 m deep. The lead ions in this experiment were accelerated in the SPS to 158 GeV per nucleon – a total of more than 32 TeV per ion. The measurements made on the collisions suggest that the first signs of quark–gluon plasma have been observed in collisions like this.

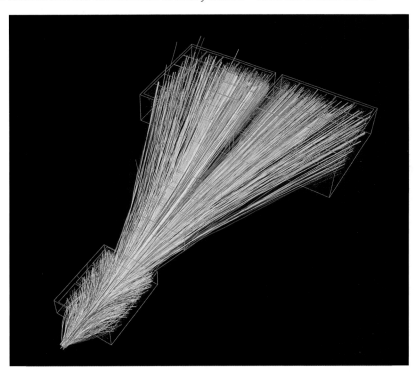

Fig. 10.12 At the Relativistic Heavy
Ion Collider (RHIC) at the Brookhaven
National Laboratory two rings of 1740
magnets guide beams of gold ions –
atoms stripped of electrons – in
opposite directions before they collide
head on. The rings sit in a tunnel
3.8 km in circumference, where the
ions are accelerated to 100 GeV per
nucleon – that is, 100 GeV for each of
the 79 protons and 118 neutrons the
gold ions contain. To bend the ions at
the highest energies requires
superconducting magnets which are
kept at their low operating
temperatures in a cryostat – in effect
a long pipe-like vacuum flask – filled
with liquid helium. Here one of the
magnet rings is visible, the other lies
behind it.

particles become easier to produce within a QGP, so more electrons and positrons should emerge unscathed. By detecting these electrons and positrons, we are effectively looking into the heart of the QGP, just as by detecting solar neutrinos we can look into the heart of the Sun (see p. 120). There are tantalizing hints that electrons and positrons *are* escaping directly from the nuclear collisions at CERN, exactly as expected if QGP has been formed.

Taken alone, each of these effects is not enough to show that QGP is made when energetic lead ions collide with heavy targets. But taken together, the evidence has a greater significance. Either nature is being unkind, producing 'background' processes that mimic the effects of QGP, or QGP has been produced at CERN. By January 2000, the various experimental teams had decided that the weight of evidence tilted in favour of QGP and CERN announced the first signs of the new state of matter.

CERN pioneered the search for QGP in the 1980s by converting its accelerators to handle beams of heavy ions, which were then directed at stationary targets. The Brookhaven National Laboratory in the USA, however, has chosen to build a dedicated machine to make beams of heavy nuclei collide head on. As with simpler particles, such as electrons and protons, the great advantage of a colliding beam machine is that all the hard-won energy gained in accelerating the particles goes into the collision. The Relativistic Heavy Ion Collider (RHIC) achieved its first collisions between beams of gold nuclei (79 protons and 118 neutrons) in June 2000, reaching energies of 130 GeV per nucleon, or a total of 50 TeV in the collision.

Fig. 10.13 This computer display
shows the tracks of as many as 1000
charged particles created in a head-on
collision of gold ions at RHIC, and
detected by the STAR experiment. In
this 'end view' the ion beams, with an
energy of 30 GeV per nucleon, have
collided at the centre of the detector.
The tracks are detected in the world's
largest time projection chamber (see
Fig. 8.9, p. 136), or TPC, which has a
volume of about 50 cubic metres.

By 2007, RHIC will have been superseded in energy by the Large Hadron Collider (LHC, see p. 207) at CERN, which will make lead ions collide at a total energy of 1300 TeV. At these extreme energies, akin to those that would have been the norm in the Universe when it was less than a trillionth of a second old, QGP should become commonplace, so that experimenters can study its properties in detail. A dedicated experiment, named ALICE for A Large Ion Collider Experiment, will search specifically for the signals of QGP – and glimpses of matter as it was a long, long time ago.

What is the Dark Matter?

By observing the heavens at wavelengths spanning the electromagnetic spectrum, from gamma rays through visible light to radio waves, astronomers have discovered a vast range of galaxies, stars, and clouds of gas and dust. Protons and the nuclei of ordinary atoms make up all the 'luminous matter' that shows up in these observations. But in spiral galaxies, to take one example, as much as 90% of the matter present remains undetected. It appears that the Universe we see by its radiations is outweighed by some mysterious 'dark matter', which does not show up at any wavelength in our telescopes.

How do we know this? By observing the movements of stars in spiral galaxies. Stars whirling round a galaxy should travel more slowly the further they are from the galactic centre. This is what happens, on a smaller scale, in the Solar System. Pluto and the outer planets move more slowly than the Earth, and the Earth moves more slowly than Mercury, the closest planet to the Sun. The speed of the planets is just right to counteract the gravitational pull of the Sun. If Mercury travelled any slower, it would spiral down into the Sun. If Pluto travelled any faster, it would escape and whirl off into outer space, flung out by the same centrifugal force that presses us outwards when a car speeds round a bend.

The problem with spiral galaxies is that the stars in their outer reaches are travelling too fast relative to those near their centres. They should by rights whirl off into intergalactic space. Astronomers can calculate the mass of a galaxy from all the luminous matter they observe in its stars and nebulae. But on the periphery stars are orbiting faster than this galactic mass says they should. It is as if some extended halo of unseen matter were contributing additional invisible mass to the galaxy's known 'luminous' mass.

This is not a small problem. In order to make the motions of the stars conform with the laws of gravity, as much as 90% of galaxies must be invisible. And the motions of individual galaxies within clusters of galaxies require even larger amounts of dark matter.

What might dark matter be? One possibility is that it is made up of massive bodies that live in the outer regions or 'halos' of galaxies but do not shine at any wavelength. These so-called 'massive compact halo objects' – MACHOs – could be bodies about the size of Jupiter, which are not big enough to become stars, or they could be black holes. Astronomers can search for objects of this kind because the gravitational field around the object can bend the light coming from stars or galaxies beyond and even create double or multiple images of the distant star or Galaxy through the effect of gravitational lensing. However, searches of this kind have not found enough MACHOs to explain the vast amount of dark matter the Universe appears to harbour. So astrophysicists have had to turn to particle physics for further ideas.

Fig. 10.14 (LEFT) M81 – a spiral galaxy in the constellation of the Great Bear (Ursa Major). The speed at which the outer stars rotate about the centre of such galaxies indicates that the galaxies contain around 90 times as much invisible 'dark matter' as the matter visible as stars.

Fig. 10.15 (RIGHT) This three-dimensional map shows the distribution of invisible or 'dark' matter in a cluster of galaxies. The more dark matter, the higher the peaks, which indicate dark matter associated with individual galaxies in the cluster. The map was made by analysing the way that light from a distant galaxy was bent by the gravitational effects of an intervening cluster of galaxies. The analysis revealed the total amount of matter in the cluster, and subtraction of the 'visible' matter left the proportions of dark matter shown here.

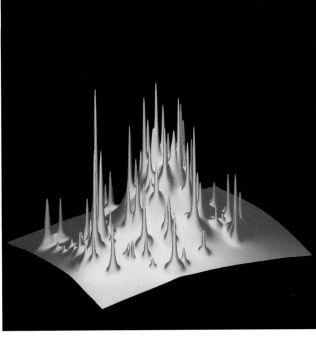

Instead of large objects such as planets or black holes, the dark matter could consist of vast quantities of subatomic particles that do not interact electromagnetically (otherwise we would be able to detect their electromagnetic radiation). One obvious candidate, with its ability to travel through light years of matter without interacting, is the neutrino. It would have to have some mass – and there are indications that this may indeed be the case – but its mass need only be very small since neutrinos are as common in the Universe as photons.

There is a problem with this low-mass neutrino hypothesis as an explanation for the dark matter, however. In the early Universe, these neutrinos would have been highly energetic, moving at almost the speed of light. They would have been 'hot', and computer simulations of galaxy evolution in a 'hot dark matter' Universe show galaxies forming in dense clusters with large voids between them. This is not what astronomers observe.

The evolution of galaxies would have been very different if the dark matter consists of slow-moving and therefore 'cold' particles. Simulations with exclusively or largely cold dark matter are better at reproducing the distribution of galaxies we see in the heavens today. But what could cold dark matter particles be? They could be slow because they are very massive, but none of the known particles that live long enough are massive enough.

So we have the situation where the motion of stars within galaxies strongly suggests the presence of invisible dark matter, and the distribution of the visible galaxies throughout the Universe seems to imply that at least some of this dark matter consists of particles that have not yet been seen in particle physics experiments.

A tantalizing possibility is that the dark matter could consist of entirely new forms of particle predicted by theories based on a 'supersymmetry' between particles of matter and force-carrying particles (see p. 205). The lightest varieties of these 'supersymmetric' particles include forms that do not respond to the electromagnetic or strong forces, but which may be hundreds of times more massive than the proton. Collisions at the highest-energy particle accelerators, in particular the Tevatron at Fermilab and the LHC under construction at CERN, may have enough energy to create them. If such a particle is found, the challenge will then be to study its properties in detail, in particular to see if it could have formed large-scale clusters of dark matter in the early Universe.

While physicists working at accelerators search for massive dark matter particles in conditions that mimic the early Universe, others are taking a complementary approach by seeking to capture the occasional fossil relic. In underground laboratories, detectors wait for particles of dark matter left over from the early Universe to play their part in a delicate game of subatomic billiards. If a heavy particle ricochets off an atomic nucleus, it may impart just enough energy to the recoiling nucleus to make it detectable. The difficulty is that there are many possible ways in which false signals can be produced in these experiments, unless they are protected from as many unwanted particles as possible.

Building laboratories underground provides a first level of protection, because the surrounding rock shields the experiments to a large extent from cosmic rays. One site that has been developed especially for dark matter experiments is at the Boulby potash mine in northern England – the deepest mine in Europe. The experiments there are installed in three disused caverns, 1100 m below the surface. Additional shielding for the detectors is

Fig. 10.16 (LEFT) A map of the distribution of galaxies in a portion of sky near the south galactic pole shows how galaxies tend to cluster together, with 'voids' between. Each small square or 'pixel' represents a patch of sky equivalent to a millimetre square held at arm's length. Black pixels contain no galaxies, dark blue pixels up to 20 galaxies, and in bright blue pixels there are more than 20 galaxies. Any viable theory of matter in the Universe must be able to reproduce a galaxy distribution like this.

Fig. 10.17 (RIGHT) Computer programs can simulate the evolution of galaxies during the history of the Universe and map out the resulting distributions. This image, which is a close match to the structure of the real Universe, was produced by a simulation that assumed that much of the matter in the Universe is in the form of cold dark matter, possibly in the form of weakly interacting massive particles.

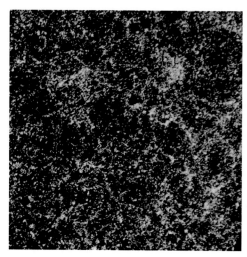
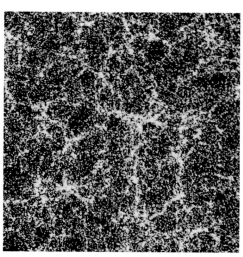

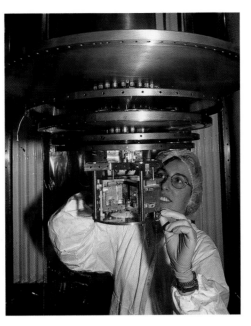

provided in two forms. Detectors are suspended in a 6 m tank of purified water, which absorbs gamma rays and neutrons from natural radioactivity in the surrounding rock. Further protection comes from special lead and copper with particularly low levels of natural radioactivity. Even within this low-radiation environment, the rate of uninteresting interactions due to residual radioactivity and cosmic rays is at least a thousand times greater than the predicted rates for massive dark matter particles, which are expected to interact less than once a week in a kilogram of detector!

The researchers at Boulby, and other places including the Gran Sasso Laboratory in Italy, are striving to develop increasingly sophisticated techniques for separating out the tiny recoil signals that dark matter would produce from the signals due to other particles. The race is on to see whether dark matter is first identified underground or at an accelerator.

Do Neutrinos have Mass?

Although it seems that neutrinos with mass cannot by themselves solve the mystery of the dark matter, there is increasing evidence that these elusive particles do have mass. We know that if neutrinos have mass, then their masses are very small, for no one has yet managed to measure them directly. The best that experiments have done is to show that the masses of the neutrinos must be less than the smallest mass the experiments can measure. In the case of the tau-neutrino, the mass must be less than 18 MeV, while the muon-neutrino's mass must be less than 0.17 MeV, neither of which is very definitive. For the electron neutrino, the limits are much better. Measurements show that it weighs in at less than 3 eV, or less than one part in 150 000 of the electron's mass of 0.51 MeV, the lightest measured mass of all particles.

Surely such small values are consistent with the idea that neutrinos are massless, so why bother to investigate further? One reason is simply the challenge to measure the Universe to the limits of technological ability. But there is another more profound reason. Every cubic centimetre in the Universe contains on average about 300 neutrinos left over from the Big Bang. They are travelling around the cosmos at or near the speed of light; and they are passing through you as you read this. Along with photons, they are the most common particles in the Universe and outnumber protons, the stuff of the stars and interstellar space, by a factor of some billion to one. So, if an individual neutrino's mass were only a few billionths of that of a proton – in other words, a few eV – neutrinos would collectively outweigh everything we see in the Universe and form part of the dark matter. They could even be major players in determining the fate of the Universe, causing it to collapse under its own weight.

Mass could also have important consequences for the neutrinos themselves. According to

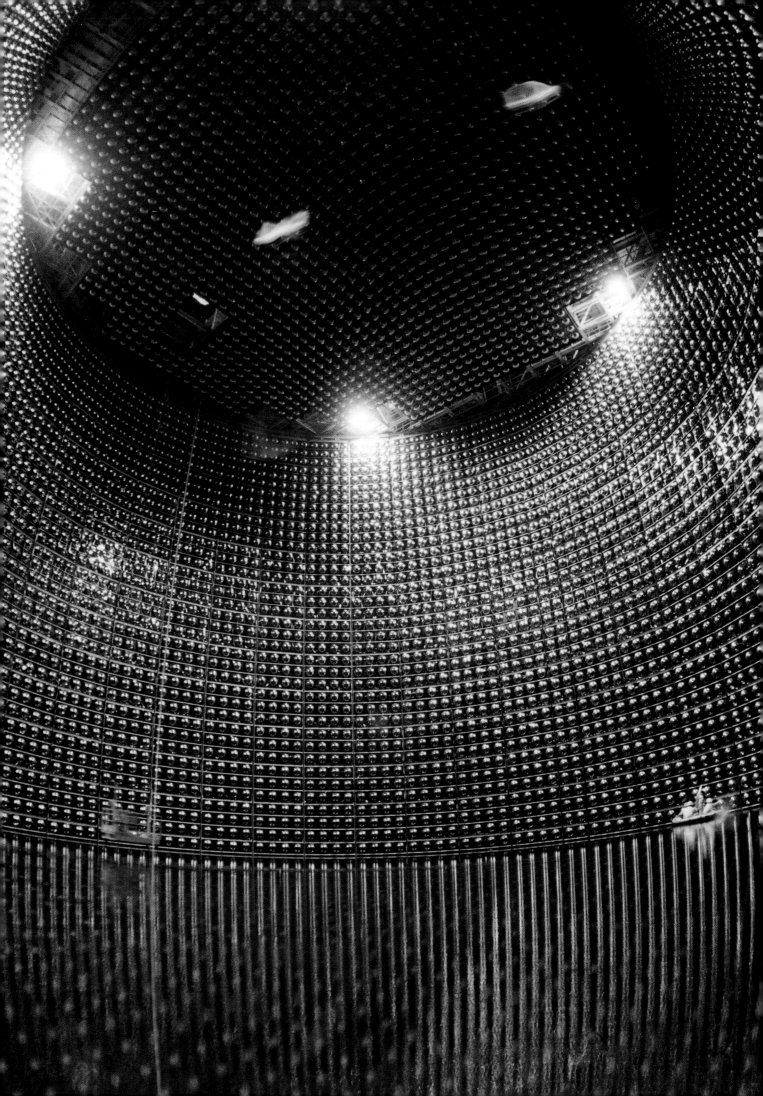

quantum theory, if neutrinos do have mass they can change their form as they travel. For example, a neutrino created as an electron-neutrino could change into a muon-neutrino as it travels, and then back again into an electron-neutrino, and so on. This regular transmutation is known as 'oscillation', and looking for evidence of such neutrino oscillations is one of the most sensitive ways to discover if neutrinos have mass.

The first hints that neutrinos might behave in this bizarre way came gradually during the 1980s and 1990s from experiments that watch for the arrival of solar neutrinos, emitted by the nuclear reactions that fuel the Sun (see pp. 120–121). What has surprised physicists is that the detectors record only a third to a half of the events predicted by detailed calculations. Astrophysicists find it difficult to explain these results by adjusting, for example, the temperature at the centre of the Sun. The alternative, more exciting proposal is that something is happening to the neutrinos en route from the centre of the Sun. A real possibility is that neutrinos produced in the Sun, which are electron-type, transmute as they travel through the matter in the Sun, oscillating into another variety. This could account for the missing solar neutrinos, because most of the detectors built so far have been sensitive to the expected electron-neutrinos, but not to muon- or tau-neutrinos, no matter how many pass through.

The case for neutrino oscillations strengthened during the 1990s when another instance of missing neutrinos became apparent, this time in detectors recording neutrinos made in the Earth's atmosphere in the interactions of cosmic rays. When high-energy cosmic rays hit the upper atmosphere they produce sprays of pions which subsequently decay. The decays of the charged pions are usually via muons to electrons (see p. 73), so they create two muon-neutrinos for every electron-neutrino. Yet in 1992 experiments measuring this ratio began to find that the numbers were nearly equal. Did this mean that there are too many electron-neutrinos or too few of the muon variety? The suspicion arose that muon-neutrinos are oscillating out of sight, presumably to tau-neutrinos.

Deep beneath the Japanese 'Alps', a detector filled with 50 million litres (12 million gallons) of ultra-clean water is studying both the neutrino mysteries by intercepting neutrinos that were created both in the Sun and in cosmic-ray showers in the atmosphere. Super-Kamiokande – or Super-K for short – detects neutrinos when they interact in the water to make either an electron or a muon, depending on the neutrino's type. These particles, unlike the neutrino, are electrically charged and can emit Čerenkov radiation (see p. 91) as they travel through the water. The Čerenkov light forms patterns of rings on the inner surface of the water tank, where it is picked up by thousands of phototubes arrayed around the walls. By carefully analysing the patterns of light that they detect the physicists at Super-K can distinguish between muons and electrons created in their detector, and hence between muon- and electron-neutrinos.

Fig. 10.20 (OPPOSITE) The Super-Kamiokande detector – a huge cylindrical tank of water, its walls lined with 11 200 phototubes to register Čerenkov radiation produced when charged particles fly through the water faster than light does. This photograph was taken while the detector was being filled with water. The huge scale of the structure is clear from the tiny boat at lower right, close to the wall with its 50 cm diameter phototubes.

Figs. 10.21–10.22 (BELOW) Computer reconstructions of rings of Čerenkov radiation in the Super-Kamiokande detector reveal the footprints of neutrinos created in the atmosphere above the detector. Colours show the time when the Čerenkov light arrives at the phototubes, violet and blue being the earliest, orange and red the latest, with a spread of about 160 nanoseconds.

Fig. 10.21 (BELOW LEFT) Here a muon-neutrino has interacted to create a muon, which travelled through the detector from upper left to lower right, and produced a clear Čerenkov ring with well-defined edges.

Fig. 10.22 (BELOW RIGHT) The fuzzier edges of this Čerenkov ring indicate that it has been produced by an electron, created in the interaction of an atmospheric electron-neutrino. Electrons, being lighter than muons, radiate more easily to produce subsidiary photons (gamma rays) and hence electron–positron pairs. These additional charged particles also produce Čerenkov radiation, which tends to blur the ring due to the original particles.

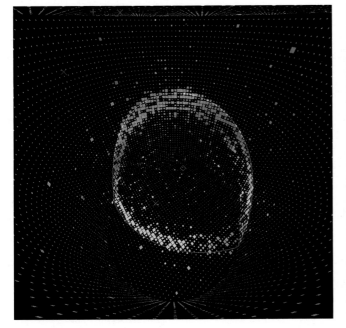

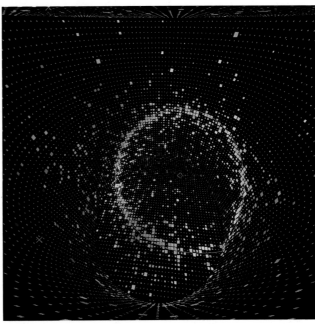

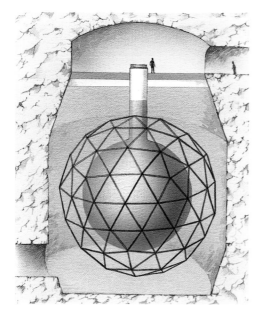

Super-K started up in 1996 and like its predecessors has found too few electron-neutrinos from the Sun. It has also found the clearest evidence so far that muon-neutrinos created in cosmic-ray decays in the atmosphere are oscillating to another type. Super-K can not only differentiate muon- and electron-neutrinos, it can also pinpoint their direction of travel. It can distinguish between 'downward-going' neutrinos, which arrive from the sky overhead, and 'upward-going' neutrinos, which arrive from below after travelling 12 000 km through the Earth.

In 1998 the Super-K team announced that the ratio of muon- to electron-neutrinos depends on whether the neutrinos come from above or below. Specifically, there seem to be too few muon-neutrinos coming from below, from the opposite side of the Earth. The best explanation is that these upward-going muon-neutrinos, having travelled through the Earth, have had more time in which to oscillate to another variety. Presumably they are turning into tau-neutrinos, because they do not appear in the detector as an increased number of electron-neutrinos.

Super-K's study of atmospheric neutrinos is a 'disappearance experiment' – expected muon-neutrinos disappear before they reach the detector. The most conclusive evidence for neutrino oscillations, on the other hand, requires an 'appearance experiment', in which the changeling neutrinos are seen in their new form. This is the philosophy behind the newest solar neutrino detector, which has been designed to be sensitive to all three types of neutrino. The Sudbury Neutrino Observatory, or SNO, is 2070 m below ground in a nickel mine in Sudbury, Ontario. It uses water to detect neutrinos, but water with a difference. The heart of SNO is an acrylic vessel filled with 1000 tonnes of 'heavy water'. This is water in which the hydrogen is in the form of its isotope, deuterium. It is heavier than ordinary water because in the nucleus of

Fig. 10.23 The Sudbury Neutrino Observatory (SNO) is 2 km below ground in a working nickel mine near Sudbury, Ontario. At its heart is a 12 m diameter transparent acrylic sphere filled with heavy water – water with a neutron as well as a proton in the hydrogen nuclei. The acrylic sphere is surrounded by a geodesic structure 18 m in diameter, which supports 10 000 phototubes that register Čerenkov radiation produced by fast charged particles created in the neutrino interactions. The detector occupies a cavern 34 m high and 22 m across, which is itself filled with ordinary – but very pure – 'light' water. This water, in which the phototubes are immersed, acts as a shield to protect the heavy water from radioactivity from the cavern walls, and to reveal charged particles entering the heavy water from outside the acrylic sphere.

Fig. 10.24 The complete geodesic structure for SNO, with its 10 000 phototubes installed. Later the cavern, which is lined with concrete and polyurethane, was filled with ultra-pure ordinary water while the acrylic sphere, hidden within the geodesic structure, was filled with heavy water.

deuterium a neutron joins the single proton of ordinary hydrogen. In SNO, electron-neutrinos interact with the neutrons in the deuterium to create protons and electrons, and the fast-moving electrons emit cones of Čerenkov radiation as they travel through the heavy water.

SNO came into operation in 1999, and in June 2001 the SNO team announced that their detector, like others, sees fewer solar neutrinos than expected. But the results from SNO revealed more than this, even from the electron-neutrino data alone. By comparing their data with solar neutrino results from ordinary, light water in the Super-K detector, the team was in a position to confirm that the solar electron-neutrinos do indeed change to a different type. This is because in light water all types of neutrino can be detected at some level, although electron-neutrinos are more readily observed. Thus if the solar electron-neutrinos change type, Super-K will still detect a few of them. However, the interactions with *neutrons* in SNO, in which a proton and an electron are produced, reveal electron-neutrinos only. Comparing the electron-neutrino data from SNO with data from Super-K shows that Super-K detects relatively more solar neutrinos. This provides the first clear evidence that some of the neutrinos arriving at Earth from the Sun really have changed from electron-neutrinos to another type.

However, the key feature for neutrino oscillations is that SNO can also detect all three types of neutrino through a reaction unique to deuterium. A neutrino of any kind can split the deuterium nucleus, freeing the neutron, which can be captured by another nucleus. The capture is detected when the newly bloated nucleus gets rid of its excess energy by emitting gamma rays, which in turn make electrons and positrons that create characteristic patterns of Čerenkov light in the surrounding water. By comparing the rates of electron-neutrino interactions with those due to all types of neutrino, SNO is in the process of demonstrating unambiguously that solar neutrinos do oscillate.

The received wisdom, which underpins the Standard Model of particles and forces, has assumed for decades that neutrinos are massless. However, as evidence accumulates that neutrinos really are oscillating back and forth from one variety to another, then according to quantum theory one or more of the neutrinos must have a small mass. This has profound implications for our understanding of particles. Measurements to reveal the masses of the neutrinos are therefore continuing, not only with detectors such as SNO and Super-K, but also with neutrinos generated at particle accelerators, as Chapter 11 will describe.

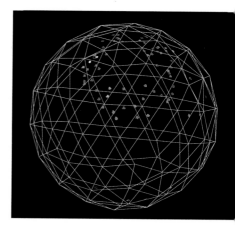

Fig. 10.25 A circle of green dots on this computer reconstruction of an event in the SNO detector shows what is most probably the 'footprint' of a solar neutrino. The electron-neutrino has interacted with a neutron in the heavy water in SNO to produce a proton and an electron. The lightweight electron moves away fast enough to emit Čerenkov radiation, which forms the ring of light picked up by the phototubes marked by the green dots.

Is there a Theory of Everything?

Technology limits our view of nature. This is particularly frustrating for theorists whose calculations can probe extreme conditions impossible here on Earth. How can their predictions ever be tested? Theories of particle physics now deal with levels of energy that existed only in the first instants of the Big Bang with which the Universe began. Yet experiments at even the largest accelerators, such as the Tevatron at Fermilab and the new LHC at CERN, reach only a million millionth of such energies. The difference between the energy of the LHC and the energies of the early Universe is as enormous as the span from the energies of processes in chemistry and molecular physics to the energy of the LHC. Yet, encouraged by the fact that we are in a sense half-way home, having spanned half the energy range, theorists can imagine a Herculean quest to bridge this gap with an all-encompassing theory.

The interest in ultra-high energies and the confidence in this quest stem from a belief that the laws of nature contain an elegant symmetry that is hidden below a certain temperature. The rich structures and variety that we experience in the cold world of chemistry and low-energy nuclear physics hide this deeper symmetry and simplicity. Thus, theory indicates that the first hints of the underlying laws and the true 'theory of everything' (or TOE) will be revealed at the highest energies now attainable at the Tevatron and eventually at the LHC. However, to realize the full symmetry within these fundamental laws will require experiments at extremely high energies and, therefore, temperatures – conditions unknown since the original hot Big Bang.

An analogy is the complete symmetry we find in a spherical drop of water in the weightless conditions of a spacecraft. It looks the same from every direction because there

Fig. 10.26 (LEFT) The perfect symmetry of water is elegantly shown in this high-speed flash photograph of a droplet formed from the column of water created by the impact of a previous droplet.

Fig. 10.27 (RIGHT) A snowflake has its own beautiful six-fold symmetry – rotate it through 60° and it will look the same, and after six identical rotations it will be returned to its original position. But this symmetry in frozen water is more restricted than the all-encompassing symmetry of the water droplet. Cooling the water to form ice crystals has 'hidden' the original symmetry.

is a symmetry in the laws that control the behaviour of water molecules. If the raindrop freezes and forms ice crystals to become a snowflake, symmetry is broken. Snowflakes have their own symmetry, but it is not complete; the snowflake does not appear the same from every direction. However, if you heat the snowflake so that it melts, the complete symmetry – hidden in the frozen form – is once again revealed.

There are signs of analogous 'melting' in the behaviour of elementary particles when they are heated. 'Heating' means colliding them at high energies, and it appears that the nature of their interactions is changing subtly as energy is increased. These observations, together with the mathematical elegance of the theories of electromagnetic, weak, and strong nuclear forces have led to the idea that these forces are intrinsically of the same strength. Only in our 'frozen' Universe is their symmetry hidden.

The mathematical formulations that predict and describe the unification of these three forces are resoundingly known as 'grand unified theories' or more prosaically as GUTs. All GUTs agree that grand unification occurs only at temperatures above 10^{28} degrees. Even the centres of stars, where elements are cooked in an inferno of some billion (10^9) degrees, are cool by comparison. At temperatures far below those of the GUTs, nature's laws and forces always appear asymmetrical, which is fortunate as far as we are concerned as our existence depends on the low-temperature disparity among the forces. The strong force grips together a compact nucleus; the less powerful electromagnetic force holds electrons remote in the periphery of atoms. Even feebler is the weak force responsible for radioactivity and the burning of the stars; it is feeble enough for the Sun to have survived long enough for life to evolve on Earth, but not so feeble that life never started at all.

Although temperatures at which grand unification occurs are still far beyond those attainable in particle accelerators, some hints of the GUT have emerged. When individual electrons and positrons collide at energies of 100 GeV, the effective temperature is some 10^{15} degrees, which is warm enough for the weak force to 'melt'. Weak radiation is then as 'liquid' as electromagnetic radiation, escaping in the form of the W and Z particles. This successful prediction of electroweak symmetry has given theorists confidence in the idea that the strong and electroweak forces become equivalent, in their turn, at a temperature of 10^{28} degrees. To realize such conditions in an accelerator would require particle collisions at an incredible 10^{15} GeV – clearly beyond the reach of present technology.

Testing the GUTs requires alternative approaches in parallel with 'low'-energy accelerator experiments. For instance we can observe cosmic rays that have been raised to extremely high energies by nature's own accelerators in the Universe at large (see p. 214). Another approach is to look for relics of the Big Bang, when the Universe was so hot that particles collided at the extreme energies meaningful to GUTs.

As GUTs have inspired particle physicists to take an interest in the early Universe, so have astrophysicists become aware of particles in the Universe at large. Cosmologists developing theories of how galaxies form and how the Universe evolved look to particle physics for ideas on how matter behaves. Measurements made by astronomers can in turn impose important constraints on theories such as GUTs.

The GUTs describe the strong and electroweak forces, but a valid TOE must also incorporate gravity. Developing such a theory is the cutting edge of theoretical research and it thrives on new ideas with exotic names such as supergravity and superstrings. Although compelling to many theorists, it is too soon to know to what extent these theories mirror the natural law of the Universe.

One important piece of the superstring theories is a new kind of symmetry – supersymmetry or SUSY. The GUTs imply that there are basically two families of particle – particles of matter (quarks and leptons) and force-carrying particles (the 'gauge bosons'). Supersymmetry links all of these particles within one 'superfamily', but it does so at the expense of predicting many more particles in the following way.

Fig. 10.28 Theodore Kaluza (1885–1945).

One feature that distinguishes the matter particles from the force carriers is the property known as 'spin'. Many particles behave like spinning tops but quantum theory dictates that they cannot spin at any arbitrary rate. Instead, they are constrained to spin only at certain 'allowed' rates, specific to each kind of particle. This spin can be measured experimentally and the table of properties at the end of this book (see pp. 230–233) gives the value for each particle expressed in units of Planck's constant, $h/2\pi = 1.055 \times 10^{-14}$ joule seconds. The electron and proton, for example, have spin 1/2 in these units whereas the photon and the W and Z particles have spin one. Indeed, the matter particles (quarks and leptons) all have spin 1/2, while the force carriers have spins of one.

In linking these particles of different spin, supersymmetry requires a host of matter particles and force carriers. It predicts 'supermatter' built from particles with integer spin (0, 1, 2, ...) rather than half-integer spin (1/2, 3/2, ...); and 'superforces' transmitted by agents with half-integer rather than integer spin. The search for such supersymmetric particles is high on the agenda at CERN and Fermilab.

What has this to do with gravity? The idea of supersymmetry grew out of detailed studies of the structure of space–time. Gravity is also intimately related to this structure and supersymmetry implies that general relativity – Einstein's theory of gravity – is but part of a richer theory. One consequence is that particles called gravitinos should exist; they would be related to the graviton, the hypothetical carrier of the gravitational force.

Fig. 10.29 Oskar Klein (1894–1977) in 1920.

The ideas of supersymmetry and supergravity may also lead theorists to understand why space has three dimensions. Einstein's theory of relativity follows from treating time as a fourth dimension. Are there further dimensions subtly intertwined with the familiar ones so that we do not perceive them? Some theories suggest that we may already be aware of the effects of additional dimensions. Over 50 years ago, Theodore Kaluza and Oskar Klein found that electromagnetism may be the effect of gravity 'spilling over' from a fifth dimension. They worked out a theory of gravity in five dimensions and then allowed one of the dimensions to 'curl up' and become imperceptibly small. The result of this 'compactification' was something that Kaluza and Klein recognized as electromagnetism. In a similar way the weak and strong forces may be the effects of gravity in higher dimensions.

More recently there has emerged the possibility of constructing a theory that contains all these bizarre ideas and more. There is the promise of a unique theory, which requires that the Universe began with ten dimensions of which only four expanded to form what we now call space and time. In this theory particles arise out of the basic mathematical structure not as point-like objects but as entities that are extended in space – albeit with dimensions of a mere 10^{-36} m. These extended objects are referred to as 'strings'. Supersymmetry is an essential ingredient of the theory, which has become known as the theory of 'superstrings'. Physicists are excited about the theory's potential because it promises to produce the long-sought marriage of gravity and quantum theory – essential in any TOE.

It has proven so difficult to formulate a theory of gravity that is consistent with quantum theory and relativity that the discovery of the superstring theory has encouraged theorists to suspect that this is indeed the unique true description. The challenge now is to make contact between this theory where the perfect symmetry is unblemished at extreme energies, and the rich families of particles and forces that emerge at the relatively 'frozen' temperatures attained at our most powerful accelerators on Earth.

11. Futureclash

In 1994, particle physicists at Fermilab in the USA caught their first glimpse of the top quark – the final piece in the quark puzzle which had begun 30 years previously with the work of Murray Gell-Mann and George Zweig. But just when this story came to a close, another began as, in the same year, the member states of CERN decided to build the Large Hadron Collider, or LHC. This would be the machine to take twentieth-first century CERN to a new high-energy frontier, where particle physicists could seek answers to fundamental questions. What is mass? What is quark–gluon plasma like? Is supersymmetry the cause of dark matter in the Universe? Where did all the antimatter go?

The LHC, which is planned to start in 2006, is being built in the 27 km tunnel that housed the Large Electron Positron (LEP) collider from 1989 to 2000. Unlike LEP, which accelerated lightweight electrons and positrons, the LHC will bring two counter-rotating beams of protons into head-on collisions. It will first accelerate the protons to energies of 7 million million electronvolts (7 TeV) per beam, so that they will collide at a total energy of 14 TeV. This is nearly 100 times greater than the energy of LEP's collisions, and nearly 10 times greater than the energy of proton–antiproton collisions at Fermilab, where the top quark was discovered.

Unlike any machine since CERN's pioneering Intersecting Storage Rings (the ISR, see p. 143), the LHC will make two beams of protons collide head on, rather than beams of particles and antiparticles. The advantage of the latter technique is that the particles and antiparticles naturally bend the same way when travelling in opposite directions through a magnetic field. So colliders such as LEP and the Tevatron have used particles and antiparticles in a single ring of magnets. But antiprotons have first to be made, while protons come simply from hydrogen gas, and it is difficult to make very intense beams of antiprotons. So to make as many particles as possible collide together, the LHC will use only protons. However to make the two beams travel in opposite directions around the ring requires opposite magnetic fields. This would normally imply two rings of magnets, as in the ISR or in RHIC at Brookhaven (see p. 196), but the LHC will use a single ring of magnets to supply adjacent magnetic fields in opposite directions. The magnets will be threaded by twin vacuum chambers to provide the channels for the two proton beams. This novel design, with two beams within a single magnet rather than two separate rings of magnets, has helped to keep down costs and make possible more efficient use of the limited space in the tunnel.

In the LHC, as with LEP, CERN will be capitalizing on existing accelerators, including the Proton Synchrotron (the PS, see p. 97), which will have been working for nearly half a century when the LHC starts. The protons will exit the PS at energies of 26 GeV, and then pass into the Super Proton Synchrotron (the SPS, see p. 102) for acceleration to 450 GeV. The SPS will feed the protons into the LHC to be whirled up to their final energy of 7 TeV. Two beam lines will link the SPS to the LHC, enabling the protons to circle the big machine in both clockwise and anticlockwise directions. Once in the LHC, the beams will be guided by 1800 superconducting bending magnets, each 14 m long.

Superconducting magnets can reach much higher magnetic fields than magnets made from a normal conductor such as copper (see p. 81). These electromagnets are among the most challenging pieces of technology in the accelerator. They are being made from a mixture of the metals niobium and titanium and, at an operating temperature of 1.9 degrees

Fig. 11.1 (OPPOSITE) A prototype section of the Large Hadron Collider (LHC) under test at CERN. The LHC, due to start up in 2006, will accelerate counter-rotating beams of protons to an energy of 7 TeV (7000 GeV), and then bring them to collide head on. Although the LHC will have a gently curving circumference of 27 km, superconducting magnets are needed to bend the high-energy proton beams at these high energies. The magnets will operate at very low temperatures, cooled by liquid helium to only 1.9 degrees above absolute zero. The 'test string' here consists of two focusing (quadrupole) magnets and six bending (dipole) magnets within a 106 m long section of the large pipe-like cryostat needed to contain the ultracold liquid helium.

Fig. 11.2 The LHC will contain two proton beams travelling in opposite directions but within a single ring of magnets. The magnets will have a novel '2-in-1' design, incorporating oppositely directed magnetic fields side by side, to control the oppositely curving beams. This engineering model shows the design of the superconducting dipoles – the bending magnets – with the two holes for the twin beam pipes for the counter-rotating beams.

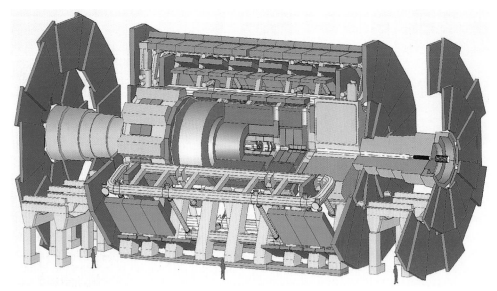

above absolute zero, they will be colder than outer space. To reach these extreme temperatures, liquid helium will be used as refrigerant. The magnets will sustain a magnetic field of 8.36 tesla – greater than anything previously used in a particle accelerator.

Huge detectors will be housed at the collision points. Two 'general purpose' detectors, called CMS (for Compact Muon Solenoid) and ATLAS (for A Toroidal LHC ApparatuS), will explore the new energy region looking for all kinds of new effects – both expected and unexpected. Two further detectors, LHC-b and ALICE (for A Large Ion Collider Experiment), will be more specialized, seeking to investigate specific puzzles in particle physics. In particular, ALICE will observe collisions of complex lead nuclei, rather than simple protons, in the hope of glimpsing signs of quark–gluon plasma (see p. 193), the high-energy state of matter that should have existed in the early Universe.

The technological challenges of these gargantuan precision 'particle cameras' are hardly less than that of constructing the 27 km ring of magnets and the associated infrastructure for the LHC. The ATLAS detector will be five stories high (20 m) and yet able to measure particle tracks to a precision of 0.01 mm. Its innermost sensors will contain nearly as many transistors as there are stars in the Milky Way – some ten thousand million. These and other detectors, together with sophisticated electronics, will measure the energies, directions, and identities of the hundreds of particles produced when the two beams of protons collide at the heart of the apparatus.

In the proton experiments, bunches of particles will pass through each other 40 million times a second and each time they cross there will be up to 25 collisions, making nearly a billion collisions per second in all. The ensuing data collection rate demanded of the detectors is equivalent to the information processing for 20 simultaneous telephone conversations by every man, woman, and child on Earth. However, only 1 in 20 million collisions is likely to produce new phenomena, and only one in a million of these, for example, might produce a Higgs boson – a particle that must exist if mass is due to the so-called Higgs mechanism (see pp. 192–193). This means that with up to a billion collisions each second, a Higgs boson would appear about once a day in each experiment at the LHC. The computers in each experiment must recognize this veritable needle in a haystack and record only selected data onto magnetic tape. The ability to do this has set new challenges for automatic data handling.

The CMS detector is more compact than ATLAS – ATLAS weighs 7000 tonnes but contains much gas and could float, whereas the more compact CMS weighing 12 500 tonnes would sink. However, CMS has its record-breaking features. Central to its design is the biggest superconducting magnet with the highest field for its size and the highest stored energy ever. The solenoid, or coil, is 6 m in diameter and 13 m long and the magnetic field of 4 tesla stores 2.5 thousand million joules of energy – enough to melt 18 tonnes of gold!

CMS and ATLAS each follow the time-honoured structure for modern particle detectors. There are three main features that a particle might meet as it moves outwards from the centre of the cylindrical structure where the collisions will occur. Each of these main layers is customized to recognize and record the different classes of particles.

First comes the logically named 'inner tracker'. This records the positions of electrically charged particles to an accuracy of about one hundredth of a millimetre, enabling computers to reconstruct their tracks. The intense magnetic fields cause the tracks of the particles to curve, and this in turn reveals the momentum of the particles.

The next layer is a two-part calorimeter, designed to capture all the energy of many types of particle. The inner part is the electromagnetic calorimeter, which traps and records the energies of electrons and photons. Strongly interacting particles – hadrons – tend to penetrate this first section of the calorimeter and escape to the outer hadron calorimeter. Here hadrons deposit most of their energy and come to a halt, leaving only muons and neutrinos to continue.

The third, outermost layer consists of special muon chambers, which track muons, the only electrically charged particles that can penetrate this far. The neutral neutrinos, on the other hand, escape from the detector entirely unseen. However, their existence can be inferred by adding up the measured momenta of all the other particles and balancing the sum, as momentum is conserved overall.

The scale of these projects is huge in terms of human endeavour. A total of some 1700 physicists from 150 universities and research institutions in 39 countries on six continents are collaborating on ATLAS alone; similar numbers are involved with CMS. With physicists and engineers around the world working on components of ATLAS and CMS, it is probably true that the Sun never sets on preparations for the LHC, which could bring its first glimpses of a new energy region in 2006.

Particle Factories

One of the major mysteries of existence is where all the antimatter has gone. To explain the dominance of matter over antimatter in the Universe requires a way to distinguish these two forms of substance. Nature provided a clue over 30 years ago when the decays of the neutral strange particles known as kaons (see pp. 74–75) revealed a subtle difference – a lack of symmetry for a combined particle–antiparticle changeover ('charge conjugation' or C) and mirror reflection ('parity' or P). This is the asymmetry known as CP violation (see p. 191).

That was in 1964, and for more than 30 years it remained the only direct observation of an asymmetry between matter and antimatter. However, the discovery that there are three pairs of quarks led theorists to realize that the effect should also occur in neutral B mesons (see pp. 166–167). These are particles analogous to kaons, but where the strange quark is replaced by the heavier bottom quark.

To test these ideas, experimenters have devised ways of producing as many kaons or B mesons as possible in particle 'factories'. These machines are custom-designed to produce large numbers of one variety of particle. The idea is to make electrons and positrons collide at specific energies, 'tuned' to the masses of neutral particles that contain one variety of quark – strange or bottom – bound with its antiquark. The decays of these neutral particles will then produce kaons or B mesons, respectively, in preference to other kinds of particles. By the end of the twentieth century three such factories had been built and begun operation: DAFNE in Italy, PEP-II at Stanford, and KEKB in Japan.

DAFNE, which stands for Double Annular Phi Factory for Nice Experiments (*sic*), has electrons in one ring and positrons in a second interlaced ring (the 'Double Annular'). The complete machine is relatively small, contained within less space than half a football field. At the two points where the rings cross, the beams mutually annihilate with a total energy of about 1 GeV. This is 200 times less than the final energy of the Large Electron Positron collider at CERN, but it coincides with the energy at rest ('rest mass') of the short-lived phi meson, a particle that consists of a strange quark bound with a strange antiquark.

The all-important feature of the phi meson is that it rapidly decays into a kaon, K, and its

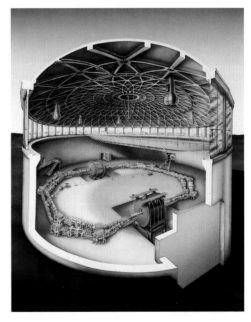

Fig. 11.5 An artist's impression of the two interlaced magnet rings of the DAFNE 'phi factory' at the Frascati Laboratory in Italy. One ring is to accelerate and store electrons, while the other operates on positrons travelling in the opposite direction. The two beams collide head on at the centre of detectors shown here on opposite sides of the ring. The beams collide with a total energy equivalent to the mass of the phi particle, which consists of a strange quark and antiquark, and soon decays to kaons.

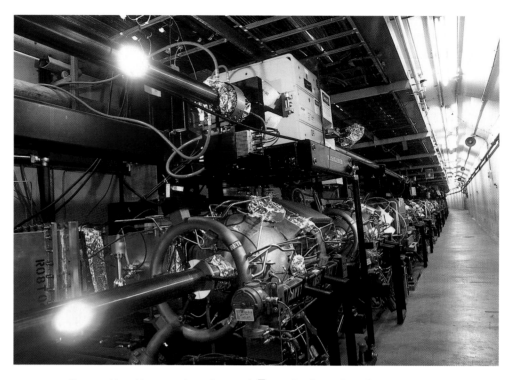

Fig. 11.6 The PEP-II machine at SLAC contains counter-rotating electron and positron beams which are accelerated separately in two rings. The upper ring here contains the positron beam, the lower ring the electron beam. With two rings, the particle beams can be accelerated to different energies before being brought to collide head on. The electrons and positrons annihilate at the correct total energy to rematerialize as a bottom quark and antiquark, bound together as an excited upsilon particle (see p. 164). However, as the initial particles have different energies, the quark–antiquark pair is born in motion, and the B mesons they form travel onwards in the direction of the highest energy beam (the electron beam). This gives the B mesons extended lifetimes due to 'time dilation', an effect of special relativity. (The bright pink and blue spots simulate the passage of bunches of positrons and electrons.)

corresponding antimatter version, denoted $\overline{\text{K}}$. And when the phi is at rest, as it is at DAFNE, the kaon and antikaon move off back to back at equal speeds and with equal energies. With its intense beams tuned in this way, DAFNE was designed to produce up to 50 billion phi mesons each year, leading to 100 billion kaons and antikaons, or about 3000 each second! With numbers like these the experiments can measure with great precision the subtle differences between the decays of the K and $\overline{\text{K}}$ first glimpsed in 1964.

The asymmetry in the kaons is very small. However, theory predicts that a similar but much larger effect should occur in B mesons, where the heavier bottom quark replaces the strange quark. The B meson is some nine times more massive than a kaon and so needs more energy to make it. For this reason it has not been possible to study B mesons precisely enough to see the breakdown of CP symmetry until recently.

A 'B factory' makes electron–positron collisions at a total energy of around 10 GeV, optimized to produce B mesons and their antiparticles ($\overline{\text{B}}$) together. So compelling is the challenge of CP violation that two machines were built in the late 1990s – PEP-II at SLAC in California and KEKB at the KEK laboratory in Japan. Each has its custom-built detector – called BaBar at SLAC and BELLE at KEKB.

Fig. 11.7 This view of the BaBar experiment at SLAC, taken during maintenance, shows its basic cylindrical structure, with layers of detectors which surround the central beam pipe when the complete apparatus is in position at the PEP-II storage ring. The grey sections provide the iron for the magnet, but also contain layers of detectors to measure the energy of hadrons (particles made from quarks). The two outer 'doors' form the end-cap, which is also made from iron interleaved with detectors.

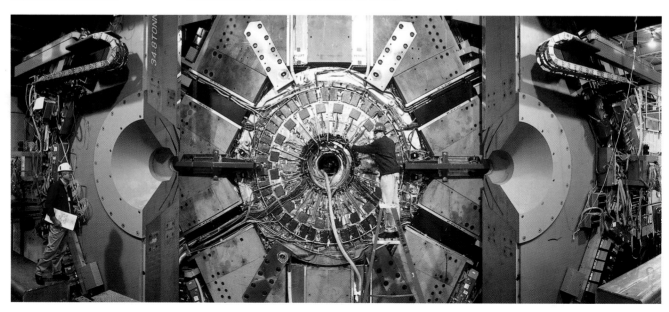

Fig. 11.8 The BELLE experiment surrounds a collision region at the B factory at KEK. The apparatus has the cylindrical layered structure typical of experiments at particle colliders. Here two endplates are left open while a technician works on one of them, revealing the central structure. The shiny circular plate marks the end of the central tracking chamber, with a radius of 90 cm. Surrounding this, the large iron yoke of the superconducting magnet is clearly visible. The iron here is interspersed with detectors to register muons and long-lived neutral kaons.

The B factories differ from previous electron–positron colliders in an intriguing way. In a standard electron–positron collider, the beams travel in opposite directions but with the same speed, so that when particles meet their motion exactly cancels out. The resulting 'explosion' when the electrons and positrons mutually annihilate is at rest, and newly created particles of matter and antimatter emerge rather uniformly in all directions. In the B factories, the colliding beams move with different speeds, so the resulting explosion is itself moving. The matter and antimatter that emerge tend to be ejected in the direction of the faster initial beam, and at higher speeds than from an annihilation at rest. This makes it easier to observe not only the particles created, but also the progeny they produce when they die – thanks to an effect of special relativity (time dilation, see p. 13) which means that particles survive longer when moving at high speed. These are essential tricks because a B meson, at rest, lives only for a picosecond, a millionth of a millionth of a second, and this is on the margins of visibility.

The strategy behind the experiments at the B factories is based on the fact that at the moment of creation a B and $\bar{\text{B}}$ are together before they fly out from the explosion. In their brief lives, however, the quirks of quantum mechanics come into play and it turns out that what starts out as a B can change, like a chameleon, into a $\bar{\text{B}}$. So the experiments compare how the B and the $\bar{\text{B}}$ evolve, a difference in their evolution patterns giving a measure of CP violation.

In July 2001 the BaBar and BELLE experiments announced clear evidence for CP violation in neutral B mesons – the first observation of this effect in particles other than kaons. Moreover the amount of CP violation measured in both experiments agreed with the theoretical predictions based on the CKM matrix – the mathematical matrix that links the weak interactions of the different quarks (see p. 192). The challenge of CP violation in B mesons goes beyond this, however. The aim is to see CP violation in the many different ways that B and $\bar{\text{B}}$ mesons can decay. By measuring the relative probabilities and other properties of the decays, the experiments at B factories will determine the values of key parameters in the CKM matrix – parameters that are directly linked to the origin of CP violation.

Fig. 11.9 A 'golden event' in the BaBar detector at SLAC, of the kind that has played a key role in the experiment's studies of CP violation. This display shows particles produced in the decays of a B meson and an anti-B meson created in an electron–positron collision at the centre of the detector. One meson has decayed into a J/psi particle and a neutral kaon, each of which has decayed to produce the orange tracks. The J/psi has decayed almost immediately into a pair of muons – the widely spaced orange tracks at right. The neutral kaon has decayed into a pair of charged pions – the closely spaced orange tracks at left. The other B meson has decayed into a negative kaon (the red track that curls up at right) and three charged pions (the other red tracks).

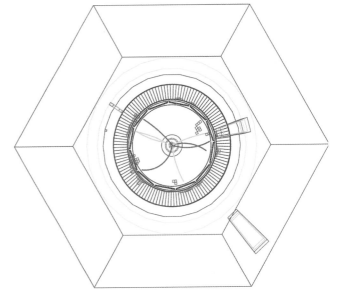

The Large Hadron Collider and the B-particle factories are the latest in a line of accelerators that began in the 1950s, stimulated by discoveries of new particles, such as pions and kaons, in cosmic-ray experiments. Now it is the turn of neutrinos from cosmic sources to stimulate ingenious new ideas for experiments using accelerators here on Earth. The aim is to measure 'neutrino oscillations' – the quantum switching between neutrino types that was initially suggested by the shortfall of electron-neutrinos from the Sun and of muon-neutrinos produced by cosmic rays (see p. 201). Physicists need to be able to test the exciting implications of these experiments with 'designer neutrino beams' provided by particle accelerator laboratories.

Theory shows that neutrinos oscillate more rapidly if the three types of neutrino have a large difference in mass. However, we know that neutrinos have small masses, so the differences in masses must at best be small too. This means that we must observe neutrinos over large distances if we are to see the oscillation develop from one type to another. To have any chance of detecting oscillations in neutrinos produced at particle accelerators, it is proving necessary to place the detector far away from the accelerator, at distances of hundreds of kilometres, perhaps even in another country!

Neutrinos are produced when pions or kaons decay. So the first step in making neutrinos is to make a beam of pions and kaons by directing a proton beam from an accelerator at a suitable target. The decays of the short-lived pions and kaons lead mainly to muon-neutrinos and one possibility is to see if this neutrino type turns into a tau-neutrino. This kind of oscillation would be revealed by charged tau particles produced when the tau-neutrinos interact with matter.

In the 1990s, two experiments at CERN, called CHORUS and NOMAD, searched for oscillations of muon- to tau-neutrinos in a neutrino beam produced about 1 km away at the Super Proton Synchrotron (SPS). The accelerator generated a beam of muon-neutrinos, so the arrival of a tau-neutrino in either experiment would be proof of oscillation. The tau-neutrino would be recognized through its production of a tau particle, although the tau would decay almost straight away, within about 10^{-13} s, to lighter particles.

The two experiments ran from 1993 to 1998, but found no examples of tau particles. So if muon-neutrinos really do oscillate to tau-neutrinos, the mass difference involved must be

Fig. 11.10 The Sun in 'neutrino light' recorded by the Super-Kamiokande neutrino detector over a period of 500 days. The image was created by plotting the difference in the Sun's position 'horizontally' (right ascension) and 'vertically' (declination) and the neutrino's direction as inferred from the electrons produced in neutrino interactions in the detector. One pixel corresponds to 1 degree, which is the size of the Sun in the sky, so the image is clearly much larger than the actual Sun. This is due to scattering of the electrons, which smears out the information on their direction.

Fig. 11.11 This computer display shows the first long-distance detection of an accelerator-produced neutrino, in the Super-Kamiokande detector. The cylindrical detector is shown as if it had been opened like a can at both ends, and its wall slit and unrolled. The two circles represent the top and bottom of the detector, the rectangle depicts the wall of the cylinder. The neutrino, in a beam of muon-neutrinos from the KEK accelerator 250 km away, has interacted in the water in the detector to produce two charged particles. These create rings of Čerenkov light on the internal surfaces of the detector. The display shows phototubes that have produced signals, the colour indicating the timing, with red corresponding to signals about 0.5 microseconds later than yellow-green. Two yellow rings are visible – a small almost solid one near the junction with the top of the detector, and a larger one to its left which spans the wall and the top. These rings are due to the two charged particles from the initial muon-neutrino interaction. A third partial ring in red, corresponding to a later time, is due to the electron produced when one of the two charged particles (a muon) decays.

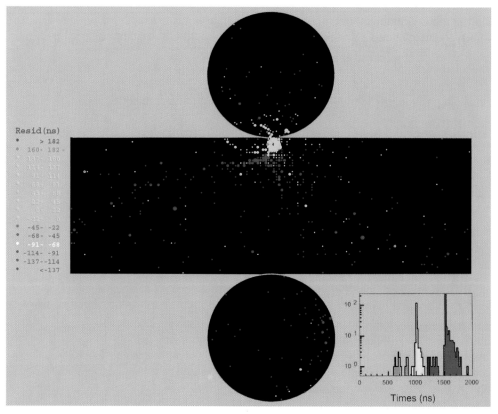

Fig. 11.12 One of the 484 giant steel plates that will form part of the MINOS long-baseline neutrino experiment in the Soudan Mine in northern Minnesota. The complete experiment will detect neutrinos in a beam from Fermilab, 730 km away in Illinois. Each steel plate is 8 m across but only 25 mm thick. With a weight of 11.25 tonnes, a challenge has been to support the sheets, as they can crumple like paper. When complete the detector will form a 5400 tonne sandwich of steel and scintillator – the steel to provide a dense target for the weakly interacting neutrinos, and the scintillator to detect the rare neutrino interactions.

too small for the oscillations to develop over the distance between the SPS and the CHORUS and NOMAD experiments. It seems that much bigger distances or 'baselines' are needed, if the high-energy neutrino beams from an accelerator are to reveal neutrino oscillations.

June 1999 saw the start of the world's first 'long-baseline' experiment with neutrinos from an accelerator. In K2K – 'KEK to Kamioka' – a neutrino beam created at the KEK laboratory travelled 250 km westwards under the Hida Sammyaku (the Japanese 'Alps') to the Super-Kamiokande detector, or Super-K (see p. 201). This huge detector was constructed principally to study neutrinos travelling 150 million km from the Sun, but it has also provided persuasive evidence for the oscillation of neutrinos made 13 000 km away in the atmosphere on the opposite side of the Earth. With the beam from KEK it was able to study much shorter 'baselines' of 250 km, until November 2001, when a major accident destroyed more than half its 11 200 phototubes.

A second long-baseline experiment, due to start in 2003, will observe man-made high-energy neutrinos at a greater distance, this time in the US. The MINOS experiment will use a high-intensity neutrino beam from Fermilab's Main Injector (see p. 82). MINOS (for Main Injector Neutrino Oscillation Search) consists of two detectors – a small one close to the source of neutrinos at Fermilab and a large one 710 km from Fermilab, in the Soudan Mine in Minnesota. To reach the far detector, the neutrinos will cross two state boundaries, starting in Illinois and traversing under Wisconsin before arriving in Minnesota.

Both K2K and MINOS are looking for a reduction in the numbers of muon-neutrinos that reach the distant detectors. An alternative technique is to look for the appearance of tau-neutrinos in the muon-neutrino beams, by detecting the tau particles they produce in their rare interactions. This is the approach being taken in two experiments at the Gran Sasso Laboratory, which is about 730 km from CERN, under the Gran Sasso massif north-east of Rome. CERN is building a new neutrino beam line that points in a south-easterly direction into Italy and which could be ready for the experiments at Gran Sasso Laboratory in 2005.

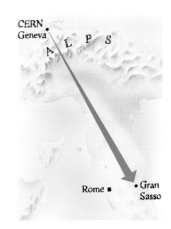

Fig. 11.13 The 730 km route of the neutrino beam planned to skim beneath the Earth's surface between CERN, near Geneva, and the Gran Sasso Laboratory in Italy.

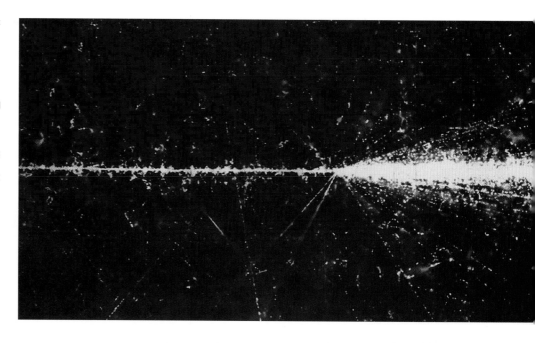

Fig. 11.14 A very high-energy cosmic ray iron nucleus shoots into some photographic emulsion (from the left) and collides with a silver or a bromine nucleus to produce a tremendous 'jet' of about 850 mesons. From the divergence of the jet, it is possible to estimate the total energy of the incoming iron nucleus as more than 15 000 GeV. But this is puny in comparison with the rarer ultra-high-energy cosmic rays. In this magnified false-colour image, the central bright core of the jet is about 0.04 mm across.

Particle Astronomy

The study of cosmic rays gave birth to particle physics in the 1930s and 1940s when the rays revealed new particles such as the positron, the muon, and the kaon. But with the development of high-energy accelerators in the 1950s particle physicists and cosmic ray physicists tended to go their separate ways. While the particle physicists concentrated on studying the products of man-made collisions, the cosmic ray experts addressed the questions of the composition and origins of the rays. Today, however, particle physicists looking to energies beyond the reach of their accelerators are once again taking an interest in the cosmic radiation.

Cosmic rays can have awesome power. In some regions of the cosmos, nature somehow manages to generate cosmic rays with energies as high as 100 billion (10^{11}) GeV. This is as much energy in one tiny particle as is carried by a tennis ball served by a top player. How nature can impart such energies to single particles is a major mystery.

Physicists believe that cosmic rays with energies up to about 100 000 (10^5) GeV are probably accelerated in the shock waves from supernovae, the explosive last acts of heavier stars. However, there is evidence that this is not the whole story, and at the highest energies

Fig. 11.15 This unusual looking telescope at the Whipple Observatory in Arizona has pioneered studies of cosmic gamma rays through the 'air Čerenkov' technique. The 10 m telescope is divided into 109 separate reflectors, which focus light onto an array of photomultiplier tubes. On dark nights, the telescope can detect the faint Čerenkov radiation emitted by the shower of energetic charged particles created when a high-energy gamma ray from outer space crashes into the atmosphere. The charged particles can travel through the air faster than light does, and so radiate cones of Čerenkov light. The time of the light's arrival at the separate reflectors gives information on the general direction of the shower, and hence of the original gamma ray.

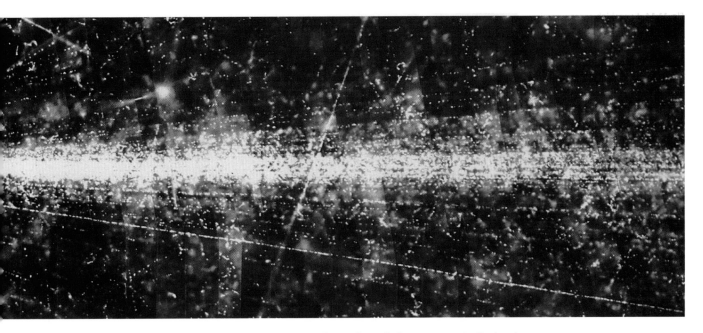

the source of the cosmic rays becomes much more puzzling. The only known way to find out what mechanisms can whip particles up to such extremely high energies is to investigate what happens when the cosmic rays impinge upon the Earth's atmosphere.

At energies below about 10 billion (10^{10}) GeV, electrically charged cosmic ray particles are deflected by the magnetic fields in our Galaxy, so the direction from which they arrive gives no indication of their source. By contrast, electrically neutral cosmic rays – mainly gamma rays and neutrinos – are unaffected by these fields. When they arrive at Earth, these neutral rays should point directly back to the place where they were created. Several teams of physicists are therefore building neutrino and gamma-ray telescopes in an effort to find out more about nature's enigmatic cosmic accelerators.

So far, the most progress has been made in identifying sources of high-energy gamma rays, with energies up to around 10 000 (10^4) GeV. When a gamma ray strikes the upper atmosphere at high energies, it generates a shower of particles, mainly electrons and positrons, which travel faster through the air than light does. The charged particles that exceed the speed of light in this way produce Čerenkov radiation (see p. 91), and they create a pool of light that travels along the direction of the shower. The total amount of light is tiny and by the time it reaches the Earth's surface it is spread over an area of about 100 000 square metres. But on a clear dark night, it can be detected.

The Whipple Observatory in Arizona pioneered this technique, studying showers produced by gamma rays with energies up to about 10^4 GeV. The 10 m 'air Čerenkov telescope', which is divided into 109 separate reflectors, could reveal the direction of the shower – and hence the gamma-ray source – by measuring the varying arrival times of light across the telescope. During the 1990s, this instrument discovered several sources of gamma rays, including the Crab Nebula (a supernova remnant) and some active galactic nuclei, including Markarian 421. Encouraged by this success, the team is planning the VERITAS project in which eight 10 m telescopes like the one at the Whipple Observatory will together act as a much larger, more sensitive telescope.

Unlike air Čerenkov telescopes, neutrino telescopes bear little resemblance to the image conjured up by the word 'telescope'. High-energy neutrinos can be 'seen' when they interact to produce muons, but their interactions are rare, so that even in the intense beams created at particle accelerators, physicists must use big detectors. However, the number of neutrinos arriving from any one distant cosmic source must be very small – just as the light from a distant star is faint compared with that from the nearest star, the Sun. The search for neutrinos from cosmic sources therefore requires extremely big detectors, so big that experimenters are incorporating vast volumes of the natural world, in the form of water or ice, into their neutrino telescopes.

Fig. 11.16 The Crab Nebula, photographed in visible light and colour enhanced to reveal its structure. The green, yellow, and red filaments are gaseous remnants of a supernova that was seen in 1054 AD. This nebula is one of several sources of high-energy cosmic gamma rays that have been detected by the gamma-ray telescope at the Whipple Observatory in Arizona.

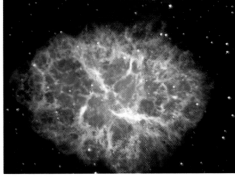

Fig. 11.17 (LEFT) A computer reconstruction reveals the path of a muon – the blue line – detected by phototubes in the AMANDA experiment at the South Pole. The display shows the first four 'strings' deployed during the Antarctic summer of 1995–96, each with 20 phototubes indicated by the short horizontal white lines (the phototubes that detected light in this event are shown by purple spots). The strings form a three-dimensional array. The numbers are the time in nanoseconds (billionths of a second) when each tube was 'hit' by light after the first 'hit'. This information is used to calculate the direction of the Čerenkov light (the purple lines) emitted by the muon. The muon is going upwards through the array, having been created when a neutrino, probably produced in the atmosphere on the other side of the Earth, interacted in the rock or ice beneath the detector.

Fig. 11.18 (RIGHT) An artist's impression of the ANTARES detector planned for deployment nearly 2.5 km deep in the Mediterranean, off the South of France. The phototubes point down towards the ocean floor to register Čerenkov light from muons moving up through the array, having been produced by neutrinos that have travelled from the other side of the Earth. The aim is to detect neutrinos from distant cosmic sources.

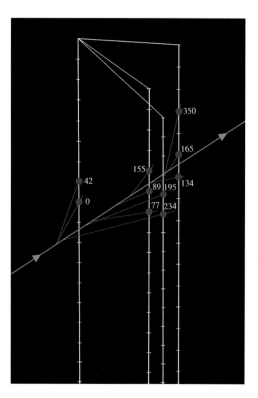

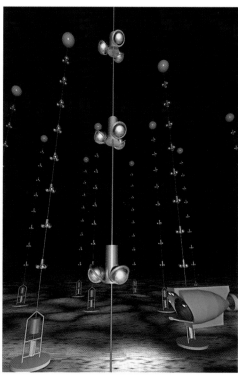

Down near the South Pole, a detector called AMANDA (described first in Chapter 1, pp. 4–5) uses the Earth as a 'target' for cosmic neutrinos, and ice to detect the muons they produce. Like anywhere on Earth, the South Pole is bathed in a constant rain of cosmic-ray muons, which are produced mainly in the decays of pions and which can penetrate deep into the ice. But AMANDA is designed to recognize muons produced by the interactions of cosmic neutrinos, and to do this the trick is to look for muons travelling up, rather than down, through the ice. Although muons are penetrating particles, they cannot traverse the Earth, so the upward-going muons can only have been created when neutrinos from sources on the opposite side of the Earth interacted in the rock or ice beneath the detector.

The high-speed muons emit Čerenkov radiation as they travel through the ice, and this is picked up by an array of phototubes. Careful timing of the signals from the tubes picks out the rare upward-going muons, derived from neutrinos that have passed through the Earth, from among the far more common downward muons in the cosmic radiation. Installing the phototubes in the ice, to depths of 2400 m, is a challenge in itself. The AMANDA team lowers 'strings' of regularly spaced phototubes down holes drilled in the ice by high-pressure hot water – before it all freezes!

The first phase of AMANDA, with 10 strings of phototubes covering a detection area of about 10 000 square metres, was completed in the Antarctic summer of 1996–97. The second phase, with an eventual area several times larger, was begun the following year.

AMANDA, at the South Pole, is searching for neutrino sources that are on the far side of the Earth, hence in the northern skies. A survey of the southern skies requires a detector in the planet's northern hemisphere, where experimenters have considered building similar detectors using water rather than ice. This at least gives them the opportunity to consider more hospitable surroundings for their experiments! A relatively small neutrino detector, with an area of a few thousand square metres, has been working since 1998 in Lake Baikal, the world's deepest freshwater lake. However, a larger detector is needed to be truly sensitive to distant neutrino sources, so plans are underway to build a big array of phototubes in deep natural trenches in the Mediterranean Sea. The ANTARES collaboration (for Astronomy with a Neutrino Telescope and Abyss environmental RESearch) is building a detector off the south coast of France, near Toulon, while NESTOR (the Institute for Deep Sea Research, Technology and Astroparticle Physics) will exploit the deepest part of the Mediterranean to the south-west of the Peloponnese in Greece. With these 'telescopes' in full operation, the study of neutrinos could become a fully-fledged branch of astronomy.

Cosmic Record-breakers

At lower energies, the paths of charged cosmic rays (by far the majority) are twisted and turned by the magnetic fields in space, so that by the time they reach Earth their directions bear no clues to their origins. But at the highest energies – above 10 billion (10^{10}) GeV – the magnetic fields are not strong enough to bend the paths significantly. So these ultra-high-energy charged cosmic rays should point back to the exotic parts of the cosmos from whence they came, just as gamma rays and neutrinos do. Unfortunately, cosmic rays at such high energies are very rare. Indeed, at energies greater than 10^{10} GeV we can expect only one cosmic ray per square kilometre per century! Yet interest in them is so great that physicists are prepared to think big, with detectors planned to cover areas as great as 3000 square kilometres – as big as greater London, or Long Island, New York.

When a high-energy cosmic ray shoots down through the upper atmosphere, it generates an avalanche of subatomic particles; a primary cosmic ray with an energy of 10^{10} GeV creates a shower containing up to 10 billion particles by the time it reaches sea level. These particles spread out sideways while preserving the direction of the main thrust. A snapshot of the shower would reveal it as a thin disc of particles moving towards the ground at nearly the speed of light. The disc can be several kilometres across, and its leading edge reaches the ground before the trailing edge. By measuring the relative arrival time of particles at several widely separated detectors, physicists can determine the direction of the shower to within two or three degrees. The total energy of the shower can be determined once its proximity and intensity are known.

One question that has been raised by studies of these 'extensive air showers' is whether there is an upper limit to the energies of cosmic rays. We know that the primary cosmic radiation consists of protons and nuclei, but these electrically charged particles should be cut off at energies above about 40 billion (4×10^{10}) GeV for a rather exotic reason.

The Universe is bathed in a 'background' of microwave radiation, which is a relic of the high temperatures of the Big Bang. But a proton rushing through this background at a velocity close to the speed of light will be confronted by gamma rays rather than the cool microwaves that we detect on Earth. This is because as far as the proton is concerned the background radiation is rushing past it, and this has the effect that the low-frequency microwaves appear as high-frequency gamma radiation. And when a gamma ray hits a proton it is absorbed, and pions are emitted: the proton slows down. This implies that any proton travelling through the microwave background with an energy above 4×10^{10} GeV will not last more than about a 100 million years without being slowed. This may sound a long time but it is brief compared with the 15 billion years of the Universe. It is surprising

Fig. 11.19 The shower of particles created when a high-energy primary cosmic ray strikes the upper atmosphere moves towards the ground at close to the speed of light. As it progresses, the shower broadens as the number of particles increases through secondary interactions. At any moment in time the shower is like a disc of particles, which becomes wider as it proceeds. When it arrives at the Earth's surface, detectors can sample the disc of particles and time their arrival in order to reconstruct the direction of the primary cosmic ray track.

Fig. 11.20 The main part of the Fly's Eye detector, in Utah. The large 'cans' contain mirrors – some appear white as they reflect the surrounding snow – which focus light on a small array of phototubes (the small rectangles). The mirrors point in different directions to cover adjacent patches of the sky, rather like the compound eye of a fly. On moonless nights the detector would register faint light from nitrogen fluorescence when a shower of particles from a very high-energy cosmic ray swept across its field of view.

that any cosmic rays of such high energy reach Earth at all. However, around 1980, the array of extensive air shower detectors at the Haverah Park experiment near Leeds in England found the first hints that the spectrum of cosmic rays might continue beyond this energy.

During the 1990s, other detectors confirmed the existence of cosmic rays with unexpectedly high energies. The highest energy so far, of 300 billion (3×10^{11}) GeV, was detected on 15 October 1991 by an instrument known as the 'Fly's Eye', which was able to track cosmic ray showers on dark, moonless nights. The Fly's Eye, in the Dugway Desert in Utah, in fact consisted of two 'eyes' four kilometres apart. The main 'eye' consisted of 67 mirrors mounted in an array of large 'cans'. Phototubes at the focus of each 1.5 m diameter mirror picked up reflected flashes of light – scintillations – produced as the cosmic rays pass through the atmosphere.

This is nothing more than an old technique put to new use. More than 90 years ago, Rutherford detected alpha particles by the faint flashes, or scintillations, emitted when the particles collided with atoms in a zinc sulphide screen. Nitrogen in the air also scintillates when electrically charged particles pass through but it does so very weakly: it gives off five photons for every metre along the track of a very high energy electron. The flashes are too faint to see, but they can be detected by modern high-quality phototubes coupled to sensitive electronics.

The Fly's Eye was so called because its view of the whole sky, built from the overlapping segments seen by each mirror, is like the scene from a fly's compound eye. On clear moonless nights the detectors could see cosmic ray showers streaking across the sky more than 20 km away. A computer recorded how much light triggered the various phototubes and in what sequence; and from this information it could reconstruct the flight of the shower and the direction of the primary cosmic ray. The Fly's Eye, which ran from 1982 to 1992, was so successful that it has now been replaced by a bigger version, known as HiRes, which has one 'eye' at the original site in western Utah, and another 12.5 km away. Together, the two sites will have 64 'mirror units', each with four glass segments that together synthesize a spherical mirror viewed by 256 phototubes.

Fig. 11.21 A map of the particle density measured for the highest-energy cosmic ray observed by the Akeno Giant Air Shower Array (AGASA) in Japan. The cosmic ray – calculated to have had an energy of 2×10^{20} eV – created a shower of particles spreading across an area nearly 6000 m square. This is the second-highest-energy cosmic ray ever recorded. (The radius of each circle represents the logarithm of the number of particles per square metre at each of the detectors in the array, with more than 20 000 at the centre of the shower.)

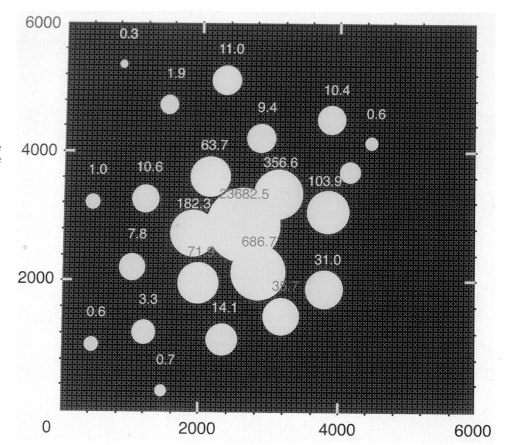

Ten years of data from the Fly's Eye showed no sign of any sharp cut-off in the cosmic ray spectrum. A similar tale has also come from data from the world's largest cosmic ray detector, which began its hunt for ultra-high-energy cosmic rays in 1991. The Akeno Giant Air Shower Array (AGASA) consists of 111 separate particle detectors spread over 100 square km near Akeno, in Japan, and it boasts detection of the second-highest-energy cosmic ray so far, with an energy of 2×10^{11} GeV.

In an attempt to capture more of these extremely elusive record-breaking cosmic rays, an international team began setting up an even bigger detector in Argentina in 1999. The Pierre Auger Project, named after the French physicist who first detected cosmic ray air showers in 1938, will combine the Fly's Eye technique with a vast array of 1600 particle detectors. When complete, the array will cover 3000 square kilometres. It will map out the energy spectrum and directions of rays in the southern heavens whose energies exceed 10^{10} GeV, and should detect about 30 cosmic rays with energies above 10^{11} GeV. The eventual plan is to complete the picture of the cosmic ray sky by building a similar observatory in Utah, to view the Northern Hemisphere skies.

The Fly's Eye and AGASA have already confirmed that cosmic rays with energies above 4×10^{10} GeV manage to reach Earth while avoiding the microwave background radiation. This implies that the particles are relatively young – less than 100 million years old. But where can they come from? The fact that the magnetic fields in our own Galaxy are too weak to contain such high-energy particles, suggests that they may have origins beyond our Galaxy. However it is also possible that they are coming from sources within our Galaxy, perhaps even from the mysterious dark matter that most astronomers believe forms a halo round the Galaxy. The high energies of these cosmic rays would be natural if they are the decay products of exotic supermassive particles, some 10 million million times more massive than a proton. If so, we would have a direct hint that the dark matter consists of exotic supermassive particles.

One possibility is that these supermassive particles are fossil remnants of the Big Bang – an idea that melds with an early theory of the source of the rays. In 1946 Georges Lemaître proposed that cosmic rays all come from the radioactive decay of the 'primaeval atom' from which the Universe began. As such they should provide a unique insight into our origins. This idea of Lemaître's was a forerunner of the now accepted Big Bang theory. If the Big Bang spawned supermassive metastable particles that have condensed to form the dark matter in and around galaxies, as some theories suggest, then the highest-energy cosmic rays could illuminate the darkness.

Fig. 11.22 In the Pampa Amarilla desert, near Malargüe in Argentina's Mendoza Province, the team pose around their first surface array detector for the Pierre Auger Project. Named after the physicist who discovered cosmic ray air showers, the project involves a 3000 sq km array of 1600 detectors like this – each one a self-contained, solar-powered tank of 12 tonnes of pure water to detect Čerenkov radiation from the charged particles in a shower. Time differences across the array will allow the direction of the primary cosmic ray to be determined.

Fig. 11.23 The HEGRA (High Energy Gamma Ray Astronomy) observatory at the Roque de Los Muchachos Observatory on La Palma in the Canary Islands. The observatory has many instruments spread over hundreds of square metres. The huts contain scintillators to detect directly the charged particles in cosmic ray showers, or phototubes to detect, on clear moonless nights, the Čerenkov radiation that the particles produce.

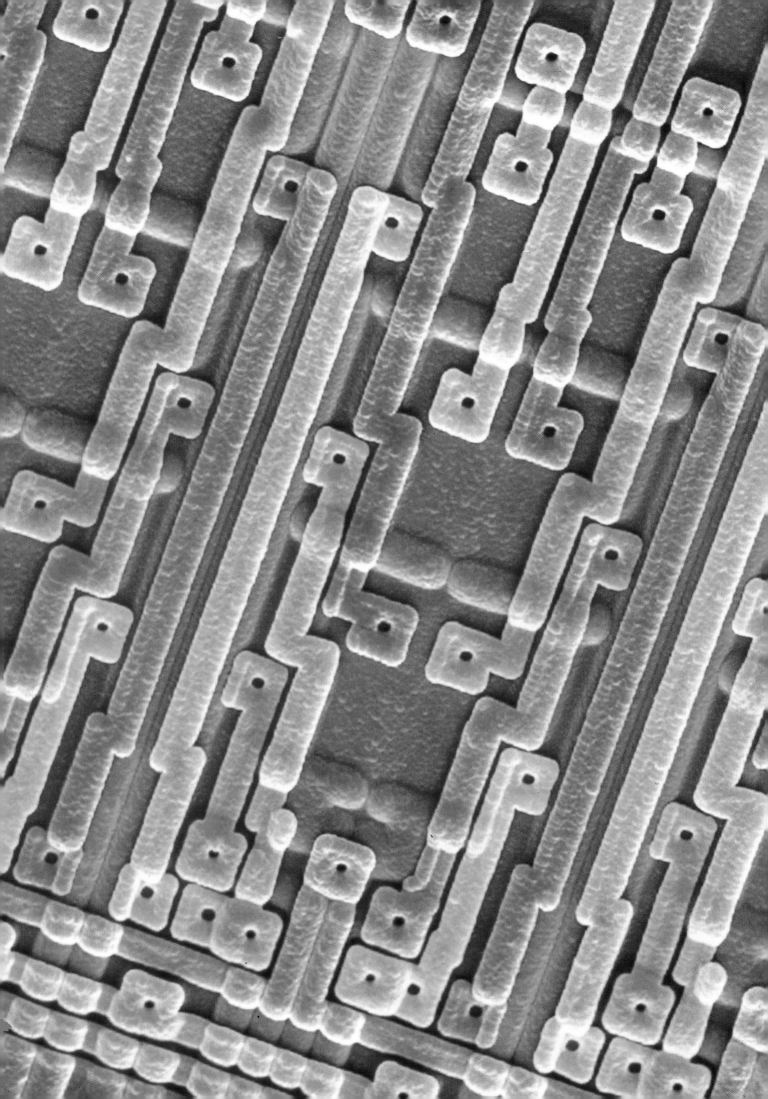

12. Particles at Work

Particle physics is an exhilarating adventure in scientific exploration. Like any branch of pure science, it is driven by curiosity. Like any exploration into new territory – such as the polar ice caps, the deep oceans, or outer space – it challenges technology to adapt to new and extreme conditions, and the results can be of lasting and widespread importance.

Some of the applications of particles and their associated technology have become so pervasive in the developed world today that we take them for granted and forget their origins. More than a century ago, visitors to J.J. Thomson's laboratory in Cambridge would advise him to put aside his bizarre-looking apparatus and to spend more time on something useful. Thomson ignored them, for he was curious about the nature of electricity, and in 1897 he was rewarded with the discovery that electricity is carried by tiny particles – electrons – which are the constituents of every atom.

Today, descendants of Thomson's apparatus sit in almost every living room – the ubiquitous television. More importantly, and with more far-reaching consequences, our understanding of the behaviour of materials in terms of the electrons they contain has led to major developments in many areas of science. Chemists have learned how to synthesize new drugs and materials; biochemists are unravelling the intricate workings of the human body and brain; and in physics the invention of the transistor and the microchip have led to revolutions in computing and information technology.

Modern experiments in particle physics are vastly more complex than Thomson's relatively simple apparatus, yet he would surely recognize the basic principles at work. First there are the beams of particles to act as probes of matter or to be investigated in their own right. Then there are the accelerators that speed the particles to their destiny, just as the electric fields in Thomson's cathode-ray tube propelled the electron beam towards the end of the tube where he observed the resulting glow.

Nowadays violent collisions of highly energetic particle beams reproduce the exotic conditions of the early Universe, and complex detectors reveal the outcome. Finally, the data must be recorded and analysed. This is a task that Thomson performed with pen, paper, and his own brain, but the vast amounts of data from modern experiments require state-of-the-art computers first to filter out and record the useful information, and then to analyse it. The passage of particles through detectors is usually completely invisible to the naked eye, but leaves electrical traces that are picked up by miniature electronic devices fabricated by the thousand on small 'chips' of silicon. Circuitry is designed to make decisions, rapidly responding in different ways to various combinations of signals from the detector, producing 'stop' or 'continue' commands to circuits further down the line. As in an industrial production line, data from the different parts of the detector are checked, sorted, and streamed together to form the final product – the 'event' that is stored for analysis. Today, particle physicists must also be computer experts, so that they can implement their calculations through sophisticated computer programs that extract the final results from the millions of events recorded.

Each of these elements of a typical experiment – particles, accelerator, detector, computing – has had some impact on our lives beyond particle physics even though most of us are completely unaware of it. In this chapter, to illustrate how particle physics works for us in medicine, industry, and commerce, as well as in other areas of science, we describe a few applications of each of these stages in an experiment.

Fig. 12.1 Electrons scattering from the surface reveal the structure of part of an EPROM (Erasable Programmable Read-Only Memory) microchip in this image from a scanning electron microscope. The electron – the first subatomic particle discovered – has led not only to new forms of microscopy, but underlies electronics, a whole new area of science and technology developed in the twentieth century, which led for example to silicon chips like this, and to the modern revolution in computing and communications.

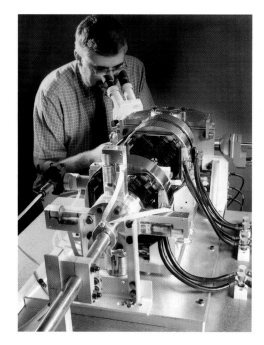

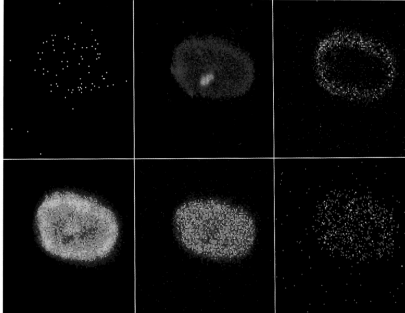

Proton Detectives and Neutron Special Agents

Fig. 12.2 (LEFT) Geoff Grime checks the target end of a beam line at the Scanning Proton Microprobe at Oxford University. Clearly visible are the two sets of precision quadrupole magnet lenses that focus the proton beam down to a diameter of less than 0.5 micrometres. This fine proton beam can probe materials through several techniques – including PIXE (see Fig. 12.3) – and reveal the distribution of tiny amounts of different elements within a specimen.

Fig. 12.3 (RIGHT) PIXE – Proton-Induced X-ray Emission – images of a strand of hair from the 'Ice Man' found in an Alpine Glacier in 1991. The technique uses a beam of protons to scan a sample and induce the emission of X-rays with energies characteristic of the different atoms. The scans here show, clockwise from top left: arsenic, calcium, copper, zinc, sulphur, and iron. The presence of arsenic suggests the Ice Man may have smelted the copper for the implements he carried. (Each scan covers an area of 0.1 by 0.1 mm.)

We see the world about us through scattered light, which is detected by our retinas and analysed by our brains. With good unaided eyesight we can see tiny fleas, only a millimetre or so long, while optical microscopes make visible the flea's legs that power the insect from host to host. But light alone tells us nothing about the composition of the flea – for example, the elements that form its hard shell. Instead, to learn about the structure of an object at the atomic level, we can use particles as probes and utilize their different properties to burrow beneath a material's surface. Protons and neutrons – the charged and neutral components of the atomic nucleus – are both brought into service in this way, providing complementary ways of investigating the small-scale structure of all kinds of materials – animal, vegetable, and mineral.

When energetic protons, with their electric charge, penetrate into materials they can interact both with the clouds of electrons and with the nuclei they encounter – and each kind of interaction yields information about the microscopic structure of the material. The protons may knock electrons out of their atomic orbitals, so that other electrons move in to take their place. However, the new incumbents must initially have been in orbitals with more energy, so they must lose energy as they fill the newly created vacancies. If the original electrons are in orbitals close to the nucleus of a relatively heavy atom, this energy is emitted as X-rays, and the precise energies of these X-rays provide a unique signature of the atom that emits them. An iron atom, for example, will produce an X-ray with an energy of 6.4 kiloelectronvolts (keV) while a calcium atom will yield 3.7 keV X-rays.

The technique of proton-induced X-ray emission – or PIXE – provides a powerful means of revealing the different elements within a specimen, even at levels as small as a few parts per million. The protons used are first accelerated to a few MeV, and then focused to form a beam only a micrometre or so across. There are about 50 facilities around the world, and they have been used to analyse elements in a wide variety of specimens, from the brain tissue of sufferers of Alzheimer's disease to the pigments used in famous paintings and in medieval manuscripts. One fascinating study, undertaken at the Scanning Proton Microprobe Facility at the University of Oxford, involved the famous 'Ice Man', the body found preserved in an Alpine glacier in 1991. Analysis of hair from the body revealed the presence of arsenic, which in turn suggested that the man had smelted the metal for his own implements, as arsenic is a common by-product of smelting.

While protons disturb a material with their electrical interactions as soon as they enter it, electrically neutral neutrons behave more stealthily. Like X-rays, they can penetrate a substance and emerge beyond, carrying messages about the matter they have passed through. But unlike X-rays, neutrons are oblivious to the clouds of electrons in atoms, and interact instead with the atomic nuclei. This makes neutrons especially effective in revealing the presence of lighter elements, which have fewer electrons, in particular hydrogen. Moreover, neutrons behave like tiny magnets – they have a 'magnetic moment' – and this means that they can tell us about the magnetic environment within materials.

In a crystalline material, the various atoms are arrayed in a regular structure, rather like soldiers on parade. When a beam of neutrons infiltrates the serried ranks of atoms in a crystal, the neutrons ricochet from the nuclei at specific angles, and the patterns of the emerging neutrons provide a unique 'fingerprint' of the structure they encountered. In the mid-1980s, this kind of neutron scattering – more properly known as neutron diffraction – provided the first clear insight into the structure of an exciting new material, yttrium-barium-copper oxide. This material had been found to be superconducting – electrically conducting with no resistance – at temperatures of 90 K. In the world of superconductivity, this is a searing temperature, far higher than the few degrees above absolute zero typical of most superconductors (see p. 81). Moreover, it is high enough for the superconducting materials to be cooled by liquid nitrogen, which is relatively easy to produce, rather than liquid helium.

Neutrons, with their ability to reveal the lightweight oxygen as well as the heavier copper, barium, and yttrium atoms, proved to be the important agents in elucidating the structure of the new material. However, understanding the exact mechanism that makes this structure superconducting has proved more difficult, and there is still no definitive theory. Neutrons continue to help in studies of these materials, however, as they also provide a picture of magnetic effects within the complex structures of the yttrium-barium-copper oxides and other materials that are 'high-temperature' superconductors. Continued research of this kind may provide the clues that will lead physicists to materials that are superconducting at room temperature. Such a discovery would trigger a new breakthrough in technology – but it would be just one of the many ways in which neutrons are put to work in modern science.

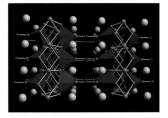

Fig. 12.4 Studies with neutron beams helped to reveal the structure of the high-temperature superconducting material, yttrium-barium-copper oxide, YBCO. In this computer rendition, oxygen atoms are red, barium is yellow, yttrium is purple, and copper is blue. The copper atoms lie at the centre of the blue copper oxide pyramids.

The Reality of Antimatter

In science fiction, antimatter is presented as a potential fuel source to propel astrocruisers to the stars. But though the components of antimatter – antiprotons, antineutrons, and positrons (antielectrons) – are routinely made at particle accelerators, 'antimatter drives' are likely to remain a fiction. Even if a means of storing antimatter could be found – to keep it isolated somehow from matter in a vacuum – the antimatter would first have to be created in bulk, one atom at a time. According to some estimates, as little as a kilogramme of antimatter could power an astrocruiser, but even making such a small amount one atom at a time is far-fetched. All the particle accelerators that have existed so far have made no more than a microgramme of antiprotons. As Chapter 7 describes (see p. 114), current experiments at CERN aim to make about one thousand atoms of antihydrogen an hour. To make a kilogramme would require more than 10^{26} atoms and to do so would take a billion billion years – far longer than the Universe has existed! Even if it were possible to have dedicated accelerators producing antihydrogen at far faster rates, the cost of providing power to run those accelerators would exceed the returns from the antimatter fuel.

The idea of antimatter as a fuel source seems destined to remain science fiction. However, one species of antiparticle – the positron – is already used routinely in applications far closer to home. The positrons used are produced not in high-energy particle collisions, but in the relatively low-energy decays of radioactive nuclei, in a form of beta decay. In solid matter, positrons emitted by an appropriate radioactive nucleus will annihilate with electrons nearby and produce gamma rays.

If the positrons annihilate with electrons in a metal they can reveal the onset of metal fatigue. Distortions in the atomic lattice of the metal provide 'resting sites' where the positrons survive slightly longer before they eventually annihilate. By observing this slight delay it is possible to detect fatigue before any cracks appear in the material. The ability to push turbine blades and other expensive components safely towards their ultimate

Fig. 12.5 An image created with a positron microscope reveals a pattern 0.11 mm across, formed by platinum on oxidized silicon. The microscope uses a beam of positrons from a radioactive source. The positrons are repelled by positive nuclei in the specimen and move towards locations where nuclei are missing – 'defects' in the regular structure of the material. In these positions there are fewer electrons, so the positrons live longer before annihilating. The microscope measures the lifetimes of the positrons. In this image, the height of the map corresponds to the positrons' lifetime, so the silicon dioxide regions with longer-lived positrons stand out against the platinum background.

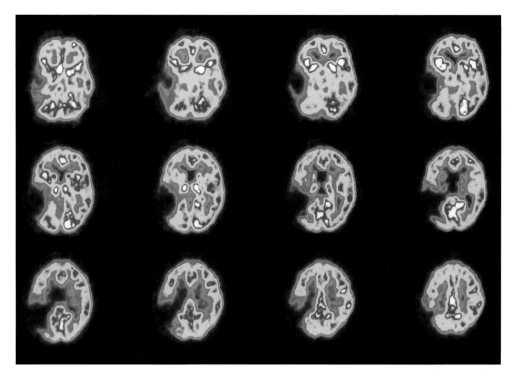

Fig. 12.6 Coloured scans produced by Positron Emission Tomography (PET) show the brain of a person who has suffered a stroke. White areas show regions of high brain activity, while blue indicates low activity. A dark blue region in the left hemisphere shows low activity, indicating an area of brain damage with reduced blood flow due to the stroke.

breaking point, thereby narrowing large safety margins, promises to be of immense economic importance.

The electrons and positrons are more or less at rest when they meet, and to conserve momentum, the annihilation produces two gamma rays that shoot off in opposite directions, so their net momentum remains close to zero. The two gamma rays can be detected in coincidence, using the kind of circuitry familiar in particle physics. Detecting *pairs* of gammas in this way yields the location of the emitting nucleus with greater precision than can be achieved with a single ray from a gamma emitter.

This coincident detection of gamma rays is the basis of the technique known as Positron Emission Tomography, or PET, which is used in many hospitals to provide detailed images of the brain. A 'halo' of gamma-ray detectors surrounds the patient's head, feeding information from gamma-ray pairs to a computer. The computer uses the information to build up images of 'slices' through the brain – hence the term tomography from the Greek *tome*, which means 'cutting'.

Accelerators at Work

Huge accelerators built to reach the highest possible energies have been a theme of this book. In LEP, the Large Electron Positron collider at CERN, a 27 km ring of magnets guided electrons (and positrons) at energies of a hundred billion electronvolts. But at the other extreme, in many homes across the world, television tubes some 10 000 times smaller than LEP accelerate electrons to about ten thousand electronvolts. While big machines like LEP may grab headlines and typify 'particle accelerator' to most people, the use of accelerators in particle physics is the exception rather than the rule. There was only one LEP, but there are billions of televisions in our homes. And in between the extremes of LEP and the television are numerous other examples, their numbers falling as their energies increase.

One of the major applications of particle accelerators is in medicine, where the machines are used both for therapy and in medical imaging. In either case, the accelerator can produce particle beams that are used directly; alternatively, its beams can be used to manufacture radioactive isotopes that emit the useful radiation. X-rays produced when energetic electrons strike a metal target are used both in imaging – where they reveal all kinds of conditions, from holes in the teeth and damaged bones to blockages and tumours – and in radiotherapy for cancer treatment. The most penetrating X-rays for radiotherapy are produced by electron beams from linear accelerators, reaching energies up to 25 million electronvolts (MeV). When the electrons encounter the intense electric fields around the heavy nuclei in a target such as tungsten, they feel a braking force. As they slow down they lose energy as 'bremsstrahlung' – braking radiation – at X-ray energies. Nowadays energetic X-rays from electron linacs are being used to destroy deep-seated tumours in patients at more than 4000 hospitals worldwide.

Radiotherapy with X-rays has been available since the early years of the twentieth century, but a century later therapy with proton beams is beginning also to play a role. Energetic protons will travel in a straight line through soft body tissue, slowing down until they come to a stop at a well-defined distance – 25 cm for protons with an energy of 200 MeV – where they do maximum damage. This ability to cause relatively little damage while travelling through a material makes proton beams an attractive option in cancer therapy. The protons should pass through healthy tissue and leave it largely undamaged, and then deliver their destructive energy mainly at the site of the tumour itself.

More than 20 000 people around the world have been treated with proton beams, mainly at centres based at accelerators that were built for research in particle and nuclear physics. However, there is increasing interest in building accelerators that are dedicated to therapy with protons – and with heavier nuclei, especially carbon ions, which are even more effective in depositing energy at the site of a tumour. By 1994, two hospitals – one at the Loma Linda University Center in California and one at Chiba in Japan – had machines dedicated to proton (or heavier ion) therapy, and three others were planned for hospitals in the USA, Japan, and Italy.

A major challenge is to make proton therapy cost-effective in comparison with more traditional X-ray treatments – and this translates into a technical challenge in terms of accelerator design. The proton therapy facility planned for the Italian National Institute of Health in Rome will use a proton linac specially designed as part of a programme for the development of inexpensive medical accelerators set up by the TERA Foundation. TERA, which stands for Terapia con Radiazioni Adroniche (Therapy with Hadronic Radiation), was set up in 1992 by Ugo Amaldi, an Italian particle physicist, and others with the express aim of promoting the development of radiotherapy with hadrons – protons and light ions such as carbon.

While the nature of protons makes them valuable destructive agents against cancer, a particular characteristic of electrons gives rise to the use of circular electron accelerators for a broad range of work. The bane of electron rings is the synchrotron radiation they emit as the electron beam whirls round the ring. This is a major problem for particle physicists who would like as much energy out of the electron beam as possible, but in other areas of research the radiation is proving to be of great value. In particular, specially designed synchrotrons can provide an intense source of ultraviolet and X-ray radiation.

In 1981, the world's first accelerator to be built purely as a source of synchrotron radiation started operation at the Daresbury Laboratory in Cheshire in the UK. The relatively compact machine – the electron beam follows a circular path only 96 m long – consists of 16 magnets to bend the electron beam, which radiates as it passes through each of the magnets. Additional specially designed magnets in a few locations bend the electron beam even more to produce X-ray beams with specific properties. 'Undulator' magnets make the

Fig. 12.7 (LEFT) Radiotherapy for cancer has been used almost since radioactivity was discovered. This photograph from 1905 shows a woman receiving treatment for breast cancer, the X-rays being produced by a movable electron tube directed towards the malignancy.

Fig. 12.8 (RIGHT) At the Loma Linda University Center in California, this facility has an accelerator dedicated to proton therapy, an effective alternative to radiotherapy for certain kinds of tumour.

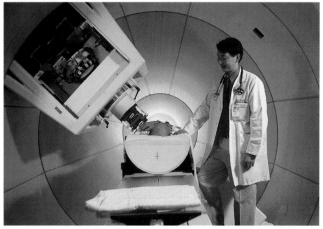

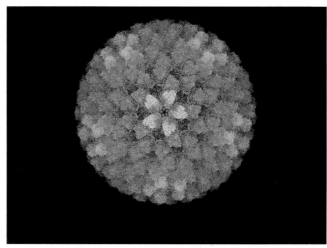

Fig. 12.9 (LEFT) The HASYLAB area at DESY in Hamburg has many experiments crowded together to receive beams of X-rays and UV radiation produced by the synchrotron, DORIS III. This 'synchrotron radiation' is emitted by electrons or positrons in the machine as they follow sharply curving paths through special magnets, called 'wigglers' and 'undulators'. Beams from DORIS III serve 42 experimental stations for research in areas from materials science and geology to medicine.

Fig. 12.10 (RIGHT) The crystal structure of the particle at the core of the Blue Tongue Virus as revealed by studies of 1000 specimens with synchrotron radiation at the Daresbury Laboratory in the UK and the European Synchrotron Radiation Facility in Grenoble. (The particle is about 70 nanometres across.)

electrons follow a regular wave-like path for a short distance, so that the X-rays they emit add together to produce a bright, narrow beam. 'Wiggler' magnets provide regions of strong magnetic field, which in effect create a hair-pin bend in the electrons' path so that they radiate 'harder' X-rays, which have higher energies or, equivalently, shorter wavelengths.

There are now dedicated synchrotron radiation facilities in several countries, and their beams probe phenomena ranging from the structure of proteins and enzymes to the behaviour of catalysts in chemical reactions. For the future, physicists are turning again to linear accelerators to build 'free-electron lasers' that operate at shorter and shorter wavelengths. The laser action requires the beam to zig-zag through a long undulator magnet in such a way that the emitted radiation interacts repeatedly with the electron beam, resulting in a huge increase in the intensity of the radiation. In 2000, a team at the DESY laboratory produced the shortest wavelength radiation to date from a free-electron laser, with wavelengths as small as 180 nanometres. The eventual goal is to build an X-ray laser, which will produce radiation with wavelengths as short as 0.1 nm.

Pixels in Medicine

The discovery of X-rays in 1895 may have marked the beginning of modern physics, as Chapter 2 described, but to most people X-rays are most readily associated with their ability to provide images of broken bones or hidden decay in teeth. 'Taking an X-ray' was one of the major advances in medical diagnosis in the first half of the twentieth century, but by the century's end, there was growing awareness of the hazards that excessive exposure to X-rays could cause. However, particle physics has made an unexpected impact in helping to solve this problem, through detectors being developed for the first major new accelerator of the twenty-first century, the Large Hadron Collider (LHC, see pp. 207–209).

X-rays produce an image on photographic film directly – recall Röntgen's famous image of his wife's hand (see Fig. 2.4, p. 19). But to create an image, the X-rays must first pass through the object, which in medicine and dentistry is live tissue, so the challenge is to minimize damage to this tissue by reducing the intensity of X-rays needed to form a useful image. This is especially a problem in mammography – the X-ray imaging of breast tissue – and in angiography, where radioactive material is injected into blood vessels to image organs such as the heart or kidneys.

One reason why mammography is not used to screen for breast cancer in young women is that the amount of X-ray radiation used is likely to cause as many tumours as it reveals. This is because breast tissue in young women is denser than it is in older women, and so younger women need a higher dose of X-rays to pass through the breast to form a useful image. Moreover, the risks involved accumulate over the years with successive mammograms, presenting yet another problem for younger women.

A key to solving these problems has been to create a detector that can record low numbers of X-rays much more efficiently than X-ray film. In 1992, when Georges Charpak received the Nobel prize for his development of wire chambers to reveal particle tracks, he said that a detector sensitive to an X-ray dose 50 times lower than normal would lead to real advances in the fight against cancer. Since then, such a detector has been found, not as a planned development for medicine, but serendipitously in the development of better particle detectors for the LHC.

At the LHC, the collisions between the beams will be so violent and so frequent that the detectors close to the interaction point will be exposed to vast numbers of particles that can damage the detectors. In the regions that are most at risk, tracking detectors are being made from gallium arsenide (GaAs), as opposed to silicon, as the gallium compound is more resistant to such 'radiation damage'. It turns out that these sensors are very good at catching almost all the X-rays that pass through them, so the team developing them at CERN realized that they might be able to create a novel X-ray detector that would meet Charpak's challenge.

The result was the 'Medipix1 chip' – a sliver of gallium arsenide that contains about 4000 pixels (picture elements) each about 0.17 mm across. The chip can count individual X-rays and respond to X-rays above a specific chosen energy. This combination of attributes can not only produce clear images at about one thirtieth the usual X-ray dose, but it also highlights the small differences in tissue density that are crucial when looking for cancerous growth.

In seeking to observe particles through their faint tracks, particle physicists often find themselves sharing the same goals as other scientists who are trying to record images in difficult circumstances. In one case, experimenters aiming to detect particles produced at the LHC have found a common interest with others interested in imaging organs in the body through the emission of gamma rays from tiny amounts of specially injected radioactive substances. Physicists at CERN have worked together with others from the Instituto Nazionale di Fisica Nucleare (INFN) in Rome to develop a fast, but very sensitive kind of camera, which can detect light at the level of one photon at a time. The device – known as the Imaging Silicon Pixel Array (ISPA) tube – is a twenty-first century development of the phototube, something that has been a common component of many particle physics experiments for half a century.

As in a standard phototube, when photons of light strike one end of the ISPA (the photocathode) they generate electrons. But in the ISPA tube the electrons are guided by an electric field so that they form an inverted image on an array of 1024 tiny pixels of silicon, each bonded directly to a microchip behind the array. The main magic occurs in the chip, which contains circuits to read out the signals from each pixel as quickly as possible. The tube has been considered for the detection of the faint Čerenkov light that charged particles produce in appropriate materials (see p. 91). However, the device is also showing promise for medical imaging with gamma rays, although in this case the gamma rays must first be converted to photons of visible light. This is achieved with a special crystal – a compound of yttrium, aluminium, and peroxide, doped with cerium – which absorbs gamma rays and rapidly emits their energy as light. The prototypes have shown a ten-fold improvement in the resolution of images compared with those from conventional gamma cameras. This could lead to better images that take less time to record and which, importantly, require smaller doses of gamma-emitting substances.

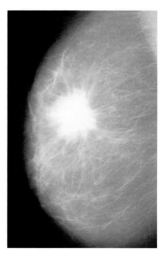

Fig. 12.11 A breast tumour appears as a white region with a ragged edge on this mammogram. The X-rays that produced this image on photographic film have the potential to damage the breast tissue, so the use of the technique is limited to older women. Sensitive detectors developed originally for particle physics experiments can create images with lower numbers of X-rays, and offer the potential for a diagnostic that could be more widely used.

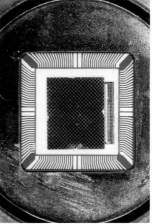

Fig. 12.12 (LEFT) The Imaging Silicon Pixel Array (ISPA) tube is a highly sensitive version of the phototube. Originally developed for particle physics, it is now being applied to medical imaging. As in a standard phototube, photons produce electrons when they strike a photosensitive cathode, but in ISPA the electrons are guided onto a silicon anode finely divided into individual 'pixels'.

Fig. 12.13 (RIGHT) The 'retina' of the ISPA tube is a silicon detector divided into pixels a few tenths of a micrometre square. The pixel array is directly bonded to a special microchip, which very quickly reads out the signals from the array.

The Final Analysis

Fig. 12.14 Modern experiments at particle colliders require banks of electronics to filter out the useful events from the millions of collisions. Here some of the 'trigger' electronics for the H1 experiment at the HERA collider is being tested. In HERA electron and proton beams meet every 96 nanoseconds – 96 billionths of a second – and the trigger's role is to decide rapidly whether a particular 'beam crossing' has produced an interesting event, before the next crossing occurs.

In any scientific experiment the apparatus is only half the story. The main aim is to collect the data, analyse the results, and tell the world what you have discovered. Thomson could observe the fluorescent spot produced by the beam at the end of his cathode-ray tube, write down the values of the electric and magnetic fields required to centralize the spot, and then calculate the ratio of the mass to electric charge for the particles in the beam. He first conveyed his preliminary results during a lecture at the Royal Institution in London in April 1897, and less than four months later he had submitted a paper describing his work, and his conclusions, for publication in *The Philosophical Magazine*.

In a large modern particle physics experiment, the sequence of events is similar but each stage is now impossible without the use of state-of-the-art electronics and computers. Nowadays the analysis often takes place on 'farms' of computers – in some cases PCs similar to those in many homes. Complex computer programs reconstruct the paths of particles, test hypotheses about particle identities, and compare the recorded events with expectations based on a detailed understanding of the behaviour of the detector. Sophisticated statistical methods give weight to the final results for the measurement of, say, the mass or lifetime of an ephemeral particle, such as the tau lepton or the top quark.

In the whole process from data acquisition to final result, particle physicists in an experimental team must become expert in electronics, computer programming, and statistical analysis. It is no wonder that one of the unseen products of modern particle physics is a steady stream of young people who are in high demand by companies dealing in information technology, computer software, and financial markets.

With the large numbers of people involved in a typical experiment dispersed across many countries, the transmission of information is of paramount importance at each stage of the experiment. By the 1980s, this involved the printing and mailing of large numbers of documents throughout the life of an experiment – from the minutes of design meetings, to the 'preprints' in which the final results appear immediately prior to publication. Then in 1984, Tim Berners-Lee, who had graduated in physics at Oxford in 1976 before turning to computing, arrived at CERN to work on software for data acquisition and systems control.

CERN, with its hundreds of visiting researchers based in different countries, proved fertile ground for a dream that Berners-Lee had been harbouring since a brief fellowship there in 1980. He envisaged a system for accessing information that works more like a human brain than a conventional computer – a system that could make *ad hoc* links between information stored in a variety of places. In 1989 he returned to this idea and proposed the project that led to the World Wide Web, and to global connectivity not only between scientists, but between millions of ordinary people.

Local network connections between computers had existed for some time, and connections between networks had become established with the Internet, a network system developed in the USA during the 1970s, which was eventually to spread throughout most of the world. The Internet allowed scientists (and others) to transfer data and communicate by email, so they could transmit electronic versions of documents and images. But the system was cumbersome. Each connection operated over a specific route, rather as a telephone calls does, and would involve the transfer of a whole document, say, before it could be viewed on the requesting computer.

With the scheme proposed by Berners-Lee, however, 'links' in a document viewed on your computer would allow you to jump in any way you liked from one document to another to find the information you wanted. It would be like browsing through an encyclopaedia or a library – but the information could be on computers literally oceans apart. Key components in the scheme were the 'browser' – a computer program to make the links – the progamming language (HTML) needed to instruct the browser, and the means of uniquely identifying information files (documents, images, etc). Berners-Lee was soon given the go-ahead to proceed – and so the World Wide Web was born, with the world's first Web site set up on a computer at CERN in 1990.

Fig. 12.15 Tim Berners-Lee (b. 1955).

For particle physicists and other scientists the Web has allowed the easy distribution of designs, minutes, preprints, and so on. It even allows us all to see 'online' events as they occur in some experiments, such as CDF and D0 at the Fermilab's Tevatron. And in the wider context of the world at large, the simplicity of Berners-Lee's scheme has meant that nowadays anyone with an Internet connection can have easy access to information on millions of computers around the world. You can choose the latest fashions, book holidays, buy stocks and shares all at the push of a button on your computer keyboard, or even via your mobile phone.

However for particle physicists, and indeed for other scientists – including meteorologists, astronomers, and biologists – with large amounts of data to analyse or complex systems to simulate, the next step is to access computers around the world to process data, as if on a vast global computer. This is the concept behind the 'Grid', so named because it is analogous to a grid for electrical power production – you plug in at the wall and immediately have access to huge amounts of 'processing power'.

Fig. 12.16 In December 1990, the world's first Web server ran on this NeXT computer at CERN, set up by Tim Berners-Lee and Robert Cailliau. In the previous months, Berners-Lee had developed the first Web browser and editor.

The key lies in making use via the Internet of spare processing power on computers at times when they are not busy with other tasks. Such 'Internet computing' already exists for specific projects, such as SETI@home, which links three million PCs around the world to help analyse signals from radio telescopes for SETI (the Search for Extraterrestrial Intelligence). However, the concept of the Grid is to develop software and network services that provide a general resource for a variety of projects.

This development involves a large number of people from a broad range of science, not only particle physics. In the USA, NASA is developing an Information Power Grid, the National Science Foundation is funding the National Technology Grid, and NEESgrid is the Network for Earthquake Engineering Simulation grid. In 2000, the European Community began funding the DataGRID project, for researchers in biological science and Earth observation as well as for particle physics. For the particle physicists the Grid offers a means – perhaps the only means – to analyse the vast quantities of data that will be produced at the LHC, and any future accelerators built in the next decade.

Few people realize that the World Wide Web was invented at CERN, but still fewer would have predicted its invention. In 1933, Rutherford famously asserted that 'Anyone who expects a source of power from the transformation of these atoms is talking moonshine'. Very few of us have the vision to see where a line of scientific investigation may lead decades from now, or whether developments like the Web will arise.

At present no one can foresee what use may be made of W and Z particles, top quarks, or tau leptons, or what discovery of the Higgs particle would lead to. But that is no reason for us to stop asking questions. Particle physics may be a pure science but it is also a practical one; the beautiful symmetric theories of matter and force now emerging are built on a solid foundation of measurement and observation. The experiments are performed – and technology extended to new limits in the process – because people ask questions. The questions that confront us today could not have been imagined a century ago. At the start of the twenty-first century, we must hope the opportunity remains to find the answers.

Table of Particles

This table includes only the major particles described in this book. In a number of cases (e.g. the muon), details of a particle and its antiparticle are given in the same entry; in other cases (e.g. the positron), antiparticles have a separate entry; but in many cases details of antiparticles are not given at all. Our criteria have been to include antiparticles mentioned separately in the book, and ones whose discovery occurred separately from their matter equivalent. Note also that antiparticles such as the positron and antiproton are described as stable, although this is true only so long as they do not meet and annihilate with an electron or proton.

Leptons

| NAME | SYMBOL | Physical Properties | | | | Discovery | |
		MASS	LIFETIME	CHARGE	SPIN	DATE	BY WHOM
ELECTRON	e^-	0.511 MeV	stable	-1	$1/2$	1897	J.J. Thomson
POSITRON	e^+	0.511 MeV	stable	$+1$	$1/2$	1932	C. Anderson
MUON ANTIMUON	μ^+ μ^-	105.6 MeV	2×10^{-6}s	-1 $+1$	$1/2$	1937	S. Neddermeyer & C. Anderson
TAU ANTITAU	τ^- τ^+	1.777 GeV	3×10^{-13}s	-1 $+1$	$1/2$	1975	M. Perl's team at SLAC
ELECTRON-NEUTRINO ELECTRON-ANTINEUTRINO	ν_e $\bar{\nu}_e$	< 3 eV	stable*	0	$1/2$	1956	C. Cowan & F. Reines
MUON-NEUTRINO MUON-ANTINEUTRINO	ν_μ $\bar{\nu}_\mu$	< 0.17 MeV	stable*	0	$1/2$	1962	M. Schwartz & team from BNL & Columbia
TAU-NEUTRINO TAU-ANTINEUTRINO	ν_τ $\bar{\nu}_\tau$	< 18 MeV	stable*	0	$1/2$	2000	DONUT team at Fermilab

*There is increasing evidence that the neutrinos are formed from a quantum superposition of 'base states', which allows the neutrinos to oscillate from one type to another.

Quarks

| NAME | SYMBOL | Physical Properties | | | | Discovery | |
		MASS	LIFETIME	CHARGE	SPIN	DATE	BY WHOM
UP ANTI-UP	u $\bar{u}$	~ 5 MeV	stable*	$+2/3$ $-2/3$	$1/2$	1964	Gell-Mann & Zweig quark model
DOWN ANTI-DOWN	d $\bar{d}$	~ 10 MeV	variable*	$-1/3$ $+1/3$	$1/2$	1964	Gell-Mann & Zweig quark model
STRANGE ANTISTRANGE	s $\bar{s}$	~ 100 MeV	variable*	$-1/3$ $+1/3$	$1/2$	1964	Gell-Mann & Zweig quark model
CHARM ANTICHARM	c $\bar{c}$	~ 1.5 GeV	variable*	$+2/3$ $-2/3$	$1/2$	1974	B. Richter & team at SLAC, S. Ting & team at BNL
BOTTOM (or BEAUTY) ANTIBOTTOM	b $\bar{b}$	~ 4.7 GeV	variable*	$-1/3$ $+1/3$	$1/2$	1977	L. Lederman & team at Fermilab
TOP (or TRUTH) ANTITOP	t $\bar{t}$	~ 170 GeV	variable*	$+2/3$ $-2/3$	$1/2$	1995	CDF & D0 teams at Fermilab

*As quarks occur only in pairs (mesons) or triplets (baryons), their lifetimes are variable, depending on the nature of the individual meson or baryon. The up quark, being the lightest, is as stable as the proton that contains it.

Gauge Bosons

| NAME | SYMBOL | Physical Properties | | | | Discovery | |
		MASS	LIFETIME	CHARGE	SPIN	DATE	BY WHOM
PHOTON	γ	0	stable	0	1	1923	A. Compton (implied: A. Einstein, 1905)
W (W-plus) (W-minus)	W^+ W^-	80.4 GeV	10^{-25}s	$+1$ -1	1	1983	UA1 & UA2 teams at CERN
Z	Z	91.19 GeV	10^{-25}s	0	1	1983	UA1 & UA2 teams at CERN
GLUON	g	0	stable	0	1	1979	TASSO & other experiments at DESY

The masses of the particles are given here, as throughout the book, in units of energy – million electron volts (MeV) or giga electron volts (GeV). This is a standard 'shorthand' for mass units of MeV/c^2 or GeV/c^2, where c is the velocity of light.

Laboratories where particles were discovered are referred to by their acronyms: BNL is Brookhaven National Laboratory, LBL is Lawrence Berkeley Laboratory, CERN is the European Organization for Nuclear Research (originally Conseil Européen pour la Recherche Nucléaire), DESY is Deutsches Elektronen Synchrotron, and SLAC is Stanford Linear Accelerator Center.

Discovery		NATURE AND ROLE	PAGES
SOURCE	DETECTOR		
cathode ray tube	cloud chamber	lepton of 1st generation; constituent of atoms; carrier of electricity	36–38
cosmic radiation	fluorescent glass	lepton of 1st generation; antiparticle of electron; formed in cosmic ray showers	66–68
cosmic radiation	electronic	leptons of 2nd generation; decay products of pions, kaons, etc.; components of cosmic rays	69–71
electron–positron annihilation	cloud chamber	leptons of 3rd generation	162–163
nuclear reactor	antineutrino capture detected by liquid scintillator	leptons of 1st generation; produced by, and probe of, weak interaction	120–123
decays of pions produced at accelerator	regenerated muon detected by spark chamber	leptons of 2nd generation; produced by, and probe of, weak interaction	120–123
high-intensity neutrino beam	regenerated tau detected in iron–emulsion layers	leptons of 3rd generation	120–123

Discovery	NATURE AND ROLE	PAGES
HOW		
direct observation in 1968–72: electron scattering at SLAC, neutrino scattering at CERN	quarks of 1st generation; up is constituent of protons, neutrons, and other particles	124–127
direct observation in 1968–72: electron scattering at SLAC, neutrino scattering at CERN	quarks of 1st generation; down is constituent of protons, neutrons, and other particles	124–127
direct observation in 1968–72: electron scattering at SLAC, neutrino scattering at CERN	quarks of 2nd generation; constituents of strange particles	124–127
inferred from J/psi (1974), charmed baryon (1975), charmed meson (1976), & charmonium spectroscopy	quarks of 2nd generation; constituents of charmed particles	158–161
inferred from upsilon (1977) & bottomonium spectroscopy	quarks of 3rd generation; constituents of bottom particles	164–167
inferred from decay into W and b particles	quarks of 3rd generation	182–185

Discovery		NATURE AND ROLE	PAGES
SOURCE	DETECTOR		
X-rays scattered from atomic electrons	crystal spectrometer	carrier of electromagnetic force; 'packet' of electromagnetic radiation	46–47
proton–antiproton annihilation	electronic	carriers of weak force (along with Z)	172–175
proton–antiproton annihilation	electronic	carriers of weak force (along with W^+ and W^-)	176–179
electron–positron annihilation	electronic	8 types of gluon; carriers of strong (colour) force	168–171

Mesons

NAME		SYMBOL	MASS	LIFETIME	CHARGE	SPIN	QUARK CONTENT	DATE	BY WHOM
				Physical Properties				**Discovery**	
PION	(pi-zero)	π^0	135 MeV	0.8×10^{-16}s	0	0	$u\bar{u}$ or $d\bar{d}$	1949	R. Bjorkland & team at LBL
PION	(pi-plus) (pi-minus)	π^+ π^-	140 MeV	2.6×10^{-8}s	+1 −1	0	$u\bar{d}$ $d\bar{u}$	1947	C. Powell & group at Bristol
KAON	(K-zero)	K^0	498 MeV	short: 10^{-10}s* long: 5×10^{-8}s*	0	0	$d\bar{s}$	1947	G. Rochester & C. Butler
KAON	(K-plus) (K-minus)	K^+ K^-	494 MeV	1.2×10^{-8}s	+1 −1	0	$u\bar{s}$ $s\bar{u}$	1947	G. Rochester & C. Butler
J/PSI		J/ψ	3.1 GeV	10^{-20}s	0	1	$c\bar{c}$	1974	B. Richter & team at SLAC, S. Ting & team at BNL
D	(D-zero) (D-plus)	D^0 D^+	1.87 GeV	10^{-12}s 4×10^{-13}s	0 +1	0	$c\bar{u}$ $c\bar{d}$	1976	G. Goldhaber & team at LBL & SLAC
UPSILON		Y	9.46 GeV	10^{-20}s	0	1	$b\bar{b}$	1977	L. Lederman & team at Fermilab
B	(B-zero) (B-minus)	B^0 B^-	5.28 GeV	1.6×10^{-12}s	0 −1	0	$b\bar{d}$ $b\bar{u}$	1983	CLEO team at Cornell

* The K^0 and $\bar{K}^0$ form a quantum system whose superposition yields two physical particles, the short-lived K_S^0 and the long-lived K_L^0, which reveal matter-antimatter asymmetry (CP violation).

Baryons

NAME	SYMBOL	MASS	LIFETIME	CHARGE	SPIN	QUARK CONTENT	DATE	BY WHOM
			Physical Properties				**Discovery**	
PROTON	p	938.3 MeV	stable (?), $>10^{32}$ years	+1	$1/2$	uud	1911–19	E. Rutherford
ANTIPROTON	$\bar{p}$	938.3 MeV	same as proton	−1	$1/2$	$\bar{u}\,\bar{u}\,\bar{d}$	1955	E. Segrè & team at LBL
NEUTRON	n	939.6 GeV	in nuclei: stable free: 15 minutes	0	$1/2$	ddu	1932	J. Chadwick
ANTINEUTRON	$\bar{n}$	939.6 GeV	same as neutron	0	$1/2$	$\bar{d}\,\bar{d}\,\bar{u}$	1956	B. Cork & team at LBL
LAMBDA	Λ	1.115 GeV	2.6×10^{-10}s	0	$1/2$	uds	1951	C. Butler & group at Manchester
ANTILAMBDA	$\bar{\Lambda}$	1.115 GeV	same as lambda	0	$1/2$	$\bar{u}\,\bar{d}\,\bar{s}$	1958	D. Prowse & M. Baldo-Ceolin at LBL
SIGMA (sigma-plus)	Σ^+	1.189 GeV	0.8×10^{-10}s	+1	$1/2$	uus	1953	G. Tomasini & Milan-Genoa team
SIGMA (sigma-minus)	Σ^-	1.197 GeV	1.5×10^{-10}s	−1	$1/2$	dds	1953	W. Fowler & team at BNL
SIGMA (sigma-zero)	Σ^0	1.192 GeV	6×10^{-20}s	0	$1/2$	uds	1956	R. Plano & team at BNL
XI (xi-minus)	Ξ^-	1.321 GeV	1.6×10^{-10}s	−1	$1/2$	dss	1952	R. Armenteros & team at Manchester
XI (xi-zero)	Ξ^0	1.315 GeV	3×10^{-10}s	0	$1/2$	uss	1959	L. Alvarez & team at LBL
OMEGA MINUS	Ω^-	1.672 GeV	0.8×10^{-10}s	− 1	$3/2$	sss	1964	V. Barnes & team at BNL
CHARMED LAMBDA	Λ_c	2.28 GeV	2×10^{-13}s	+1	$1/2$	udc	1975	N. Samios & team at BNL
LAMBDA-B	Λ_b	5.62 GeV	1.2×10^{-12}s	0	$1/2$	udb	1991	R422 team at CERN ISR

Discovery			
SOURCE	DETECTOR	NATURE AND ROLE	PAGES
interaction of protons from accelerator	tantalum converter and proportional counters	involved in nuclear binding; decays into photons; a source of cosmic gamma rays	108–109
cosmic radiation	emulsion	involved in nuclear binding	72–73
cosmic radiation	cloud chamber	strange meson; shows matter–antimatter asymmetry (CP violation)	74–75
cosmic radiation	cloud chamber	strange meson	74–75
interactions of protons from accelerator (Ting), electron–positron annihilation (Richter)	electronic	first known member of charmonium family	158–159
electron–positron annihilation	electronic	charmed mesons	158–161
interactions of protons from accelerator	electronic	first known member of bottomonium family	164–165
electron–positron annihilation	electronic	bottom mesons; B^0 shows matter–antimatter asymmetry (CP violation)	164–167

Discovery			
SOURCE	DETECTOR	NATURE AND ROLE	PAGES
alpha scattering from atomic nuclei	scintillator	charged constituent of atomic nuclei	42–45
interactions of protons from accelerator	scintillation & Čerenkov counters	antiparticle of proton	112–114
beryllium bombarded with alpha particles	ionization chamber	neutral constituent of atomic nuclei	42–45
interactions of protons from accelerator	liquid scintillator	antiparticle of neutron	112–114
cosmic radiation	cloud chamber	strange baryon; replaces neutron in nuclei to make hypernuclei	76–77
interactions of pion beam produced from accelerator	emulsion	antiparticle of lambda	112–114
cosmic radiation	emulsion	strange baryon	78–79
interactions of kaon beam produced from accelerator	diffusion cloud chamber	strange baryon	78–79
interactions of kaon beam produced from accelerator	bubble chamber	strange baryon	78–79
cosmic radiation	cloud chamber	strange baryon	78–79
interactions of kaon beam produced from accelerator	bubble chamber	strange baryon	110–111
interactions of kaon beam produced from accelerator	bubble chamber	strange baryon; confirmed theory of Eightfold Way	118–119
interactions of neutrino beam produced from accelerator	bubble chamber	charmed baryon	158–161
proton–proton collisions	electronic	bottom baryon	164

Further Reading

The following is a selection of generally non-technical books that cover the same subject area as *The Particle Odyssey*. It is not intended to be a comprehensive guide to the literature on particle physics. It includes some 'classics' that are out of print but which should be available through good libraries or second-hand bookshops on the ground or on the internet (such as www.abe.com).

General Interest Books

The Cosmic Onion: Quarks and the Nature of the Universe, Frank Close (Heinemann Educational, 1983; AIP Press, 1986). An account of particle physics in the twentieth century for the general reader.

Lucifer's Legacy, Frank Close (Oxford University Press, 2000). An interesting introduction to the meaning of asymmetry in matter and antimatter and other current and future areas of particle physics.

Spaceship Neutrino, Christine Sutton (Cambridge University Press, 1992). All about the elusive neutrino.

From X-rays to Quarks: Modern Physicists and their Discoveries, Emilio Segrè (W.H. Freeman, 1980). From radioactivity to charm, a detailed account by a leading experimenter.

The Particle Garden: Our Universe as Understood by Particle Physicists, Gordon Kane (Perseus Books, 1996). An introduction to particle physics and a look at where it is heading.

The Elegant Universe: Superstrings, Hidden Dimensions, and the Quest for the Ultimate Theory, Brian Greene (Jonathan Cape, 1999). A prize-winning introduction to the 'superstrings' of modern theoretical particle physics.

Pioneers of Science, Robert Weber (Institute of Physics, 1980). Brief biographies of physics Nobel prize winners from 1901 to 1979.

Marie Curie: A Life, Susan Quinn (Heinemann, 1995).

Rutherford: Simple Genius, David Wilson (Hodder & Stoughton, 1983). An authoritative biography.

The Neutron and the Bomb: A Biography of Sir James Chadwick, Andrew Brown (Clarendon Press, 1997).

Lawrence and his Laboratory, J.L. Heilbron (University of California Press, 1989). The Lawrence Berkeley lab and its founder.

Strange Beauty: Murray Gell-Mann and the Revolution in Twentieth-century Physics, George Johnson (Jonathan Cape, 2000). A biography of Murray Gell-Mann, the 'father' of quarks.

The First Three Minutes, Steven Weinberg (Andre Deutsch, 1977; Basic Books, 1993). The first three minutes after the Big Bang, described in non-technical detail by a leading theorist.

Dreams of a Final Theory, Steven Weinberg (Pantheon Books, 1992; Vintage, 1993). A 'classic' on modern ideas in theoretical particle physics.

More Specialist Books

The Birth of Particle Physics, ed. by Laurie Brown and Lillian Hoddeson (Cambridge University Press, 1983). Proceedings of a symposium on the history of particle physics in 1930–1950, with contributions from many individuals active at the time.

Pions to Quarks: Particle Physics in the 1950s, ed. by Laurie Brown, Max Dresden, and Lillian Hoddeson (Cambridge University Press, 1989).

The Rise of the Standard Model: Particle Physics in the 1960s and 1970s, ed. by Lillian Hoddeson, Laurie Brown, Michael Riordan, and Max Dresden (Cambridge University Press, 1997).

The Particle Century, ed. by Gordon Fraser (Institute of Physics, 1998). The progress of particle physics through the twentieth century.

QED: The Strange Theory of Light and Matter, Richard Feynman (Princeton University Press, 1985). The theory of quantum electrodynamics, explained by one of the theorists who developed it.

An Atlas of Typical Expansion Chamber Photographs, W. Gentner, H. Maier-Leibnitz, and W. Bothe (Pergamon Press, 1953). Out of print, but a comprehensive collection of cloud chamber photographs.

The Study of Elementary Particles by the Photographic Method, C.F. Powell, P.H. Fowler, and D.H. Perkins (Pergamon Press, 1959). Also out of print, but the authoritative compilation of emulsion photographs.

Cambridge Physics in the Thirties, ed. by John Hendry (Adam Hilger, 1984). Accounts of the Cavendish Laboratory by physicists who worked there.

The Discovery of Subatomic Particles, Steven Weinberg (Scientific American Books, 1983). A detailed introduction to the discoveries of the electron, proton, and neutron.

Acknowledgements

This book would not have been possible without the generous contributions of the many individuals who have supplied pictures, given advice, checked the manuscript, and spent time helping us during our visits to CERN, DESY, Fermilab, and SLAC. Not to mention the personal and professional friends who have put up with our demands. We have tried to remember everyone in the list below, but we apologize to anyone who has inadvertently been omitted. A special personal thanks to Caroline, Gill, and Terry.

Franz Aussenegg, Trina Baker, Dave Barney, Tomasz Barszczak, Gianni Battimelli, Stuart Bebb, Franco Bedeschi, Steve Bello, Doug Benjamin, Bob Bernstein, Steve Biller, Renilde Vanden Broeck, Chuck Broy, Volker Burkert, Bobby Byers, Robert Cailliau, Neil Calder, Ian Campbell, Larry Cardman, Philippe Charpentier, Sergio Cittolin, Darren Crawford, John Dainton, Jean Deken, Mick Draper, Hans Drevermann.

David Evans, Petra Folkerts, Paolo Franzini, Gordon Fraser, Stuart Fuess, James Gillies, Silvia Giromini, Paul Gleave, Joel Goldstein, Norman Graf, Michael Green, Geoff Grime, Laurent Guiraud, Dee Hahn, Reidar Hahn, Michael Herren, Rodney Hillier, David Hitlin, Paul Huf, Joe Incandela, Judy Jackson, Mary Janosi, Gron Tudor Jones, Charles Jui, Eric Kajfasz, Ed Kinney, Jane Koropsak, Heinz Krenn, Mark Kruse, Walter Kutschera, Masahiro Kuze, Kimberley Kuzma, Simon Kwan.

Patrice Loiez, Bonnie Ludt, Byron Lundberg, Aki Maki, Paul Mantsch, Bob Mau, Robin Marshall, Kevin McDonough, Adrian McKerney, Curtis Meyer, Joachim Meyer, Ada Molkenboer, Bob Morse, Barbara Moss, Marty Murphy, Gerald Myatt, Sheryl Nonnenberg, Luann O'Boyle, Tokio Ohska, Jaap Panman, Luc Pape, Keith Papworth, David Parker, Ritchie Patterson, Venita Paul, Cesar Pava, Joseph Perl, Brian Pollard, Felicity Pors, Paul Preuss, Kurt Riesselmann, Michael Riordan, Rob Roser.

Robert Schwarz, Jack Scott, Andrew Simmen, Gordon Squires, Jim Strait, Bob Svoboda, Lucas Taylor, Masahiro Teshima, W. Triftshäuser, Jon Trux, Fred Ullrich, Martin Veltman, Stefano Villa, Bob Wagner, David Ward, Alan Watson, Heiner Westermann, Ute Wilhelmsen, Ralf Wischnewski, Thomas Zoufal.

Thank You

Useful Web Sites

The list of Web sites is mainly composed of the major particle physics laboratories, many of which provide links to other research centres and experiments.

Particle physics "Picture of the Week"
 http://hepweb.rl.ac.uk/ppUKpics/pr_pow.html

Brookhaven National Laboratory
 http://www.bnl.gov

CERN
 http://welcome.cern.ch

DESY
 http://www.desy.de

Fermi National Accelerator Laboratory (Fermilab)
 http://www.fnal.gov

KEK
 http://www.kek.jp

Lawrence Berkeley National Laboratory (LBNL)
 http://www.lbl.gov

Stanford Linear Accelerator Center (SLAC)
 http://www.slac.stanford.edu

Picture Credits

Most of the pictures and other illustrations in this book, including pictures from the major laboratories, are available from Science Photo Library (SPL), London. They can be searched and ordered in either photographic or digital form via SPL's online service at www.sciencephoto.com.

We have attempted to obtain permission to reproduce all the pictures in the book, but in the case of some of the older photographs the copyright owner is unknown or could not be located.

Front cover: CMS collaboration, CERN/Science Photo Library
Back cover: David Parker/SPL
Back cover flap: author portraits by Stuart Bebb, Physics Photographic Unit, Oxford
Frontispiece: Tomasz Barszczak, University of California, Irvine. For the Super-Kamiokande collaboration.

1.1 David Parker/SPL
1.2, 1.3 Cavendish Laboratory, University of Cambridge
1.4, 1.5 David Parker/SPL
1.6 D-Zero collaboration, Fermilab
1.7 Ralf Wischnewski, DESY-Zeuthen, Germany
1.8, 1.10 AMANDA collaboration
1.9, 1.11, 1.12 Robert Schwarz, South Pole, Antarctica, 2000
1.13 V. Hess
1.14 CERN/SPL
1.15 David Parker/SPL
1.16 J-L. Charmet/SPL
1.17 Peter Ginter/Bilderberg
1.18, 1.19 CERN/SPL
1.20 Sergio Cittolin, CMS collaboration, CERN/SPL
1.21 CMS collaboration, CERN/SPL

2.1 IBM Corporation, Research Division, Almaden Research Center
2.2 Science Museum/Science & Society Picture Library
2.3 Jean-Loup Charmet/SPL
2.4 SPL
2.5 Jean-Loup Charmet/SPL
2.6, 2.7, 2.8 SPL
2.9 C. Powell, P. Fowler & D. Perkins/SPL. Originally published in Powell, Fowler, Perkins (1959) *The study of elementary particles by the photographic method*. Oxford: Pergamon Press.
2.10 Professor Peter Fowler/SPL
2.11 Illustration by Gary Hincks
2.12, 2.13, 2.14 C.T.R. Wilson/Science Museum/ Science & Society Picture Library (SSPL)
2.15 Cavendish Laboratory, University of Cambridge
2.16 SPL
2.17 Professor Peter Fowler/SPL
2.18 The Living Archive
2.19 By permission of the Syndics of Cambridge University Library
2.20 N. Feather/Science Museum/SSPL
2.21 Ernest Orlando Lawrence Berkeley National Laboratory
2.22 Photographed by Elliot & Fry/Hulton Getty
2.23 Cavendish Laboratory, University of Cambridge
2.24, 2.25 Science Museum/SSPL
2.26 C.T.R. Wilson/Science Museum/SSPL
2.27 Cambridge University, American Institute of Physics/SPL
2.28 P.M.S. Blackett/ Science Museum/SSPL
2.29, 2.30 Cavendish Laboratory, University of Cambridge

3.1 Steve Horrell/SPL
3.2 Niels Bohr Archive, Copenhagen
3.3 P. Auger/Science Museum/SSPL; colouring by D. Parker/SPL
3.4 J.G. Wilson, *Proc. Roy. Soc.*, London (A), 166, 482 (1938); colouring by D. Parker/SPL
3.5 Ernest Orlando Lawrence Berkeley National Laboratory
3.6 Simon Terrey/SPL
3.7, 3.8 Professor Peter Fowler/SPL
3.9, 3.10, 3.11 P.M.S. Blackett & D.S. Lees/ Science Museum/SSPL; colouring by D. Parker/SPL
3.12 Ernest Orlando Lawrence Berkeley National Laboratory; colouring by SPL
3.13 K. Brueckner & W.M. Powell, Ernest Orlando Lawrence Berkeley National Laboratory; colouring by D. Parker/SPL
3.14 I. Joliot-Curie & F. Joliot/Science Museum/SSPL
3.15 Ernest Orlando Lawrence Berkeley National Laboratory; colouring by D. Parker/SPL
3.16 I.K. Bøggild/SPL
3.17 Illustration by Gary Hincks
3.18 National Optical Astronomy Observatories/SPL
3.19 L3 experiment, CERN/SPL

4.1 Illustration by Gary Hincks
4.2 V. Hess
4.3 Courtesy of the Archives, California Institute of Technology
4.4 Illustration by Gary Hincks
4.5 Archiv zur Geschichte der Max-Planck-Gesellschaft, Berlin-Dahlem
4.6 Ullstein Bild
4.7 Courtesy of Prof. Bruno B. Rossi
4.8, 4.9 Courtesy of Prof. D. Skobeltzyn/SPL
4.10 Courtesy of the Archives, California Institute of Technology
4.11 C.D. Anderson/Science Museum/SSPL
4.12 Courtesy of George Rochester
4.13 Cavendish Laboratory, University of Cambridge
4.14 Courtesy of the Archives, California Institute of Technology
4.15 Emilio Segre Archives, American Institute of Physics
4.16 The Living Archive
4.17, 4.18, 4.19 Courtesy of George Rochester
4.20 Dept. of Physics, University of Bristol; courtesy of Rodney Hillier
4.21 C. Powell, P. Fowler & D. Perkins/SPL
4.22 Courtesy of Ilford Limited
4.23 C. Powell, P. Fowler & D. Perkins/SPL
4.24, 4.25 Dept. of Physics, University of Bristol/SPL
4.26 C. Powell, P. Fowler & D. Perkins/SPL

5.1 Goronwy Tudor Jones, University of Birmingham/SPL
5.2 Ernest Orlando Lawrence Berkeley National Laboratory
5.3 Goronwy Tudor Jones, University of Birmingham/SPL
5.4 ALEPH experiment, CERN/SPL
5.5 ZEUS collaboration, DESY
5.6 Tomasz Barszczak, University of California, Irvine. For the Super-Kamiokande collaboration
5.7 S.H. Neddermeyer & C.D. Anderson/Science Museum/SSPL
5.8 G. Piragino, experiment PS 179, CERN/SPL
5.9 C. Powell, P. Fowler & D. Perkins/SPL
5.10 C. Butler & G. Rochester, Manchester University
5.11 OPAL collaboration, CERN/SPL
5.12 CERN/SPL
5.13, 5.14, 5.15, 5.16 Ernest Orlando Lawrence Berkeley National Laboratory; colouring by David Parker/SPL (5.13, 5.14) and SPL (5.15, 5.16)

6.1 Image courtesy of AERO-METRIC, INC., Sheboygan, Wisconsin, U.S.A.
6.2 David Parker/SPL
6.3, 6.4 Fermilab
6.5, 6.6 David Parker/SPL
6.7 Ernest Orlando Lawrence Berkeley National Laboratory
6.8 American Institute of Physics/SPL
6.9 Ernest Orlando Lawrence Berkeley National Laboratory
6.10 Illustration by Gary Hincks
6.11, 6.12, 6.13, 6.14, 6.15 Ernest Orlando Lawrence Berkeley National Laboratory
6.16 Courtesy of Brookhaven National Laboratory
6.17 Ed Young/SPL
6.18 Ernest Orlando Lawrence Berkeley National Laboratory
6.19 CERN/SPL
6.20 Sandia National Laboratories/SPL
6.21 Ernest Orlando Lawrence Berkeley National Laboratory
6.22 Courtesy of Prof. Donald Glaser
6.23, 6.24 Ernest Orlando Lawrence Berkeley National Laboratory
6.25 Courtesy of Brookhaven National Laboratory
6.26 David Parker/SPL
6.27 Novosti/SPL
6.28, 6.29 CERN/SPL
6.30 Courtesy of British Oxygen Company Limited
6.31, 6.32 CERN/SPL
6.33 Courtesy of Prof. Y. Ne'eman
6.34 Photo by Francis Bello, © Bello Estate/SPL
6.35 Fermilab
6.36 CERN/SPL
6.37 David Parker/SPL
6.38 Stanford University Archives
6.39 AIP Emilio Segrè Visual Archives/Physics Today Collection/SPL
6.40 Photo by Thomas Nakashima/SLAC
6.41 David Parker/SPL
7.1 Fermilab
7.2 Ernest Orlando Lawrence Berkeley National Laboratory
7.3 R.D. Leighton/SPL
7.4 Ernest Orlando Lawrence Berkeley National Laboratory; colouring by SPL
7.5, 7.6 Ernest Orlando Lawrence Berkeley National Laboratory
7.7 Ernest Orlando Lawrence Berkeley National Laboratory; colouring by D. Parker/SPL
7.8 Experiment PS210, CERN/colouring by SPL
7.9 Courtesy of Brookhaven National Laboratory
7.10 Volker Burkert and the CLAS collaboration at Jefferson Lab
7.11 Courtesy of the Crystal Barrel collaboration
7.12 Courtesy of Brookhaven National Laboratory
7.13 Illustration by Gary Hincks
7.14 Tomasz Barszczak, University of California, Irvine. For the Super-Kamiokande collaboration
7.15, 7.16 Courtesy of Dr F. Reines
7.17 NOMAD experiment, CERN/SPL
7.18 Fermilab
7.19 IMB collaboration/SPL
7.20 UA1 experiment, CERN/David Parker/SPL
7.21 Illustration by Gary Hincks
7.22 ZEUS collaboration, DESY
7.23 H1 experiment, DESY

8.1 David Parker/SPL
8.2 CERN/SPL
8.3 Philippe Plailly/SPL
8.4 CERN/SPL
8.5 David Parker/SPL
8.6 Illustration by Gary Hincks
8.7 Stanford Linear Accelerator Center
8.8 Ernest Orlando Lawrence Berkeley National Laboratory
8.9 Courtesy of Brookhaven National Laboratory
8.10 AIP Niels Bohr Library/Orren Jack Turner
8.11 David Parker/SPL
8.12 Courtesy of Laboratori Nazionali di Frascati dell' INFN
8.13 David Parker/SPL
8.14 Ernest Orlando Lawrence Berkeley National Laboratory
8.15 Courtesy of Brookhaven National Laboratory
8.16 Stanford Linear Accelerator Center
8.17 David Parker/SPL
8.18 CLEO experiment, Cornell University; colouring by D. Parker/SPL
8.19 Cornell University
8.20 DESY
8.21, 8.22, 8.23 CERN/SPL
8.24 Cavendish Laboratory, UA5 experiment, CERN/David Parker/SPL
8.25, 8.26 David Parker/SPL
8.27 UA1 experiment, CERN/David Parker/SPL
8.28 David Parker/SPL
8.29 CERN/SPL
8.30 David Parker/SPL
8.31 CERN/SPL
8.32 Illustration by Gary Hincks
8.33, 8.34 David Parker/SPL
8.35 Stuart Bebb, Physics Photographic Unit, Oxford
8.36, 8.37 SLD collaboration, SLAC
8.38, 8.39 David Parker/SPL
8.40, 8.41 DESY

9.1 L3 experiment, CERN/SPL
9.2 Mark 1 experiment, SLAC; redrawn by Gary Hincks
9.3 S. Ting & team's experiment, Brookhaven National Laboratory; redrawn by Gary Hincks
9.4 Mark 1 experiment, SLAC
9.5 H1 experiment, DESY
9.6 Courtesy of Brookhaven National Laboratory
9.7 SLAC Hybrid Facility Photon collaboration
9.8 TASSO experiment, DESY
9.9 ALEPH experiment, CERN/SPL
9.10 Fermilab
9.11 DELPHI experiment, CERN/SPL
9.12, 9.13 CDF collaboration, Fermilab
9.14, 9.15 ALEPH experiment, CERN/SPL
9.16 JADE experiment, DESY
9.17 SLD collaboration, SLAC
9.18 Courtesy of the Crystal Barrel collaboration
9.19 ZEUS collaboration, DESY
9.20 UA1 experiment, CERN/SPL
9.21 D-Zero collaboration, Fermilab
9.22 ZEUS collaboration, DESY
9.23 DELPHI experiment, CERN
9.24 OPAL collaboration, CERN/SPL
9.25 Gargamelle experiment, CERN/SPL
9.26 UA1 experiment, CERN/SPL
9.27 DELPHI experiment, CERN/SPL
9.28 ALEPH experiment, CERN/SPL
9.29 L3 experiment, CERN/SPL
9.30, 9.31, 9.32, 9.33 ALEPH experiment, CERN/SPL
9.34, 9.35 CDF collaboration, Fermilab
9.36 D-Zero collaboration, Fermilab

10.1 Space Telescope Science Institute/NASA/SPL
10.2 NASA/SPL
10.3 UA1 experiment, CERN/David Parker/SPL
10.4 Allan Morton/Dennis Milon/SPL
10.5 AIP Emilio Segrè Visual Archives/Physics Today Collection/SPL
10.6 Courtesy of Brookhaven National Laboratory
10.7 The KLOE collaboration
10.8 CERN/SPL
10.9 David Parker/SPL
10.10 NA35 experiment, CERN/SPL
10.11 NA49 experiment, CERN/SPL
10.12, 10.13 Courtesy of Brookhaven National Laboratory
10.14 Jean-Charles Cuillandre, Canada-France-Hawaii Telescope/SPL
10.15 Lucent Technologies/Bell Labs/SPL
10.16 Maddox, Sutherland, Efstathiou & Loveday/SPL
10.17 Dr George Efstathiou/SPL
10.18 David Parker/SPL
10.19 Stuart Bebb, Physics Photographic Unit, Oxford
10.20 ICRR (Institute for Cosmic Ray Research), The University of Tokyo
10.21, 10.22 Tomasz Barszczak, University of California, Irvine. For the Super-Kamiokande collaboration
10.23 Illustration by Gary Hincks
10.24 Ernest Orlando Lawrence Berkeley National Laboratory
10.25 SNO collaboration
10.26 Martin Dohrn/SPL
10.27 Claude Nuridsany & Marie Perennou/SPL
10.28 SPL
10.29 Niels Bohr Archive, Copenhagen

11.1, 11.2, 11.3 CERN/SPL
11.4 CERN/SPL
11.5 Courtesy of Laboratori Nazionali di Frascati dell' INFN
11.6, 11.7 David Parker/SPL
11.8 High Energy Accelerator Research Organization (KEK), Japan
11.9 BaBar experiment, SLAC
11.10 R. Svoboda at Louisiana State University
11.11 K2K collaboration
11.12 David Parker/SPL
11.13 Illustration by Gary Hincks
11.14 Prof. P. Fowler, University of Bristol/SPL
11.15 Roger Ressmeyer/Corbis
11.16 Jeff Hester and Paul Scowen, Arizona State University/SPL
11.17 AMANDA collaboration
11.18 ANTARES – F. Montanet, CPPM/IN2P3/CNRS-Univ. Mediterranée
11.19 Illustration by Gary Hincks
11.20 Fly's Eye experiment
11.21 AGASA collaboration
11.22 Pierre Auger Observatory
11.23 David Parker/SPL

12.1 Andrew Syred/SPL
12.2 Stuart Bebb, Physics Photographic Unit, Oxford
12.3 Physics Photographic Unit, Oxford
12.4 Alfred Pasieka/SPL
12.5 From 'Lifetime measurements with a Scanning Positron Microscope' by A. David, G. Kögel, P. Sperr, and W. Triftshäuser, *Phys. Rev. Lett.* 87 (2001) 067402. Courtesy of Prof. Dr. W. Triftshäuser
12.6 Wellcome Department of Cognitive Neurology/SPL
12.7 SPL
12.8 Robert B. Rearick, Loma Linda University Medical Center, Media Services
12.9, 12.10 DESY
12.11 Breast Screening Unit, King's College Hospital, London/SPL
12.12, 12.13 CERN/SPL
12.14 David Parker/SPL
12.15 Hank Morgan/SPL
12.16 CERN/SPL

Index

sigma particles 78–9, *79*, 232–3
  charmed sigma 160
silicon chips *220*, 221
silicon strip detectors 150–1, *150*, *151*
Skobeltzyn, Dmitry 54, *54*
SLAC Large Detector (SLD) 152, *152*, 169, *169*
SLAC Linear Collider (SLC) 153–4, 177
Sloan, David 85
Snyder, Hartland 96
Sochet, Melvyn 183
Soddy, Frederick 25, *25*
solar neutrinos 120–1, 201, 202–3, *203*
solar plasma 195
solar system model 35
solar wind 189
spark chamber 98–100, *99*, 133
SPEAR 138–9, *138*, *139*, 158, 160
spin 205
spiral galaxies 197
Standard Model 157, 193, 203
standing wave 82
Stanford Linear Accelerator Center (SLAC)
    8, 103–5, *103*, *104*, *105*, 107
  BaBar detector 210, *210*, 211, *211*
  charged-couple device 152, *152*
  Hybrid facility bubble chamber *161*
  Mark I detector 138–9, *139*
  Mark II detector *134*, 154, 177
  neutrino detection 123, 179
  PEP (Positron Electron Project) 141, 142
  PEP-II 210, *210*
  Program Advisory Committee 158
  quark studies 125
  silicon strip detector *151*
  SLC (SLAC Linear Collider) 153–4, 177
  SLD (SLAC Large Detector) 152, *152*, 169, *169*
  SPEAR 138–9, *138*, *139*, 158, 160
  time projection chamber 135, *135*
Stanford University 103, 105, 137, 138
STAR experiment 136, *136*, *196*
stochastic cooling 144
strange particles 58–9, 74–5, 76–7, 78–9, 88, 89,
    110–1, 119
strange quark 124, 125, 230–1
strangeness 110–1, 119, *119*
Strassman, Fritz 30
streamer chamber *72*, 73, 78, *78*
strong focusing 96–8
strong force 7, 39, 40
subatomic particles, naming 7
Sudbury Neutrino Observatory (SNO) 202–3, *202*,
    203
sun 172, *212*
  photons from 47
  see also solar headings
superconducting magnets 81, 102, *154*, 155, *155*,
    207–8, *207*
superconductivity 81, 223
superforces 205
supergravity 205
superheated liquid 92
Super-Kamiokande *48*, *70*, 71, *120*, 121, *200*,
    201–2, *201*, *212*, 213
supernovae 123, *123*, 214
Super Proton Synchrotron (SPS) 102–3, *102*, 144,
    145, 194, *195*, 207
superstring theory 205
supersymmetry (SUSY) 198, 205
supersynchrotrons 101–5
symmetry 203–4
synchrocyclotron 87, 96, 97, 109, 115
synchrophasotron 98
synchrotron 88
synchrotron radiation 103, 225–6

tantalum 109
TASSO *155*, *162*, 163
tau 71, 139, *139*, 162–3, 230–1
tau-neutrino 123, *123*, 162, 199, 230–1
television 38
TERA foundation 225
Tevatron *viii*, *3*, 81, 83–4, *83*, 102, 103, 147, 151,
    153, 157, 173, *173*, 177, 182, *184–5*
theorists 8
theory of everything 203–5
thermalization 194
Thomson, Joseph John (J.J.) 1, 2, *2*, 22, 24, *24*
thorium 21, 22, 24–5, 41
time dilation 13
time projection chambers 135–6, *135*, *136*, 195,
    196
Ting, Sam 139, *139*, 158
tomography 224
Tomonaga, Sin-Itiro *7*, 46
top quark 1, 2, *3*, 142, 147, 153, 182–3, *182*, *184–5*,
    230–1
Trilling, George 138
Tsouchek, Bruno 138
Turlay, Rene 191

UA1 detector *124*, 145, *146*, 147, *172*, 173, 177,
    177, 188
UA2 detector 145, 147, 173
UA5 detector 145, *145*
Uncertainty Principle 17
undulator magnets 225–6
UNESCO 97
unification of forces 8
universe, expanding *186*, 187
upsilon 141, *141*, 164, *165*, 232–3
uranium 17, 41
uranium salt 19–20, *20*

Van der Meer, Simon 144, *144*, 147
Varian, Russell 104, *104*
Varian, Sigurd 104, *104*
Veksler, Vladimir 87
VERITAS 215
virtual photon 46
V particle 59, *59*, 74, *74*, 89

Walton, Ernest 32, *33*
Wambacher, H. 60
water droplet symmetry 203–4, *204*
weak force 7, 39, 172
Weinberg, Steven 176, 177
Whipple Observatory *214*, 215
White Mountain, California 59
Wideröe, Rolf 84–5, *85*, 137
Wiegand, Clyde 90, *90*
wiggler magnets 226
Wiik, Bjorn 155
Wilson, Charles (C.T.R.) 30–1, *30*, 60
Wilson, Robert *3*, 97, 101, *101*, 102
Wojcicki, Stan 116
Wood, John 93
World Wide Web 228–9, *229*
Wouthuysen, S.A. 109
W particles 46, 147, 172–5, *172*, *173*, *174*, *175*, *188*,
    230–1
Wulf, Theodor *22*, *48*, 50, 51

xenon *92*
xi particles 78–9, *78*, 232–3
xi-star 119
xi-zero 110–11, *110*, 232–3
X-ray laser 226
X-rays 19, *19*, 22, 60, 222, 224–5
  detectors 226–7

Ypsilantis, Thomas 90, *90*
Y-star 116
yttrium-barium-copper oxide 223, *223*
Yuan, Luke *115*
Yukawa, Hideki 57, *57*, 61, 71, 73
Yukawa, Mrs 57

ZEUS detector 69, *69*, 126, *126*, *151*, 155, 171, *171*,
    173, *173*
Z particle 46, 129, 147, *147*, 153–4, 176–9, *176*,
    177, *178*, *179*, 230–1
Zweig, George 125